Songling Huang, Wei Zhao
Magnetic Flux Leakage

Advances in Electrical and Electronic Engineering

Volume 1

Songling Huang, Wei Zhao

Magnetic Flux Leakage

Theories and Imaging Technologies

DE GRUYTER

清華大學出版社
TSINGHUA UNIVERSITY PRESS

Authors

Prof. Songling Huang
Tsinghua University
Dept. of Electrical Engineering
Haidian District
30 Shuangqing Rod
100084 Beijing
Republic of China

Prof. Wei Zhao
Tsinghua University
Dept. of Electrical Engineering
Haidian District
100084 Beijing
Republic of China

ISBN 978-3-11-047701-6
e-ISBN (PDF) 978-3-11-048015-3
e-ISBN (EPUB) 978-3-11-047971-3
Set-ISBN 978-3-11-048016-0
ISSN 2509-582X

Library of Congress Cataloging-in-Publication Data
A CIP catalog record for this book has been applied for at the Library of Congress.

Bibliographic information published by the Deutsche Nationalbibliothek
The Deutsche Nationalbibliothek lists this publication in the Deutsche Nationalbibliografie; detailed bibliographic data are available on the Internet at http://dnb.dnb.de.

Cover image: Ingram Publishing/thinkstock
Typesetting: Integra Software Services Pvt. Ltd.
Printing and binding: CPI books GmbH, Leck
♾ Printed on acid-free paper
Printed in Germany

www.degruyter.com

Preface

Magnetic flux leakage (MFL) testing has many advantages, such as the simplicity of principle, the convenience for application and no treatment for the detected surface. Thus, in the past 100 years from the beginning of its birth, MFL detection technology has been developed rapidly. Its application fields are extended to almost all industries involved in the production and life of people, such as petroleum, chemicals, automobile, aerospace, railway, ship and so on, including not only the inspection of raw materials but also the factory production line of the workpiece, semi-finished product quality control, service facilities defect detection and safety assessment.

For a long time, defect retrieval quantization has been a difficult and hot research topic in the field of MFL detection. In recent years, the quantitative inversion technology of MFL testing has experienced from simple to complex, from qualitative identification to quantitative analysis, and toward the direction of defect contour recognition and MFL imaging detection.

This book introduces basic principles of MFL detection, identification and quantification of defects and its implementation techniques, mainly including MFL detection influence factors, signal processing method, defect equivalent size quantization method, defect profile inversion, three-dimensional MFL imaging and so on.

During the last more than 10 years, the author has been conducting continuous research and practical applications in MFL testing, and much of the content of this book comes from the doctor and master theses guided by the author. These students include Wei Cui, Junjie Chen, Xinmeng Liu, Zhiyi Su, and Xinyi Wu. In the implementations of related technologies, we obtain considerable support and assistance from the leaders and engineers of Sinopec, PetroChina and CNOOC to ensure continuous improvement of the related technologies in practice. The author thanks those people sincerely!

With the growth of demand for MFL testing, research in MFL testing is drawing more attention and the industrial applications are increasing more widely. The author hopes that the publication of the book can provide references for the research and development staffs, the students and teachers in universities, and engineers and technicians. If any error is found in the book, criticism and correctness from the readers are welcomed.

Author
May 2016
At Tsinghua Yuan

Contents

1 Introduction

1.1 Outline

The principle of magnetic flux leakage (MFL) detection is shown in Figure 1.1.

The MFL testing method is based on the high permeability of ferromagnetic materials [1]. When ferromagnetic materials are magnetized by the external magnetic field, and if the ferromagnetic material is continuous and uniform, then the magnetic flux in the material will be constrained in the interior of the materials, thus almost no flux leaks from the material surface. However, when there are defects in the surface or inside of the ferromagnetic materials, due to the small permeability of the defects and the large magnetic resistance, the magnetic lines of force will change the way. When the magnetic induction intensity in the material or the defect is large enough, the magnetic flux will overflow from the workpiece near the defect, pass through the air above the defect and then enter into the workpiece; thus, the MFL of defects is created. Here, the magnitude of MFL is directly related to the magnetic field strength of ferromagnetic material. In the external magnetizing field, the relationship between magnetic induction $\vec{B}$ and magnetic field strength $\vec{H}$ of the ferromagnetic materials is

$$\vec{B} = \mu\vec{H}. \tag{1.1}$$

Because the permeability of the material varies with the magnetic field intensity $\vec{H}$, the relationship between $\vec{B}$ and $\vec{H}$ is not linear, and the magnetization curve exhibits a nonlinear variation. The magnetic characteristics of the ferromagnetic material are in accordance with the law of the magnetization if the tested material is magnetized by a permanent magnet or an exciting coil.

A steel plate with a surface defect is taken as an example to illustrate the MFL phenomenon [2]. Figure 1.2 shows the profile of the steel plate defect. Figure 1.3 shows the magnetic characteristic curve of the steel plate.

The sectional area of a defect in a steel plate is S_a. The sectional area of the steel plate is S. The remaining sectional area of the steel plate in the defect area is $S - S_a$. If the magnetizing field is uniform with a magnetic field strength H, the magnetic induction intensity in the plate with defects is $B1$, which is corresponding to the point Q on the B–H curve and to the point P on the permeability curve.

Due to the existence of defects, the magnetic induction intensity of the remaining section increases. So the operating point moves from Q to Q' on the magnetization curve. However, the permeability of point Q' is smaller than that of point Q, which changes from point P to point P' in curve 2. That is to say, due to the existence of defects, the magnetic induction intensity increases with the decreasing cross-sectional

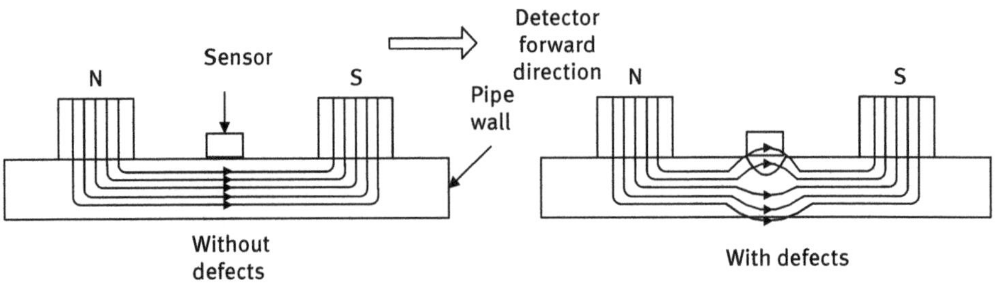

Figure 1.1: The principle of magnetic flux leakage detection.

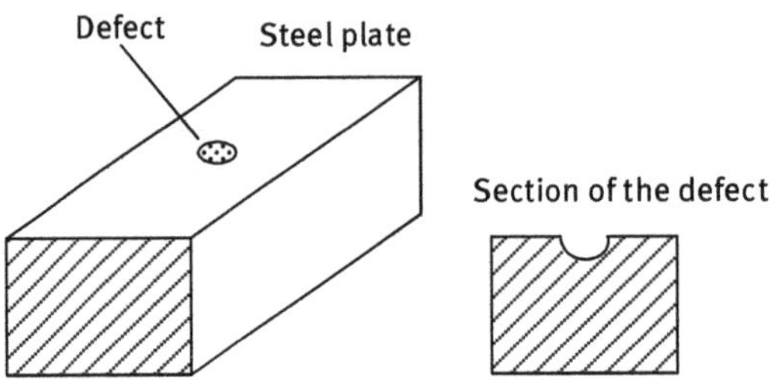

Figure 1.2: The sketch map of a defect in steel plate.

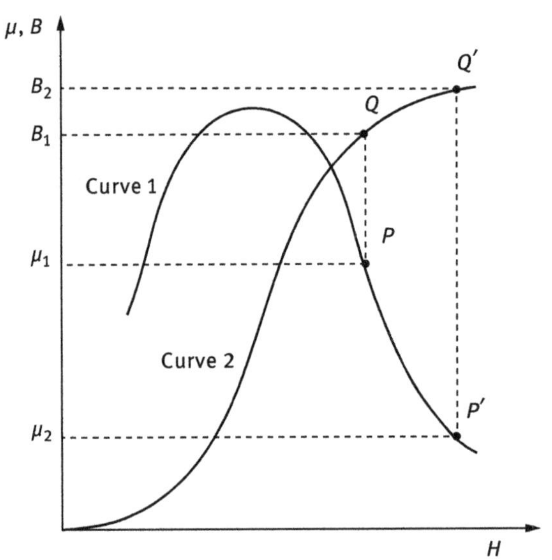

Figure 1.3: The magnetizing characteristic curve of the steel plate.

parts; at the same time, the permeability change is smaller, which shows that the plate with defects cannot pass through the original magnetic flux, so a portion of the magnetic field scatters and leaks into the surrounding medium, and forms the leakage of magnetic field.

1.2 Influence Factors of Magnetic Flux Leakage Testing

Magnetic leakage fields of defects will be affected by many factors. These factors include the magnetizing magnitude, remanence, electrical conductivity, magnetic permeability, magnetic coupling loop, the distance between the magnetic poles, the moving speed of the detector, internal stress, liftoff of the probe and so on [3, 4].

1.2.1 The Influence of Magnetizing

The function of the magnetizing system of the detector is to make the tested part to be magnetized, and then the MFL is generated around the defect. So the magnetizing field strength is the most important influencing factor on the magnetizing level. External magnetizing field strength is determined by the relationship between the magnetizing system and the tested material. Thickness of the tested material, components of material, scanning speed, remanence and other factors will affect the magnetizing system.

The relationship between the applied magnetizing field intensity and the magnetic induction intensity in the tested part is nonlinear, as shown in Figure 1.4.

After the tested material is magnetized by the permanent magnet, the MFL field can be generated near the defect, which is mainly determined by the magnetizing intensity. If the magnetizing intensity is not enough, then the decreasing wall thickness induced by the defects still can carry all of the magnetic flux, so there will be no MFL to the surface of the tested piece. Therefore, only when the two conditions of the defect

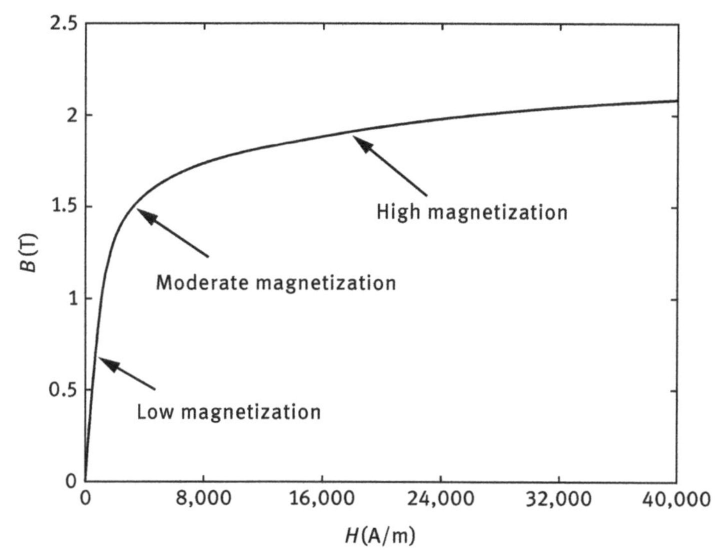

Figure 1.4: The *B*–*H* curve of a steel pipe wall.

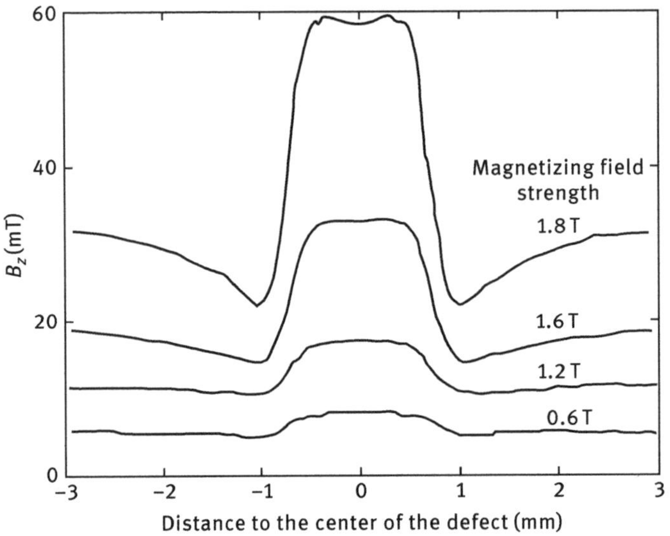

Figure 1.5: Effect of magnetization on the magnetic flux leakage signal of the same defect.

and permeability decrease with the increase of the applied magnetizing field strength are both satisfied, and the MFL field can be produced.

The strength of the applied magnetic field is mainly determined by the magnetic properties of the magnet. Figure 1.5 shows the magnetic leakage field of a defect calculated by the finite element with different magnetizing field strengths, which are 1.8, 1.6, 1.2 and 0.6 T, respectively. The length, width and depth of the defects are 14.6, 14.6 and 7.3 mm (50 % wall thickness), respectively. A value of 0.6 T corresponds to a relatively low magnetization; 1.2 and 1.6 T correspond to moderate magnetization; and 1.8 T corresponds to a higher magnetization.

As shown in Figure 1.5, the signal tested by the probe actually contains two parts: the background magnetic field signal and the defect MFL signal. When the magnetic induction intensity of the tested material is 1.8 T, the amplitude of the background magnetic induction intensity is 32 mT and the total testing magnetic induction intensity amplitude of the probe is 58.5 mT. So the magnetic induction intensity amplitude of the defect is 58.5–32 = 26.5 mT. The ratio of the MFL signal of the defect to the background magnetic field signal is about 0.8. When the magnetic induction intensity of the tested material is 0.6 T, the magnetic induction intensity amplitude of the magnetic leakage signal is only 2 mT, and the ratio of the MFL signal of the defect to the background magnetic field signal is 0.3. Here, the background magnetic field is the magnetic induction intensity direct current signal tested by the probe when there are no defects, and this is the air-coupled magnetic field mainly from the magnetic poles (abbreviating as "air-coupled magnetic field") as shown in Figure 1.6.

In this way, the effect of external magnetic field intensity on the MFL testing can be seen by comparing the ratio of the MFL signal to the background magnetic field

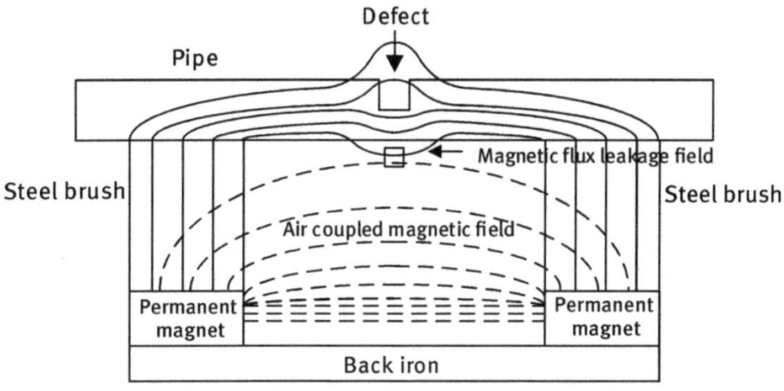

Figure 1.6: Schematic diagram of the air-coupled magnetic field.

under different external magnetic field strength. As shown in Figure 1.6, the magnetic induction intensity of the tested part is 1.8 T, and this ratio is the highest. And with the decrease of the magnetic field intensity, the proportion is gradually reduced. When the applied magnetic field strength is lower than 1.2 T, this ratio will be reduced to about fixed value. However, if the magnetization is improved on the basis of 1.8 T, the ratio of the magnetic leakage signal to the background magnetic field shows a downward trend. The main reason is that MFL signal amplitude will not change when the magnetic induction intensity of the tested material reaches saturation; at the same time, the background magnetic field will continue to improve with the increasing external magnetic field, which leads to lower the ratio of the two.

To sum up, the applied magnetizing field that can reach or moderately exceed the saturation magnetization will generate large enough leakage flux, which can ensure the effective detection of corrosion defects. Therefore, in order to improve the accuracy of corrosion defect magnetic leakage detection and quantification, a higher external magnetizing field strength should be chosen, to ensure that the tested material is magnetized to moderate saturation and to get enough high ratio of the magnetic leakage field of defects to the background magnetic field. It should be noted that the external magnetizing field strength cannot be too high, because the high magnetizing field intensity will lead to reduction of the signal-to-noise ratio (SNR).

Because the magnetizing field intensity will affect the detection and quantification, the influencing factors of the magnetizing field intensity also affect the detection and quantification. The main factors that affect the detection and quantification are as follows.

1.2.1.1 The Thickness of Magnetized Material

The greater the thickness of the material is, the stronger the external magnetizing field is required to reach saturation magnetization. In the case of constant external magnetizing field, the change of the thickness is inversely proportional to the magnetic field intensity and the magnetic induction intensity of the tested material. Let the thickness

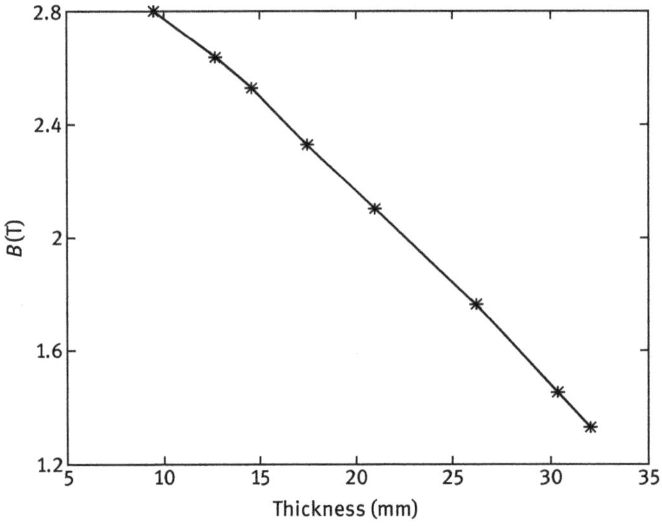

Figure 1.7: Relationship between the thickness and the magnetic induction intensity.

T of the tested materials be 9.5, 12.7, 14.6, 17.5, 21, 26.2, 30.4 and 32 mm, then the magnetizing field strength in the same situation, using finite element method, and the change trend of the magnetic induction intensity above the surface 4 mm are shown in Figure 1.7.

As shown in Figure 1.7, the thickness of the tested material and the magnetic induction intensity show a negative linear relationship, that is, the thicker the tested piece, the smaller the magnetic induction intensity. Therefore, when designing the magnetic parameters of the permanent magnet, the thickness of the specimen must be considered.

1.2.1.2 The Component of Material

Because of the change of carbon content, alloy component and impurities, the magnetic permeability of the material is changed, and the magnetization of the material is affected. By using a strong external magnetic field, the influence can be eliminated by saturation magnetization. At this time, the change of the magnetic permeability will lead to inconsistency of the leakage magnetic signal. So it is difficult to effectively evaluate and analyze the testing signal.

1.2.1.3 The Coupling Loop

In a magnetizing system, steel brushes are used to couple the magnetic flux to the measured material. The coupling efficiency between the magnetizer and the measured material will have a certain effect on the magnetic induction intensity in the material. Short steel brush can provide higher coupling efficiency and higher magnetic induction strength, but longer steel brush can improve the pass-through ability of the inner inspector of pipes. If the applied magnetic field strength is only slightly higher than the saturation level, then the reducing coupling efficiency may decrease the magnetization of the material under saturation.

1.2.1.4 The Space between Magnetic Poles

The shorter distance between the magnetizing poles can produce a higher magnetization, but this will lead to a decrease in the uniformity of the magnetic field. A longer distance between the poles can provide a relatively uniform magnetic field, but it will reduce the magnetization effect because the magnetic reluctance between the two magnetizing poles will increase with the increase of the magnetic pole spacing, so stronger magnets are needed to make the material to reach saturation. Therefore, it is necessary to select the appropriate magnetic pole spacing aiming at the properties and the thickness of the material.

1.2.1.5 The Remanence

Remanence refers to the residual magnetic field in the material that has been tested by MFL detection in the past. The remanence will affect the current level of magnetization, especially when the magnetizing level is low or moderate.

The remanence will reduce the magnetic induction intensity of the material being tested, thus affecting the detection and quantification of defects. Ferromaterial in magnetizing will show a hysteresis effect. That is, after the magnetic field is removed, the material will remain within a certain magnetic induction intensity. When the material is magnetized again, the magnetization process will start from the remanence, thus a new magnetization curve is produced. The magnetization curve is not only nonlinear but also different from the magnetization each time.

The results of MFL testing for the defects of the same length and width, and different depths are shown in Figure 1.8, which shows two important changes. First, after each testing, the amplitude of the MFL signal is reduced. Second, with the increase of the number of detection, the amplitude of the MFL signal will be smaller and smaller. Because after many times magnetizing, the tested material is saturated by the action of the remanent, so even if the external magnetizing field changes, the magnetic induction intensity in the tested material and the MFL of the defect do not obviously change.

When the magnetization of the detector is low, the remanence effect on magnetic induction intensity in the pipeline is larger. After an MFL testing, the magnetic field will remain for a long time in the pipeline. For the next MFL testing, if the magnetization of the detector is low, the remanence will be superimposed on the new magnetizing field and affects the amplitude of magnetic leakage field of defects. If the pipe is magnetized to saturation level, the influence of remanence will be smaller. Therefore, if the detector magnetizes the pipe to saturation, the effect of the remanence can be neglected.

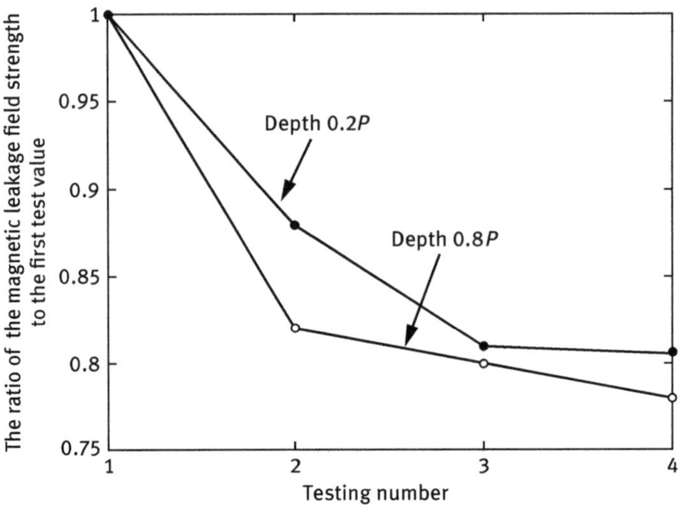

Figure 1.8: Influence of remanence on magnetic leakage field.

1.2.1.6 The Remanence

The internal stress of materials will affect its magnetic conductivity and magnetic induction intensity of the materials and defects. The magnetic field distribution will distort and the testing results will be seriously affected. At the lower magnetizing level, the influence of stress on the magnetic induction intensity of the material is very serious, as shown in Figure 1.9. If the material is magnetized to saturation, the variation of stress can be ignored.

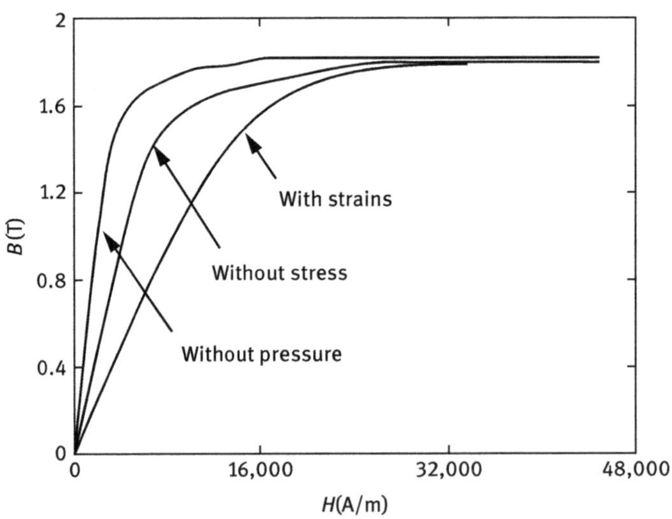

Figure 1.9: Influence of stress on magnetic field.

Based on the above conclusions, for corrosion detection, higher levels of magnetization should be used to let the tested material magnetized to saturation, so as to effectively reduce the influence of the material changes, pole spacing, sweeping of the speed and stress on the magnetizing intensity and magnetic leakage field signal.

1.2.2 The Influence of Testing

1.2.2.1 The Scanning Speed

The increase of scanning speed of the MFL detector will cause the leakage flux of the defect to be reduced obviously, and the degree of influence depends on the scanning speed and the shape parameter of the defect. When the requirement of the defect detection accuracy is higher (e. g. the depth is less than 10 % wall thickness), the change of the speed should be considered. At different speeds, the MFL of defects of different depths are simulated, and the results are shown in Figure 1.10.

The effect of speed on the MFL signal in Figure 1.10 is given as follows.

1. Speed increase will reduce the amplitude of the MFL signal.
2. The speed increase can change the shape of the front and back end of the MFL signal, and their bottom values of the waveform are different.
3. The MFL waveform is moved to the reverse direction of the scanning.
4. Speed has little effect on the magnetic leakage field of large defects whose depths are more than 80% wall thickness, but it has a great influence on the magnetic leakage field of the smaller defects whose depths are less than 20% wall thickness.

In the field of image processing technology, due to the vague images of measured objects that are caused by objects moving or image acquisition instrument movement,

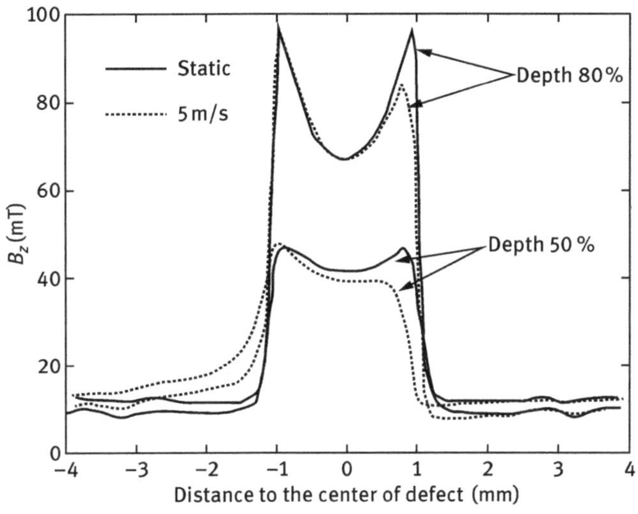

Figure 1.10: Influence of speed on MFL.

two-dimensional (2D) Fourier transform is commonly used to obtain image invariant features, thereby eliminating the effect of moving. Because the MFL signal is one-dimensional time series, we can use one-dimensional Fourier transform to design the signal reconstruction filter and carry out the inversion of the signal waveform so as to eliminate the influence of the speed.

1.2.2.2 The Liftoff of Probes

When the probe is scanning on the surface of the material, the surface roughness of the material, such as the weld seams, defects and so on, will cause the change of the liftoff of probes. Changes of liftoff of the probe induce fluctuations of output signal. It can be seen as an interference signal and it will adversely affect the results of the testing, so it should be suppressed.

In order to study the liftoff effect of the probe on the MFL testing for corrosion defects, a typical corrosion defect was simulated. Simulation model of the structure parameters include the following: the pipe diameter is 1,016 mm, the pipe wall thickness is 14.6 mm, the distance of magnetizing poles is 1,000 mm, the steel brush length is 500 mm, the defect depth is 7.3 mm, the width and length are both 10 mm, the probe liftoff values varies from 0.5 to 16 mm. Figure 1.11 shows the relationship between the value of the liftoff and the peak–valley value B_{zp-p} of the MFL.

From Figure 1.11, the peak–valley value of MFL decreases rapidly with the increase of the liftoff value of the probe. Therefore, in order to improve the detection sensitivity, the value should be small. However, if the detection sensitivity of the probe meets the requirement of need, the selection of the liftoff value should be to minimize the detection error.

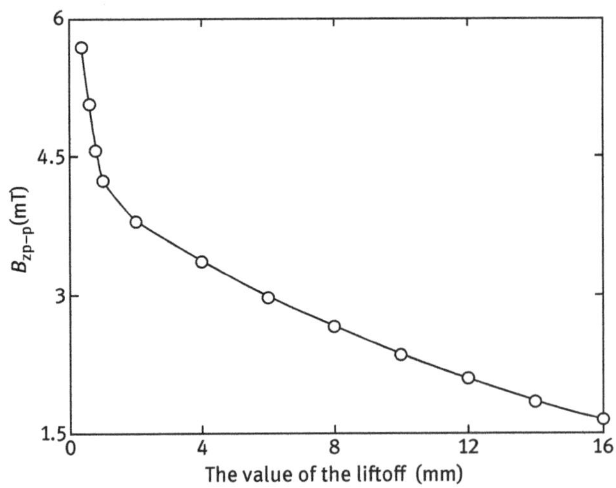

Figure 1.11: The relationship between the value of the liftoff and the peak–valley value B_{zp-p} of the magnetic flux leakage.

1.3 Research Status of Defect Quantification Method of Magnetic Flux Leakage Testing

1.3.1 MFL Detection Theory and Technology Development

In 1934, the world's first MFL testing equipment named Magnaflux A-300 with multidirection magnetizing was born. MFL testing technology started the industrial application. After more than 80 years of continuous development, MFL testing has been applied from defects detection of steel material and gradually extended to the oil and gas pipeline defects detection, rail defects detection, tanks floor inspection, cables of bridges detection and wire rope detection [5–16].

1.3.1.1 Magnetic Flux Leakage Theory and Calculation Method

Fundamental theory research of MFL testing based on analytical method has been more perfect. Mandache and Clapham [17] established an analytical model of MFL for the detection of surface crack defects, which can accurately predict the MFL signals. Melikhov et al. [18] deduced the analytical expression of the surface defect of any angle, which can realize the fast calculation of MFL detection signal. Kang et al. studied the magnetic mechanism of MFL detection and analyzed the complete process of leakage signal from the ferromagnetic material. Based on the principle of magnetic dipole, researchers of the group established the MFL signal calculation model under arbitrary direction of defects and the scanning direction of the sensor and analyzed the influence of the direction of defects and scanning direction on the MFL signal [19, 20]. Dutta et al. [21] established an MFL analytical model based on the magnetic dipole. Based on the model, the spatial-scale invariance of the MFL signal, the influence of liftoff to the signal received and the sensitivity of the signal to the defect parameters were studied. Based on the theoretical calculation model, Mandal et al. [22–25] studied the signal characteristics of MFL in 50 % depth electrochemical corrosion pits of 9 mm steel pipe wall, and the influence of the circumferential stress caused by the pipeline pressure on MFL signal. The analytical method has high computational efficiency. However, it is only applicable to the defects with regular shape, which is difficult to be applied to the MFL signal calculation of irregular shape defects in the actual oil and gas pipeline.

Since the 1970s, the numerical calculation method has been widely used in the field of electromagnetic field, which has greatly promoted the development of MFL testing technology. A lot of research works by finite element method were developed for MFL static and transient simulation model of oil pipeline defects; the applicable conditions were derived by comparing the calculation error of two static simulation models; and a three-dimensional (3D) integral equation of the MFL defects testing was established, which can calculate the defect MFL signal effectively. Kovács [26] obtained the MFL simulation signals by using the finite element method and compared them with the testing signal. Altschuler and Pignotti [27] established a nonlinear numerical model for MFL testing of pipeline defects and studied the influence of the wall thickness, the defect size and the magnetizing field strength on the MFL testing signal. Al-Naemi et al. [28] summed up various finite element model method for MFL testing system and compared the signals of 2D model and 3D model, confirmed the reason of the 2D model error is that the defective area is more likely to be supersaturated. These numerical methods can realize the accurate calculation of MFL signals of defects with arbitrary shapes. However, these methods have the disadvantage of large computation load. Therefore, Zhang et al. [29] set up an equivalent linear model for prediction of MFL signal according to the level of magnetization, which classified the nonlinear ferromagnetic material into different regions, given suitable permeability values. The equivalent model can improve the computational efficiency, but it reduces the original accuracy of the numerical calculation method.

The application of numerical calculation method has also promoted the study of the influence of the magnetization, the defect sizes and the residual stress on the MFL signal. Usarek et al. [30] studied the influence of the geometric size of the defect and the elastic deformation caused by the residual stress on the MFL testing signal by the finite element simulation. Leonard and Atherton [31, 32] found that the changes of oil and gas pipeline pressure, bending stress and residual stress will lead to the changes of the MFL testing signal, and through the finite element calculation confirmed the stresses would result in anisotropic characteristics of the pipe wall material. Babbar et al. [33, 34] studied the effect of defect size and residual stress on the MFL of steel plate defect using 3D finite element simulation. Ivanov et al. [35] proposed a two-phase finite element modeling method of mechanical damage to the pipeline MFL testing, which first derived the stress distribution from the pipeline structure model and established the mapping relationship between the stress levels to permeability, eventually embedding the pipeline stress distribution into the finite element calculation model. These studies reveal the influence of magnetization, residual stress and other factors on the MFL testing signal. However, the corresponding correction method for the quantitative inversion of MFL testing has not been put forward.

In the process of MFL testing, the eddy current caused by the velocity effect was also studied by numerical calculation method. Three-dimensional transient finite element simulation model of MFL was established, and the characteristics of the eddy current generated by high-speed MFL were studied. The change of magnetic strength and permeability of ferromagnetic material in high-speed MFL were analyzed [36]. The maximum effective testing speed of MFL detection is proved to be 200 km/h [37, 38]. Through the finite element simulation of different lengths of magnetic yokes, Norouzi and Ravanbod [39] proposed a method to reduce the eddy current caused by moving detector through the reasonable design of magnetic yoke. Based on the 3D finite element simulation, Park and Park [40] proposed a compensation method of the distortion of MFL induced by the eddy currents. On the whole, the research of the present stage is mostly focused on the principle and characteristics of eddy current, and the effect of eddy current on the MFL testing signal and the quantitative inversion results of defects have not been further studied.

1.3.1.2 MFL Signal Processing Technology

The theoretical research level has been improved; at the same time, the processing technology of MFL testing signal, such as compression, noise reduction and correction, has also been developed rapidly. A tree-weighted MFL testing data lossless compression algorithm based on context, which had higher compression efficiency compared with the traditional Lempel-Ziv-Welch compression algorithm [41] was proposed. Afzai and Udpa [42] proposed a method based on adaptive filtering and wavelet technology for MFL testing signal denoising of seamless steel pipes, which has high computational efficiency and data independence. Park and Park [43] proposed a new method to optimize the size of the permanent magnet and developed a new type of sensing system that can test small-depth defects and improve the sensitivity of MFL testing. A 3D magnetic sensor model with nonlinear proportional coefficient was established, which could effectively restrain the nonlinearity of the magnetic sensor and then the calibration method of the 3D magnetic sensor was put forward [44]. The influence of mechanical boundary conditions and material parameters on the sensitivity of the magnetic sensor was studied, and a new method for the calibration and optimization of magnetic sensor was presented [45, 46]. A compensation method of the liftoff of the MFL probes was provided, which was based on distributed Hall element structure, and the Hall device is divided into several Hall blocks distributed on the surface of a silicon wafer, and then adjusted the Hall block number and the reference current level to meet the needs of the liftoff value, the offset of the liftoff, the SNR and other aspects of the request [47]. These signal processing technologies have a great impact on MFL testing signal integrity and SNR, which can be used to improve the quantitative inversion results of MFL testing.

1.3.1.3 Novel Sensor and Testing Method

The development of high-performance magnetic sensor has promoted the development of MFL testing technology. Pelkner et al. [48, 49] used the three-axis giant magnetoresistance effect (GMR) sensor with high sensitivity for MFL testing, and the minimum testing depth of the defect on the specimen was 44 μm. Tehranchi et al. used the garnet magneto-optical films as sensors for surface defect MFL testing and researched the sensor output changes with the size of defects. The experimental results showed that the sensor has application prospect in the high-precision magnetic imaging of micro-defects [50]. At the same time, the group used binuclear giant magnetoimpedance (GMI) sensor based on Co amorphous wire to test surface defect and found that there was a good correlation between the output of the sensor and the size of the defect [51]. A static axial magnetizing field was established to satisfy both the MFL and the magnetostrictive guided wave testing requirements. A compound sensor that could be used in MFL testing and magnetostrictive guided wave testing was developed, and it was successfully applied to the bridge cable disconnection testing [52].

Angani et al. [53] used linearly integrated Hall and GMR sensor array to measure the MFL field and tested the internal impurities in the cold rolled steel.

At the same time, some researchers have proposed a number of new MFL testing methods and carried out the test and verification. Nara provided the ferromagnetic pipeline MFL nondestructive testing method based on discrete Fourier coefficients. MFL signal discrete Fourier sine and cosine coefficients were calculated, then the defects were evaluated by the ratio and the square root of the two kinds of coefficients [54, 55]. Jun et al. [56] detected the defects of austenitic stainless steel using eddy current and MFL testing methods by linearly integrated Hall sensor array. Sun et al. proposed a unidirectional magnetization MFL testing method for omnidirectional defect that could effectively detect defects parallel to the magnetizing direction. Compared with the traditional multi-magnetization method, the method of magnetizing structure has been simplified significantly [57]. A high-speed MFL testing method based on the orthogonal magnetization was proposed, which is only needed to scan the steel pipe along the axial line, without the need of complicated spiral scanning [58]. A new MFL testing method based on near-zero reference magnetic field was proposed, which could effectively test the MFL of defects under the large liftoff, thus reducing the noise and wearing of the probe [59]. Sophian et al. [60] proposed a pulsed magnetic leakage testing method, using the frequency characteristics of the square wave to get the different depth information inside the material, which had a good testing effect of near-surface micro-defect.

1.3.2 Research on Quantitative Inversion Technique of Defects

With the rapid progress of theory and corresponding technology, the quantitative inversion technique of MFL testing has been developed rapidly [61, 62]. In general, the quantitative inversion technique of MFL testing has experienced the development process from simple to complex, from qualitative analysis to quantitative analysis. The research on the quantitative inversion technique of MFL testing can be divided into three stages.

1.3.2.1 Simple Quantitative Judgment

In the early MFL testing, a simple threshold was carried out through the MFL testing signal, in order to quantitatively determine whether the defect was beyond the limit [63–65]. General magnetic potential equation of the MFL static magnetic field was deduced, 3D finite element simulation model of typical defects was established, and then the relationship between defect parameters and MFL testing signal was studied [66]. The characteristic of circumferential signal in MFL testing of pipeline was analyzed, and the mapping relationship between the parameters and the defect size was established, which could quantitatively identify the geometric parameters of the defect [67, 68]. Saranya et al. [69] compared the MFL defects' recognition methods of region growing algorithm, minimum error thresholding algorithm and morphological segmentation algorithm, and found the region growing algorithm was the strongest on different defects.

These early quantitative methods, although they can determine the relative size of defects, however, the recognition accuracy is low, and the dependence on artificial experience is large.

1.3.2.2 Equivalent Size Quantization

Since the 1990s, researchers have begun to study the methods of quantifying the length, width and depth of defects. In this stage, the statistical recognition and artificial neural network technology in the field of quantitative inversion of the MFL testing has been widely used.

Qi [70], based on the finite element method, researched the influence of pipeline MFL defect size, pipe material, testing speed, residual stress on the testing signal, extracted MFL signal feature value using multivariate nonlinear regression and statistical recognition method, and quantitatively identified defect length, width and depth of equivalent size in the allowable error range. Saha et al. [71] studied the analysis method of the radial MFL testing signal and summed up the empirical formula of the size quantization of the metal loss defects by the radial MFL testing signal. Schifini and Bruno [72] established a nonlinear 3D finite element model, which could inverse the MFL signal to the length, width and depth parameters of the surface rupture defects, and its reliability has been verified by the test results. Mukhopadhyay and Srivastava [73] defined defect characterization parameters and MFL signal characteristic parameters, using wavelet decomposition and reconstruction to reduce the noise of the original signal, and then put forward the complete classification and quantification method of metal loss defects of buried pipeline MFL testing. Reimund et al. proposed a quick estimation method for defect equivalent length, width and depth, which used magnetic dipole model of the semianalytical calculation of magnetic leakage field, and optimized defect equivalent parameters of iterative least square based on trust region reflective algorithm. The relative error of the assessment of defect depth was between 0.6 % and 15.9 % [74].

Ma and Liu [75] proposed a new improved radial basis function neural network (IRBFNN) algorithm, which could be used to quantify the size of pipeline corrosion defects by using the local approximation ability of radial basis function neural network (RBFNN) and the self-organizing ability of the immune algorithm. Abe et al. [76] used amorphous resistive sensor for testing pipeline circumferential and radial MFL components and then trained neural network to predict the cross-sectional width and depth parameters of the pipe hexahedron defects with better prediction accuracy. Amineh et al. [77] tested the circumferential MFL signal of steel pipe defects and quantitatively evaluated the shape parameters of the rectangular and cylindrical defects using the spatial mapping method. The method was extended to the application of quantifying the cracks of the pipe wall; the angle, equivalent length and depth of the crack could be obtained; and the error was very small [78].

The relative size of the defect can be assessed by the method of equivalent sizing of MFL testing signal. However, the profile of the defects and the developing trend still cannot be given.

1.3.2.3 Defect Profile Reconstruction

The inversion method of the defect profile of oil and gas pipeline MFL testing has been studied since 2000, in order to reconstruct the complex shape of the defect. At the present stage, the defect profile inversion technique is mainly focused on the reconstruction of the defect 2D section profile, and a series of technical problems still exist in the inversion of the defect 3D profile.

At present, the inversion method of defect profile can be divided into two categories: direct method and iterative method. Direct method establishes the mapping relationship between the MFL testing signal and the defect size directly, and then used to solve the inverse problem. Iterative method is on the basis of solving the forward problem and feedback to solve the inverse problem. The method starts with the initial defect shape, predicts the MFL signal by solving the forward problem and constantly revises defect shape to reduce the prediction error between MFL signal and the tested MFL signal.

The common tools to realize the direct method include neural network, support vector machine, affine projection algorithm and so on. The mapping relationship between the MFL testing signal and the defect 2D section profile was constructed using support vector machine method. And based on the simulation and the tested MFL data, it has a better mapping accuracy [79]. Based on the support vector machine, the classification and the parameters of the defects of oil and gas pipeline MFL testing were classified and the classification accuracy of the defect was 92.7 % [80]. The state space model between the defect profile parameters and the MFL signal was established, and then the improved particle filter algorithm was used to reconstruct the defect profile [81]. An inverse mapping method from the MFL signal to the defect profile was proposed. The defect length and width were realized, respectively, through selectively applying time-varying or non-time-varying wavelet mapping model in different subregions [82]. A defect profile inversion method based on multipower affine projection algorithm was proposed. After the weight vector of the algorithm was amended and adjusted in the database of defect MFL signals, defect profiles could be predicted quickly [83]. Based on the direct method, such as neural network, support vector machine and so on, rapid prediction of defect profile parameters can be achieved. However, it has great dependence on the training samples.

In the iterative method, the forward problem is solved by numerical method. The iterative correction of the defect profile is realized by the gradient optimization algorithm, genetic algorithm and so on. Priewald et al. proposed an iterative inversion method based on finite element method. The defect model was updated by the Newton Gauss optimization algorithm, and the 2D profile of the defect section could be obtained by the MFL testing signal [84]. Hari et al. [85] proposed a simplified finite element forward model, which was embedded genetic algorithm to MFL 2D profiles of defects-iterative inversion process and improved the speed of finite element calculation and defect profile inversion. A defect profile inversion method of MFL testing based on efficient management of particle swarm optimization algorithm was also put forward. By using three correction models to speed up the algorithm correction speed, and by automatic particle choice way to improve search probability optimal profiles of defects, the computing time of the method could be reduced 20–30 % compared to the conventional adaptive particle group optimization algorithm [86]. Also, an improved finite element neural network was proposed, which could be solved by using the finite element calculation model in the neural network structure [87]. These numerical methods are used to solve the problem of MFL forward problem, and the problem of low computing efficiency is inevitable. It is still difficult to be applied to the 3D profile inversion of MFL testing.

In the iterative method, the forward problem can be solved by using neural network. Ramuhalli et al. [88] obtained the mapping relationship between the MFL signal and the defect profile by training the neural networks, which was embedded in the iterative algorithm to obtain the 2D section profile of the defect. Since then, the group has improved the original iterative algorithm and embedded the forward and reverse neural networks in order to improve the efficiency of the algorithm [89]. Researchers have proposed an iterative inversion algorithm of the defect 2D contour, which uses RBFNN as the forward model in prediction defect MFL signal and uses genetic algorithm to solve the optimization problem in the iterative inversion process and to get rid of the local minimum problem in the gradient descent method [90]. An MFL signal calculation model based on the finite element neural network was proposed. By embedding the finite element model into the neural network structure, the finite element calculating speed of the MFL signal could be accelerated [91]. Adaptive wavelet was used to reduce the dimensionality of the MFL data, and then the iterative inversion algorithm based on RBFNN was constructed for 3D contour shape of defects [92, 93].

In addition to the direct method and the iterative method, the researchers also put forward some equivalent or simplified methods for the reconstruction of the defect profile. A method of reconstructing the defect shape by successive approximation of MFL testing signal was adopted, and it was applied to the reconstruction of three regular-shaped defects of the semicircle, triangle and rectangle [94]. Li and Lowther proposed an MFL testing defect recognition based on the topological gradient method. Topological gradient image of defects was obtained by 2D and 3D finite element models, then the optimization process of defect shapes could be accelerated by the topological gradient, and the shape and size of defects could be identified [95]. Ravan et al. proposed an MFL defect profile inversion method. First, the defective opening shape was estimated based on Canny edge detection algorithm, and then the defect depth contour was calculated by space mapping method to realize the equivalent contour evaluation of metal loss defects [96]. The inversion method was also used to evaluate the simplified target parameters of long and narrow opening cracks, including the number, position, angle, length and depth of the crack [97]. In general, these equivalent or simplified methods can be used to obtain the equivalent contour of some specific kinds of defects, but not the real contour.

1.3.3 The Problems of Defect Quantitative Inversion in MFL Testing

At present, the quantitative inversion technique of MFL testing can realize the quantization of the equivalent length, width and depth of the defect and can reconstruct the 2D profile of the defect section. However, there still exist a series of difficult problems in the application of defect reconstruction of 3D contour reconstruction. The causes of this dilemma can be summarized as two aspects: testing signal source and quantitative inversion method.

1.3.3.1 The Problem of Testing Signal Source

MFL testing at this stage is mostly based on axial or radial component of the MFL field, and there is a serious lack of defect information. The MFL detector mostly magnetizes the pipe along the axial direction of the pipeline, so the axial or radial MFL signal component with large amplitude is generally used as the testing signal. However, the circumferential component of the MFL signal is often neglected because of its relatively small amplitude, which leads to the lack of MFL testing signal.

In fact, the theoretical study of MFL testing shows that the circumferential component of the MFL field contains useful information that reflects the specific profile of the defect. Siebert and Sutherland [98] studied the circumferential component of the MFL testing signal, and it was proved that the circumferential component could be used in conjunction with the radial and axial components to predict the length and width of the defect. Wilson et al. [99, 100], based on finite element simulation of 3D MFL signal, confirmed that the circumferential component of the MFL testing signal contained the useful information of the defect shape and direction, which could improve the quantitative results of the narrow axial defects and irregular defects. Dutta et al. [101] studied the characteristics of 3D MFL signal, confirmed the circumferential sizes information of defects contained in the circumferential component of the MFL field rather than in the radial and axial components, and then put forward to determine the location of the defects using the circumferential component of the MFL field.

The experiments of MFL testing also proved that the 3D MFL testing, fully utilizing the 3D component of the MFL signal, could significantly improve the quantitative inversion results of the defect. Based on finite element simulation, the relationship between the MFL signal and parameters of pipeline defects was researched and artificial defect parameters were identified using the back-propagation (BP) neural network. When the axial and radial MFL signals were used at the same time, defect parameter identification results of BP neural network would be improved significantly [102]. The characteristics of 3D pipeline MFL signal were studied. It was found that the radial, axial and circumferential components of MFL signal, each having advantages and disadvantages on defect recognition of width, length and depth, confirmed that the 3D MFL testing could test nonconventional defect which was difficult to be tested in one-dimensional MFL testing [103]. Three-dimensional MFL testing could significantly improve the quantization precision of pipeline metal loss defects. Kim et al. used integrated Hall and GMR sensor array to test 3D magnetic field component in high resolution and realized the real-time visualization of the magnetic field in the 2D plane. It was proved that the 3D MFL testing in the field of defect inversion had great potential application value [104].

The results of both theoretical research and physical experiment show that the 3D MFL testing signal contains more sufficient defect profile information than one-dimensional MFL signal.

1.3.3.2 Problems Existing in the Inversion Method of Defect Quantification

The existing MFL testing method has a serious shortage in accuracy, speed and practicality, and it is difficult to realize the reconstruction of the real 3D profile of defects. There are low precision, slow speeds or low practicability in the two categories of existing direct method and iterative method of the defect profile reconstruction, which is difficult for the 3D contour inversion of defects and needs further optimization and improvement.

In the existing defect profile inversion method, the mapping relationship between the MFL testing signal and the shape of the defect is established by the direct method based on the defect sample library. These methods require a large number of defect samples for training, and there are overreliance on the consistency of the training samples and accuracy, lack of expansion and low quantitative accuracy of the complicated shape of defects. In actual use, the effectiveness of the mapping method can be greatly reduced if the actual defect appearance and the shape are different from the training defects. In order to overcome this shortcoming, it is needed to increase the variety and quantity of training samples, which are difficult to be realized in practice.

In the iterative method, the forward problem of the analytical method, numerical method and neural network method have their own shortcomings. Analytical method is simple and computationally efficient. However, because of the complexity of the shape of the defects and the arbitrary nature of the electromagnetic characteristics of the pipe material, such as the nonlinear and so on, its application in practice is not realistic. Numerical solution method is effective and accurate, but its calculation efficiency is not high, especially in the calculation of small defects. Compared with the analytical and numerical methods, the neural network method is faster. However, it has a large dependence on the training defect samples and cannot effectively predict the leakage field distribution of arbitrary complex contour defects.

At the same time, the optimization algorithm that is commonly used in the iterative method is the gradient optimization algorithm. Although the gradient algorithm can be fast iterative optimization, there are several shortcomings to be overcomed, which cannot be applied to the quantitative inversion of defects. In convergence, due to the fact that MFL defect profile inversion problems usually have multiple local optimal solutions, without sufficient prior information and appropriate initial guess solution, the gradient optimization algorithm is easy to fall into local optimum and difficult to converge to the global optimal solution. At the same time, the gradient optimization algorithm needs to solve the first- or second-order derivative of the objective function, in the application of the contour iterative solution of MFL testing; it will bring an additional amount of computation. In addition, in order to reduce the calculation error caused by the ill-posed inversion defect contour, additional regularization processing should be processed in the gradient optimization algorithm. And this will significantly increase the computation workload. In fact, defect inversion results of the practical MFL testing have shown that the gradient optimization algorithm of simple defects inversion is more effective and it is not ideal in the stress corrosion defects with complex contours [105].

2 Testing Signal Processing Method

2.1 Data Acquisition and Storage

2.1.1 Data Acquisition of Magnetic Flux Leakage Testing

In magnetic flux leakage (MFL) testing [1], the data collected includes magnetic data and nonmagnetic data. Magnetic data refers to the data tested by the magnetic sensor, which contains defect information, accounting for the majority of the data obtained by the detector. According to the different application, the number of magnetic sensors will have a corresponding change. Nonmagnetic data refers to the scanning speed, pressure, temperature and other parameters. The sampling frequency and accuracy of nonmagnetic data are much lower than those of the magnetic data.

Due to the large number of defect testing sensors, not each sensor is equipped with a separate ADC (analog-to-digital converter). Multiple switches are usually used for signal acquisition of different sensors. To 48″ diameter oil gas pipeline defect detector, for example, the data acquisition module connects 600 ID/OD (inside diameter/outside diameter) sensors through 10 signal concentrators, which produces one analog output with every 60 ID/OD sensors. The 60 ID/OD signals and the 600 MFL signals compose 660 analog signals to be collected through the analog-to-digital (A/D) conversion of the detector. Detector is connected with 660 outputs of the sensors by 10 signal concentrators, in which each signal concentrator can connect the analog output of the 66 magnetic sensors through internal analog multiplexer switch. The system relies on the field programmable gate array (FPGA) to give the address of the multichannel analog gating switch to select the 66 channel signals.

Assuming the maximum scanning speed of the detector is 6.67 m/s and the sampling space is 3.33 mm, the data of 660 channels is sampled by 10 parallel ADCs, so an ADC conversion must be completed in 7.5 μs. The stability time of multiswitch is 4 μs, so A/D conversion must be completed within 3.5 μs. Taking into account the need for the operation of the system, it is recommended to control the A/D conversion completed in 1 μs and sampling frequency above 1 m/s.

The output voltage range of Hall sensor is 0–5 V, which corresponds to 0–500 Gauss of the defect MFL induction intensity. High-resolution oil and gas pipeline detector requires to test a minimum of 0.5 Gauss magnetic induction intensity changes, so the bits of A/D conversion must be greater than (or equal to) 10. Considering the sampling precision and economy, the 12-bit A/D conversion module is more suitable.

A 48″ diameter oil and gas pipeline MFL detector has 10 parallel sampling channels. The parallel sampling rate of the 10 channels is up to 400 kHz. In the case of 660 channels per sample, the sampling trigger frequency is 6 kHz. Interface control logic is completed by FPGA. The schematic diagram of 660 channel of data acquisition

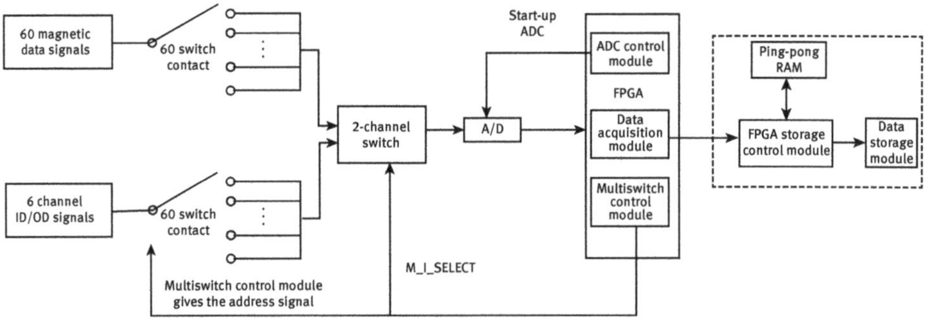

Figure 2.1: Block diagram of data acquisition for oil and gas pipeline detector. ID/OD, inside diameter/outside diameter; ADC, analog-to-digital converter; A/D, analog to digital.

is shown in Figure 2.1, which shows the schematic diagram of the single-channel sampling. Multiswitch control module gives the address signal. The address range of the 60 magnetic data signals is 0–59. The address range of the six channel ID/OD signals is 0–5. Two-choose-one analog switch is used to select signals between the magnetic data and the ID/OD signal.

2.1.2 Organization and Storage of Data

2.1.2.1 Data Storage Formats

In the storing procedure, the data is divided into a fixed size block, and the end of each block has a block tail record [3]. Each data block is composed of several scans of main magnetic channel and ID/OD channel magnetic data and some necessary non-magnetic data. The data block size is determined by the number of the main probe, the number of the ID/OD probe and the number of scans per block.

For 12-bit ADC, in order to save hard disk space, four 12-bit data needs to be combined into three 16-bit data. When the number of the main channels and ID/OD channels is a multiple of 4, the memory edge is aligned. When the number of the main channels and ID/OD channels is not multiples of 4, zero should be filled in the remaining space.

The end of the block is a total of 32 bytes, which is the nonmagnetic data including distance, time, speed, pressure difference, temperature, the system state and mark information. The system state is one of the flag bytes that represents the current state of the detector, including the state information, such as start, stop, overspeed, reset and so on.

2.1.2.2 Data Flow Organization

Data acquisition uses the 12-bit ADC and has10 channel parallel input channels. In the data storage, there needs 16-bit format, continuous serial storage. This requires a reasonable organization of the collected data. The data is organized in FPGA, whose

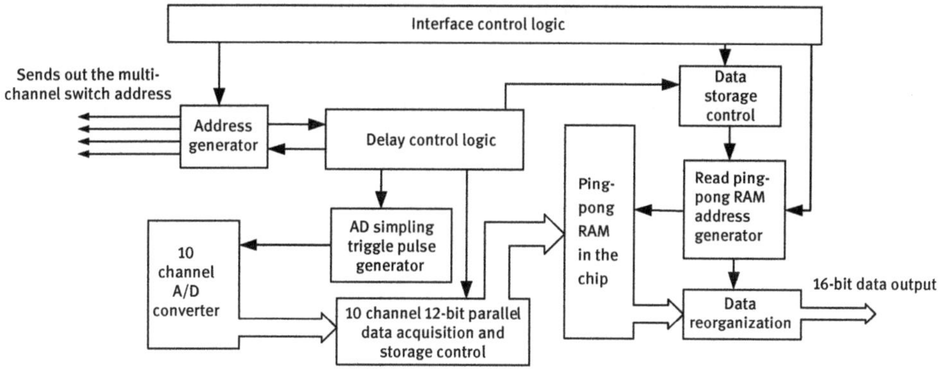

Figure 2.2: Data organization logic control chart.

biggest advantage is the parallel implementation, so that the processing speed is very fast. The logical control of the data flow organization is shown in Figure 2.2.

The multichannel switch address generator inside the FPGA sends out the multichannel switch address, which is used to select different channels. The sampled data is successively stored in the ping-pong RAM in the chip. The reason why the ping-pong RAM is used is to avoid the coverage before the previous data has time to organize the storage. Ping-pong RAM working principle is shown in Figure 2.3. After the writing pointer whites A, data storage control module will start data organization and start to read A data. In the next sampling, the writing pointer will jump to B, so reading and writing ping-pong RAM can simultaneously proceed without data confusion, which improves the processing speed of the system.

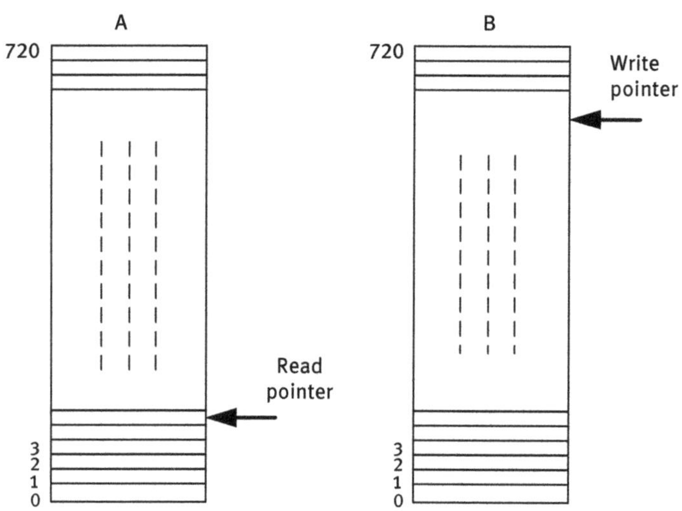

Figure 2.3: Schematic diagram of ping-pong RAM working principle.

In the single sampling process, each ADC will be carried out 66 times. The collected data of the first ADC is sequentially stored into the RAM addressed 0–65 position. The collected data of the second ADC is sequentially stored into the RAM addressed 66–131 position and so on. The collected data of the 10th ADC is sequentially stored into the RAM addressed 594–659 position.

2.2 Data Compression and Noise Reduction Method

2.2.1 Testing Data Compression

The storage hardware of the MFL detector can only provide limited storage capacity and speed [106]. For the defect inspection of large diameter and long distance oil and gas pipeline, whether general or the instantaneous data size is beyond the capacity of the embedded storage system, it is not possible to store the original testing data without any treatment. It is necessary to take corresponding online testing data compression method. Although there are many methods of data compression, it is difficult to meet the special needs of real-time compression in the online operation of the detector. The embedded system requires that the compression algorithm can achieve high compression ratio and cannot be too complex. So it is necessary to find a simple and efficient algorithm for the characteristics of the data.

2.2.1.1 Segmented Threshold Compression Algorithm

Normally, without the presence of any defects, most of the pipeline inspection data is small-amplitude signal that does not contain defects information. MFL features with larger amplitudes only appear near the defect. The core idea of this compression algorithm is to divide original data into successive subsections. Less or no compression is done to subsections with large amplitudes, and more compression is done to subsections with small amplitudes. The dynamic processing of compression ensures high overall compression ratio and preserves ideal signal quality.

The proposed compression process is as follows. First, the original sample data is divided into subsections. The number of data points of each subsection is assumed to be N. Such data set is acquired: $X_1 = \{x_0, \ldots, x_{(N-1)}\}, \ldots, X_i = \{x_{(i-1)N}, \ldots, x_{iN-1}\}, \ldots$. Then a compression threshold TH is chosen. The maximum, minimum and average values of each subsection are calculated. If $\max\{X_i\} - \min\{X_i\} > \mathrm{TH}$, a 4:1 small compression ratio is used (preserve one point for every four points), or else the average value is used as the compressed data, which means an N:1 large compression ratio.

This compression algorithm faces the problem of how to distinguish small compression ratio subsection and large compression ratio subsection during the recovering process. The solution is to insert flags between the compressed subsections. For example, 1-bit flag is used. “0” stands for data compressed using small ratio and “1” stands for data compressed using large ratio. Because the overall compression ratio

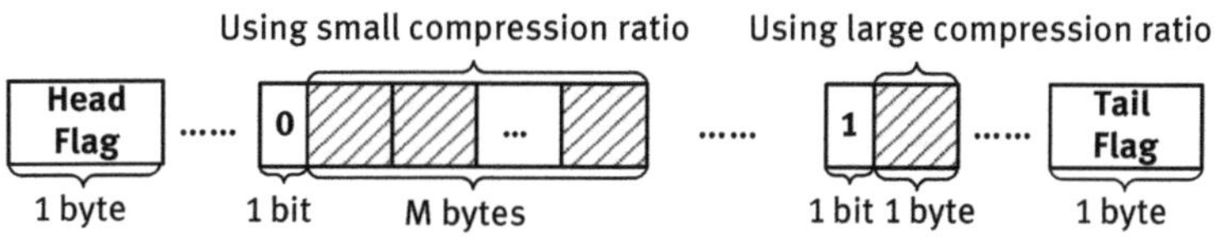

Figure 2.4: Structure of compressed data frame.

cannot be decided in advance, it is necessary to add head flag and tail flag when storing compressed data. The structure of compressed data frame is shown in Figure 2.4. One data point occupies 1 byte storing space in this figure, that is an 8-bit ADC is assumed. Signal reconstruction can be accomplished according to this structure.

Compared with wavelet decomposition and other types of transformation algorithms in frequency domain, the proposed subsection-adaptive compression algorithm has its obvious advantages. The algorithm is simple without complicated multiplication and division operations, and both compression and reconstruction are carried out in order. It can easily be applied to embedded pipeline inspection hardware system due to its good real-time performance.

Because the possibility of finding defects during the process of pipeline inspection is often very small, this algorithm can achieve very high compression ratio on most inspection data. While for useful signal containing defects or abnormal signal with large magnitude, it can store them with high quality, providing enough information for data analysis afterward. The dynamic compression process ensures the validity and quality of compressed data.

Noises in the original signal will affect the effectiveness of the compression. At places where noise magnitude exceeds compression threshold, even without defects, data will be stored as useful signal, resulting in lower compression ratio. So before A/D sampling, the inspection signal needs to go through band-pass filter, to reduce the magnitude of noise.

Another problem is the flags frequently used to divide the compressed data. They are easily confused with the data itself, bringing difficulty to data analysis and system debugging. Especially when data loss occurs at storage module, it will lead to data confusion. The length of the flag can be increased to 4 bit or even 8 bit to enable the data analysis software to better identify a flag and determine the places of data loss and errors. This will also result in lower compression ratio.

2.2.1.2 Experiments and Discussion

For the data compression algorithm, the ultrasonic testing signal is more persuasive than the MFL signal, so the effect of the compression algorithm is explained by the electromagnetic ultrasonic data. The compression algorithm is tested by the typical experimental data of the electromagnetic ultrasonic guided wave of 1 MHz and electromagnetic ultrasonic surface wave of 250 kHz, respectively. The experimental data is

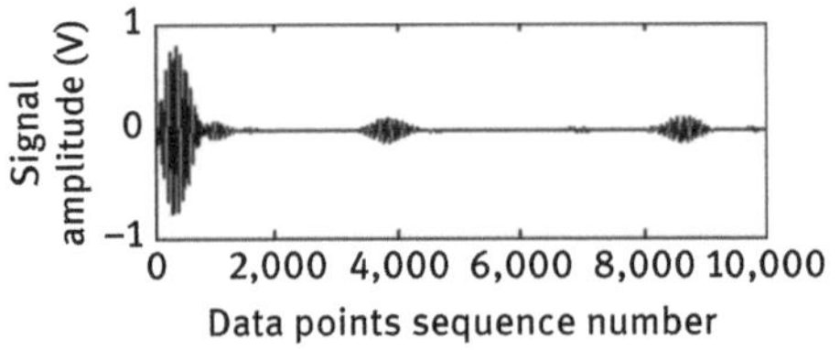

Figure 2.5: Original signal.

obtained through 20 MHz, 12-bit precision data acquisition card, and the compression program is written by MATLAB platform.

First, the 250 kHz guided wave data is compressed. The original waveform is shown in Figure 2.5. The amplitude of the original data is normalized before processing. Whether the choices for subsection points number of original data and compression threshold are suitable or not need experimental comparison and verification. If 20 data points form a subsection, then the extreme compression ratio is 20:1. In the following examples, compression thresholds of 0.05, 0.1 and 0.15 are investigated, respectively, and the reconstructed signals for these different thresholds are shown in Figures 2.6–2.8. If 40 data points form a subsection, then the extreme compression ratio becomes 40:1. Using the same thresholds, the reconstructed signals for this new subsection length are shown in Figures 2.9–2.11. Apparently, the number of subsection points or subsection length cannot be too small, or the compression ratio will be too small. Multiple experiments show that 40 subsection points might be a proper choice. If the number of subsection points is increased further, though

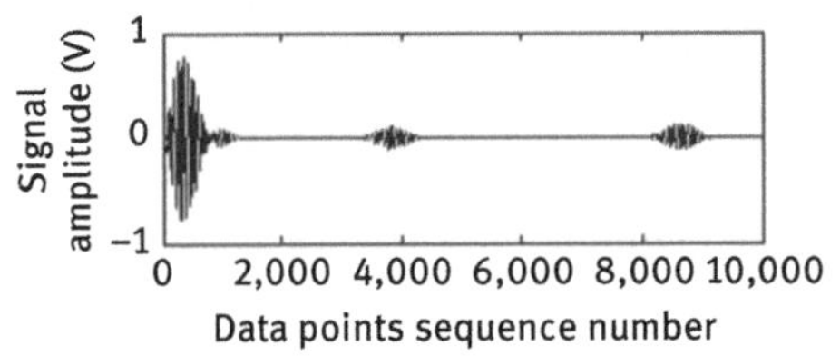

Figure 2.6: Reconstructed signal at $N = 20$, TH = 0.05, compression ratio $R = 11.21$.

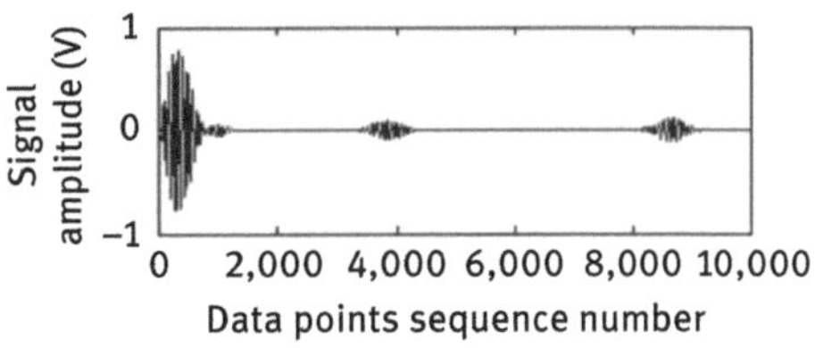

Figure 2.7: Reconstructed signal at $N = 20$, TH = 0.1, compression ratio $R = 14.12$.

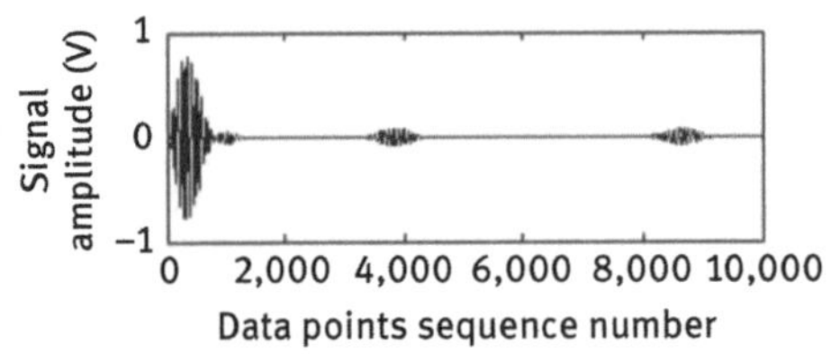

Figure 2.8: Reconstructed signal at $N = 20$, TH = 0.15, compression ratio $R = 16.23$.

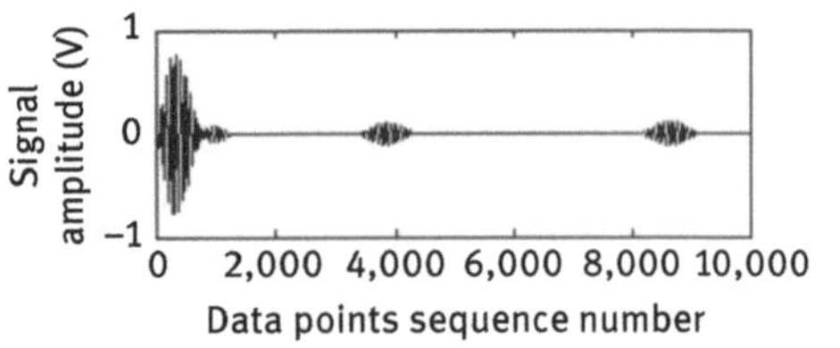

Figure 2.9: Reconstructed signal at $N = 40$, TH = 0.05, compression ratio $R = 18.59$.

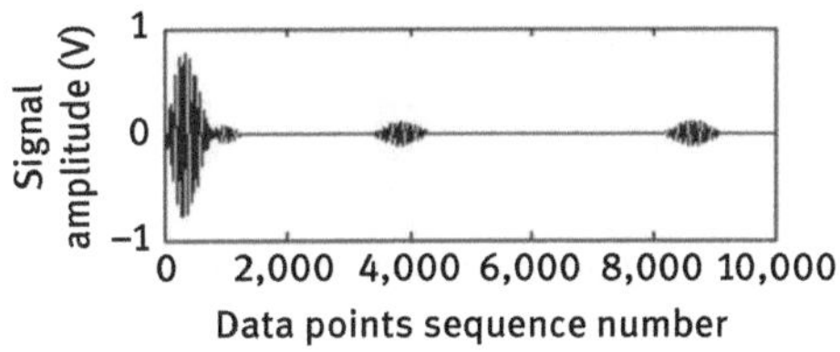

Figure 2.10: Reconstructed signal at $N = 40$, TH = 0.1, compression ratio $R = 22.42$.

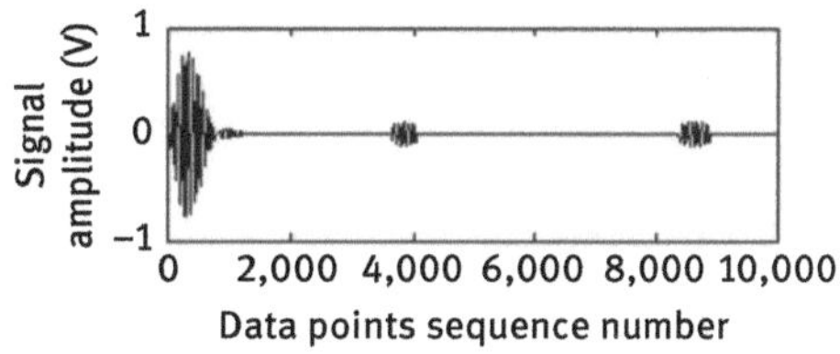

Figure 2.11: Reconstructed signal at $N = 40$, TH = 0.15, compression ratio $R = 24.39$.

the extreme compression ratio is increased, the continuity of reconstructed signals weakens. Obvious data missing and sudden changes appear in some places.

Clearly, the number of subsection points cannot be too small, otherwise the compression ratio is very low. After many experiments, it is found that, for this group of guided wave data, the number of subsection points should be 40. Increasing the compression threshold can improve the compression ratio, but it will damage more details. Increasing the number of subsection points can also improve the compression ratio at the same compression threshold. However, it makes the continuity of the compression waveform change worse.

The surface wave data of 1 MHz is compressed, and the original waveform is shown in Figure 2.12. A number of different subsection points and the compression threshold are also used. The reconstructed waveform of decompression data is shown in Figure 2.13(a)–2.13(f).

From the above experimental results, we can see that the effect of the segmentation threshold compression algorithm for different test data is different. The original waveform of 250 kHz guided wave is relatively clean and there is no large clutter, so it is easy to achieve a higher compression ratio. The experimental data of 1 MHz wave, compared to 250 kHz data, whose amplitude of the signal and noise are larger, contains a greater amount of information, so the compression ratio at the same parameters is smaller.

2.2.2 Noise Reduction Method of the Testing Signal

The typical ultrasonic testing signal is still taken as an example to illustrate the method of the signal denoising. Electromagnetic acoustic transducer (EMAT) is usually excited by high-energy single pulse or sinusoidal burst sequence. Defect

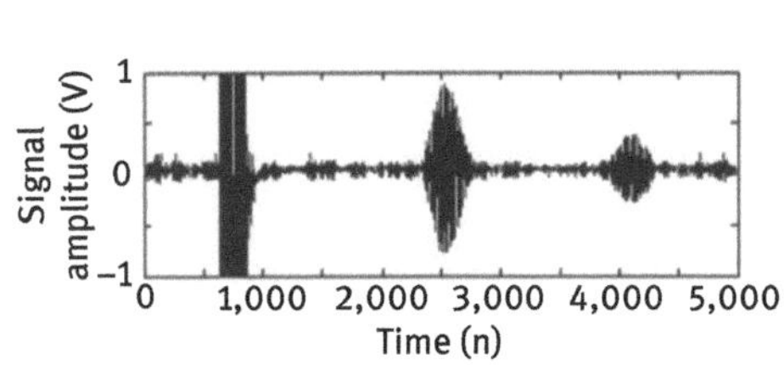

Figure 2.12: Original signal of 1 MHz electromagnetic ultrasonic surface acoustic wave.

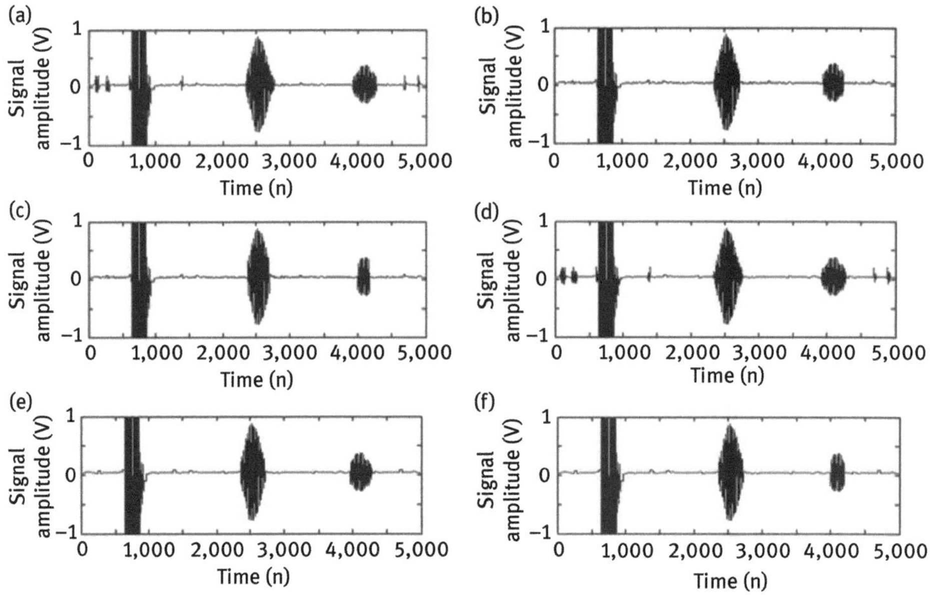

Figure 2.13: Compression waveform of 1 MHz electromagnetic ultrasonic surface wave: (a) $N = 20$, TH = 0.2, compression ratio $R = 4.28$; (b) $N = 20$, TH = 0.35, compression ratio $R = 5.31$; (c) $N = 20$, TH = 0.5, compression ratio $R = 6.19$; (d) $N = 40$, TH = 0.2, compression ratio $R = 6.18$; (e) $N = 40$, TH = 0.35, compression ratio $R = 8.61$; and (f) $N = 40$, TH = 0.5, compression ratio $R = 9.54$.

echo signals are amplitude modulation signals, whose frequency is the same as that of excitation frequency. Due to the low efficiency of the electromagnetic ultrasonic transducer, the receiving signal is usually highly contaminated with noise. The signal-to-noise ratio is not high even after the band-pass filter of hardware circuit. It is necessary to reduce the noise of the data, improve the quality of the signal and increase the accuracy of the data analysis.

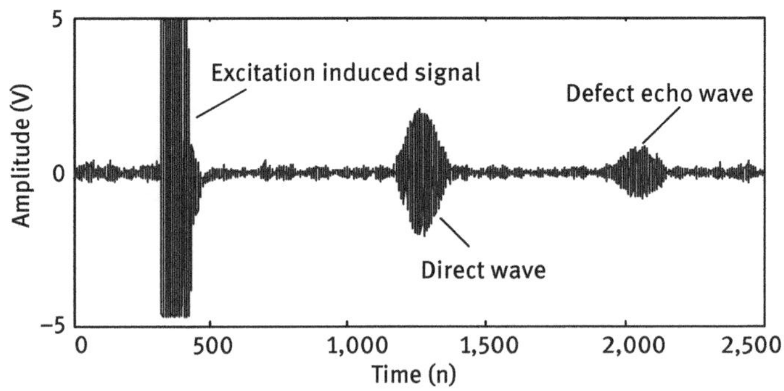

Figure 2.14: Experimental EMAT surface wave data.

Figure 2.14 shows a typical waveform data of 1 MHz experimental EMAT surface wave. A pitch-catch arrangement is used. A 1 MHz sinusoidal burst current drives the transmitter coil to generate an ultrasonic surface wave in an 8 mm thick steel plate. Weak voltage signal induced by the ultrasonic surface wave on the receiver coil is amplified and sampled by ADC. The A/D sampling frequency is 10 MHz, sampling resolution is 12 bits and sampling range is ±5 V.

The characteristic frequency of EMAT is decided by the configuration of the transducer and the frequency of driving current. In this case, it's 1 MHz. The received signal includes useful information on this frequency and noises at multiple frequencies. Generally, the composition of inspection signal $s(t)$ can be given by means of the following expression:

$$s(t) = y(t) + n(t), \tag{2.1}$$

where $y(t)$ is the wanted signal and $n(t)$ represents the noise component:

$$y(t) = u(t) \bullet v(t) = u(t) \bullet \sin(2\pi f_0 t + \varphi), \tag{2.2}$$

where f_0 is the characteristic frequency and φ is the excitation phase angle. From the signal analysis point of view, $u(t)$ is the envelope of the wanted signal and $v(t)$ is the carrier wave. $v(t)$ shows the high-frequency characteristics of the EMAT surface wave. And $u(t)$ contains critical information about the size and position of the defect.

2.2.2.1 Denoising Algorithm based on Envelope Filtering

Due to the fact that the practical testing signal contains a lot of noise, the curve of discrete sampling data is plotted, which presents many burr. It seriously affects the key information identification. The noise has a wide frequency band, and the conventional noise reduction method is often difficult to achieve the desired results.

The envelope regulation technique is good at preserving the main characteristics of the signal. The overall outline of the signal is revealed by the signal envelope, which reflects the tendency of the signal change on a large scale. We use the upper and lower envelopes to regulate the signal itself so as to suppress small-amplitude fluctuations in the signal, maintaining important signal edges at the same time.

The algorithm has two crucial steps. First is getting a smooth signal envelope. Second is using this envelope to modify the original signal properly. A schematic representation of the processing steps performed in the proposed algorithm is presented in Figure 2.15.

A good envelope extraction algorithm must be able to adapt to signals that include a variety of frequency components. Because the original signal frequency components are not pure, conventional envelope extraction algorithms such as Hilbert transform and wavelet transform usually cannot reach the desired results.

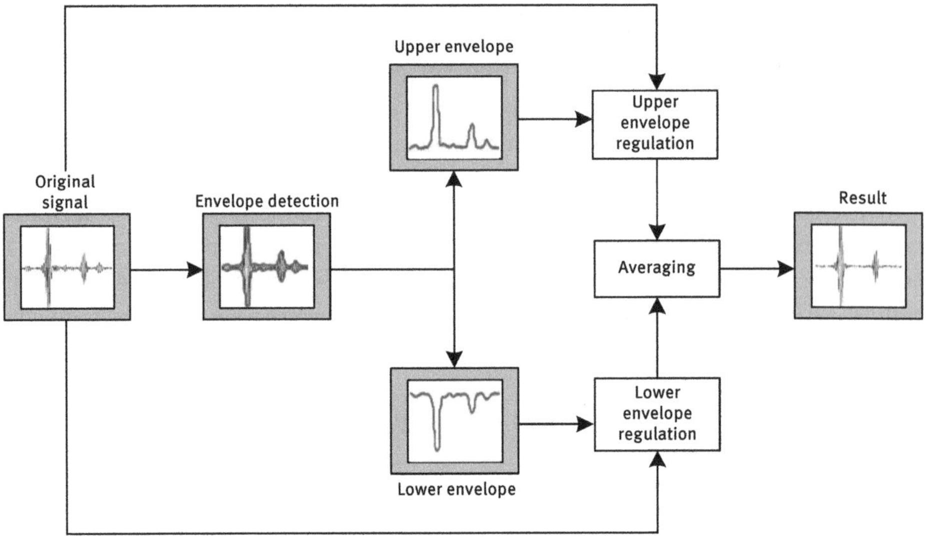

Figure 2.15: The block diagram of the denoising algorithm using envelope regulation.

Interpolation is a classical mathematical method. Compared with other polynomial interpolation methods such as Taylor interpolation and Lagrange interpolation, Hermite interpolation requires not only through the interpolation nodes but also the derivative of the interpolation curve at interpolation node equals to the derivative of the original curve. Hermite interpolation can be used to smoothly interpolate between key points of a specified curve.

We use piecewise cubic Hermite interpolation method to obtain the envelope of the inspection signal. The steps are as follows:

1. Obtain the extreme points (local maximum points or local minimum points) sequence $P = \{x_0, x_1, \ldots, x_n\}$. The extreme points are interpolation nodes. The corresponding function values are $y_0, y_1, \ldots, y_n$, and the corresponding first derivative values are $y'_0, y'_1, \ldots, y'_n$.
2. In the interval $[x_{i-1}, x_i]$, the cubic Hermite polynomial $H_3(x)$, which satisfies $H_3(x_i) = y_i$, $H'_3(x_i) = y'_i$, can be expressed as

$$H_3(x) = y_{i-1}F_0(x) + y_iF_1(x) + y'_{i-1}G_0(x) + y'_iG_1(x), \tag{2.3}$$

where

$$\begin{cases} F_0(t) = \left(1 - 2\frac{x-x_{i-1}}{x_{i-1}-x_i}\right)\left(\frac{x-x_i}{x_{i-1}-x_i}\right)^2, \\ F_1(t) = \left(1 - 2\frac{x-x_i}{x_i-x_{i-1}}\right)\left(\frac{x-x_{i-1}}{x_i-x_{i-1}}\right)^2, \\ G_0(t) = (x - x_{i-1})\left(\frac{x-x_i}{x_{i-1}-x_i}\right)^2, \\ G_1(t) = (x - x_i)\left(\frac{x-x_{i-1}}{x_i-x_{i-1}}\right)^2, \end{cases} \quad x \in [x_{i-1}, x_i]. \tag{2.4}$$

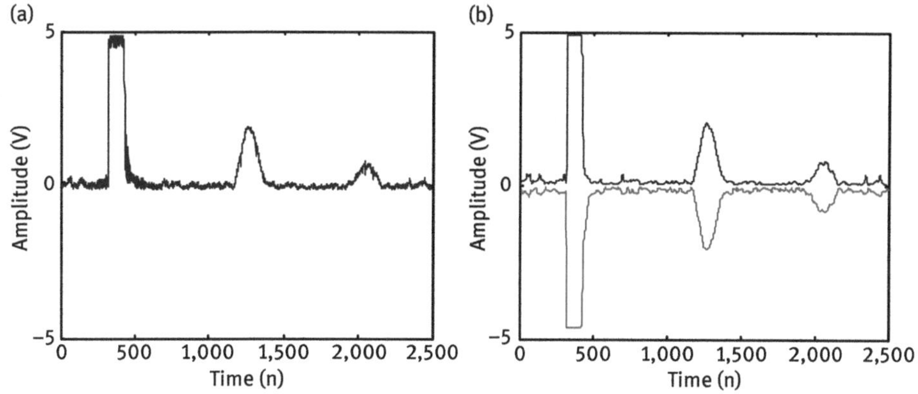

Figure 2.16: Results of two envelope extraction methods. (a) Upper envelope acquired by Hilbert transform. (b) Upper and lower envelopes acquired by piecewise cubic Hermite polynomial interpolation.

Substitute x in the cubic Hermite polynomial with discrete coordinate sequence at intervals divided by extreme points; thus, the interpolation curve is obtained.

Figure 2.16(a) shows the signal envelope acquired by Hilbert transform. Figure 2.16(b) shows the envelope acquired by piecewise cubic Hermite polynomial interpolation. The latter is obviously smoother.

2.2.2.2 Envelope Regulations

The idea of envelope regulation is demonstrated in Figure 2.17. The processing steps are as follows:

1. Obtain the extreme points and the envelope of original signal, as shown in Figure 2.17(a) and 2.17(b). S is the original signal. p_i is the local maximum point, n_i is the local minimum point. Eu is the upper envelope, and Ed is the lower envelope.
2. It is shown in Figure 2.17(b) that the difference between the envelope and original signal is great during some sections such as p_2 to p_3 and n_3 to n_4. The difference is small during some sections such as p_1 to p_2 and n_2 to n_3. Replace the original signal by upper and lower envelopes, respectively, during small difference sections as in eqs (2.5) and (2.6):

$$S_{\mathrm{u}} = \begin{cases} S, & \max\{|E_{\mathrm{u}}(j) - S(j)|\} > \mathrm{TH} \\ E_{\mathrm{u}}, & \max\{|E_{\mathrm{u}}(j) - S(j)|\} < \mathrm{TH} \end{cases}, \qquad j \in [p_i, p_{i+1}], \tag{2.5}$$

$$S_{\mathrm{d}} = \begin{cases} S, & \max\{|S(j) - E_{\mathrm{d}}(j)|\} > \mathrm{TH} \\ E_{\mathrm{d}}, & \max\{|S(j) - E_{\mathrm{d}}(j)|\} < \mathrm{TH} \end{cases}, \qquad j \in [n_i, n_{i+1}], \tag{2.6}$$

where TH is the regulation threshold, S_{u} is the upper modified signal and S_{d} is the lower modified signal, as shown in Figure 2.17(c) and 2.17(d).

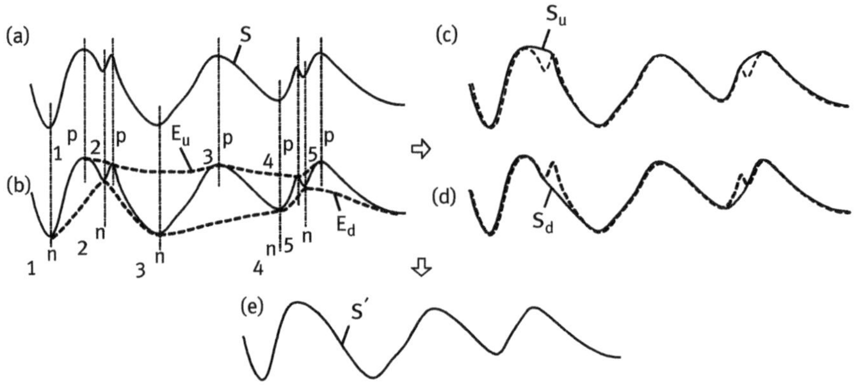

Figure 2.17: Schematic diagram of envelope regulation algorithm: (a) original signal waveform; (b) local extreme points and signal envelope; (c) upper envelope signal regulation; (d) lower envelope signal regulation; and (e) result signal waveform.

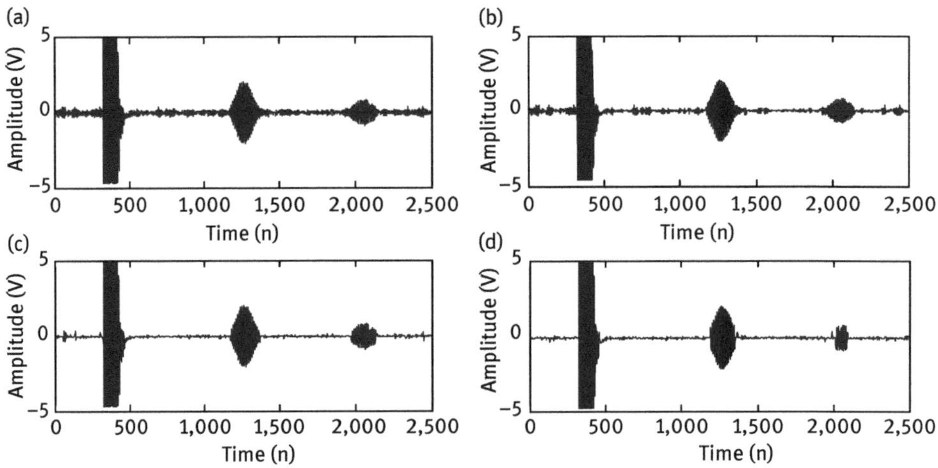

Figure 2.18: Comparison of overall signal waveforms before and after denoising: (a) overall original signal waveform; (b) result signal waveform, threshold = 0.3 V; (c) result signal waveform, threshold = 0.6 V; and (d) result signal waveform, threshold = 1.2 V.

3. The result of envelope regulation is obtained by averaging S_u and S_d, as shown in Figure 2.17(e):

$$S' = (S_u + S_d)/2. \tag{2.7}$$

2.2.2.3 Experimental Results

The algorithm is tested on 1 MHz experimental EMAT surface wave data. Choosing a proper threshold is vital. Little effort is made by the denoising process if the threshold

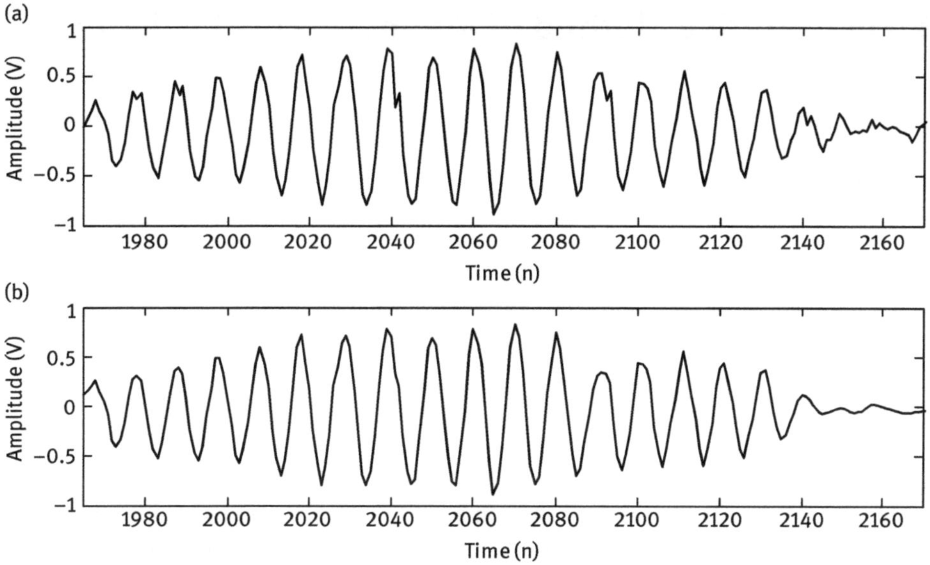

Figure 2.19: Comparison of defect echo waveforms before and after denoising: (a) original defect echo waveform and (b) result defect echo waveform, threshold = 0.6 V.

is too small. But the useful signal is damaged if the threshold is too large. Larger threshold results in more modification to the original signal. Typically, the threshold can be set as 40 % of the useful signal amplitude. The amplitude of defect echo waveform in the original signal is about ±0.8 V. Therefore, the empirically proper regulation threshold is chosen around 0.6 V.

Figure 2.18 shows the denoising results of overall signal waveform at certain regulation thresholds. Figure 2.19 shows the zoom-in results of defect echo waveform at 0.6 V threshold. It can be seen that the overall noise in the signal is reduced. And the sawtooth-shaped noises on the defect echo waveform are also eliminated.

3 Quantitative Method of Magnetic Flux Leakage Testing

3.1 Outline

Defect appearance inversion from the magnetic flux leakage (MFL) signal, including the distinction of defect types, the characteristic analysis and the shape parameters of the defect quantitative analysis, can be attributed to a typical inverse problem of electromagnetic field, and the process is shown in Figure 3.1.

A formula can be obtained from Figure 3.1:

$$y(t) = x(t) \bullet h(t), \tag{3.1}$$

where $x(t)$ represents the input excitation source; $y(t)$ represents the output, that is, the tested signal of the sensor; and $h(t)$ represents the transfer function, that is, the interaction between the magnetizing field and the defect.

The forward problem in the electromagnetic field is the excitation source input and the transfer function as known and the output to be solved. The inverse problem is the output and input as known, to solve the transfer function, or known output and transfer function, solving the input. As for the defect quantification, the forward problem means that the defect parameters and external exciting source are known, and MFL signal needs to be estimated; the inverse problem is according to the tested magnetic leakage field and external magnetizing information to solve the defect parameters.

The forward problem of electromagnetic field can usually be solved by the analytical or numerical method and obtains the only solution. But inverse problems are often ill-conditioned, which is mainly due to the lack of uniqueness and continuity with respect to the defect characteristics.

For an ill-conditioned problem, it can be processed by increasing the information, so the number of the information is very important for solving the inverse problem. The more the information is, the more accurate the solution is. The less the pseudo-solution is, the smaller the possible solution set is. But in the actual testing, the signal can only be obtained by the discrete, finite observation, which is likely to further exacerbate the problem of the inverse problem.

In view of the characteristics of the inverse problem of MFL detection, the existing methods can be classified as follows:

The direct method is to obtain the solution of the inverse problem by solving the Green equation. Because it is difficult to solve the direct method, the indirect method is proposed to solve the defect parameters according to the testing signal. These methods can be divided into three kinds: mapping method, iterative method and signal classification method (Figure 3.2).

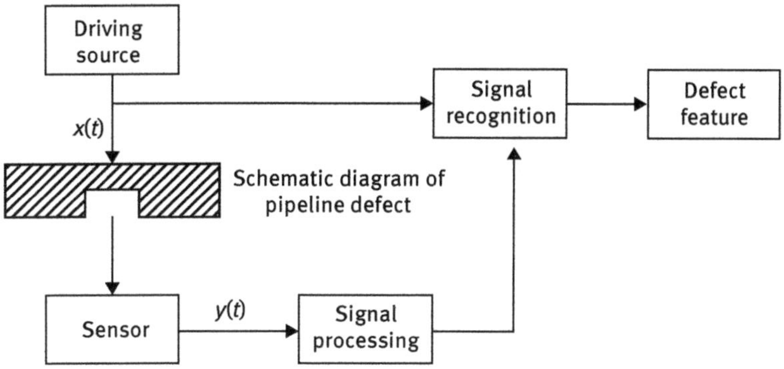

Figure 3.1: Defect quantification process.

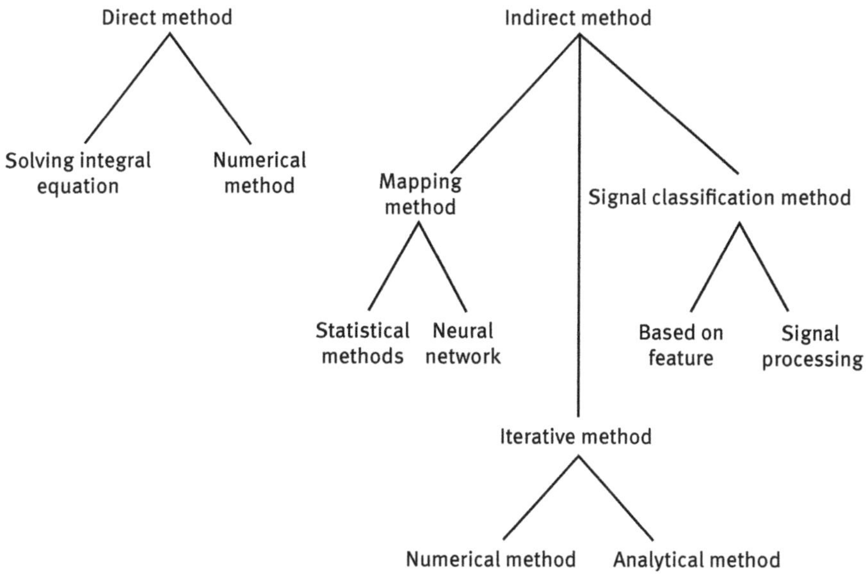

Figure 3.2: Classification of defect quantification method.

The principle of mapping method, also known as pattern matching method, is shown in Figure 3.3. Mapping method needs to establish standard defect feature library X. The standard library Y of the defects signal is established using magnetic dipole model, the finite element model or physical testing to obtain magnetic leakage signal. Based on statistical method or neural network method, the mapping relationship between Y and X is established. By using the mapping relationship, the types of defects and related parameters can be determined from the characteristics of the magnetic leakage field detected.

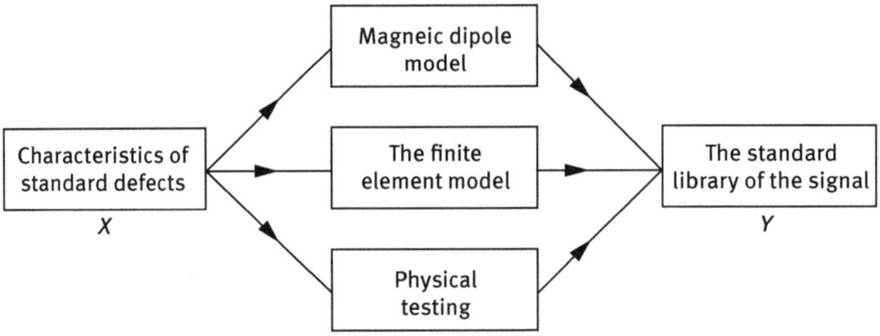

Figure 3.3: Mapping principle.

According to the specific simulation matching algorithm, the mapping method can be divided into statistical recognition-based method and neural network-based method.

At present, the theory and implementation techniques of analog matching algorithm based on statistical methods have been more mature, such as Bayesian modeling method for quantitative identification of wire rope defects. This method is mainly based on the feature extraction algorithm that extracts the most important characteristics of the defects from a large number of signal data and predicts the different types and some characteristic parameters of defects using statistical methods such as threshold method, the minimum hazard rate method, mini/max method, Neyman–Pearson method and so on. It is more successful in the application of standard defects (such as rectangular groove). However, statistical methods do not have adaptive processing capabilities, such as the self-organization of the pattern information, the ability of self-learning, pattern recognition capabilities from signals containing noise and so on. At the same time, this method requires a large amount of testing or simulation data, hence the statistical process is more complex.

At present, the analog matching algorithm based on neural network has been applied in the defect classification of ultrasonic testing and quantitative analysis of eddy current testing. As one of the methods of pattern recognition, artificial neural network is a mode distribution model established on the basis of model transformation, which has the characters of nonlinear mapping and parallel computing ability, and provides a very effective way for the identification of dynamic signal. However, the neural network prediction method of defect shape parameters is still in the research stage, especially in the prediction of real complex defects. Using the above method, the training speed, the prediction accuracy and the consistency of the prediction results cannot meet the actual demand.

The two mapping methods mentioned earlier has a common defect, which is overreliance on the consistency and accuracy of the statistical sample or the training samples. The quantization accuracy of complex structure and the actual defects is low because of its lack of expansion. If the difference between the shape of the actual defect and the defect of the calibration is larger, the effect of these methods will be greatly reduced. In order to overcome these shortcomings, there needs a substantial increase in the number of samples and categories, but this is more difficult to achieve in practice.

Another important issue is that the existing mapping methods are established on the basis of analysis of two-dimensional (2D) problems, for example, the width of the defect is assumed as the same, looking for a mapping relation between the magnetic leakage field and the length or depth of the defect. Two-dimensional method can simplify the research of the magnetic leakage field and help to understand the law of the defect signal changes. In fact, the relationship between the magnetic leakage signals and defect dimensions is nonlinear, and in the actual testing, three-dimensional (3D) shape parameters of the defect are unknown, so it is difficult to predict whether defects rely only on these independent conclusions to judge defect feature.

The process of solving the inverse problem with the iterative method is shown in Figure 3.4. The essence of this method is to solve the inverse problem in the way of solving the positive problem. The positive problem can be solved by analytical and numerical methods.

Third type of indirect method is signal classification, also known as the pattern classification method. The inverse problem is divided into a number of solutions to a limited number of categories. The classification method of MFL signal of corrosion defects is mainly divided into two types. The first method is to use a matched filter, that is, to construct a filter array, and each of which corresponds to a known type of defect. According to the response of each filter in the array, the category of defects is determined. Because the testing signal can fluctuate due to various factors, the matched filter method is easy to get the wrong classification result. Moreover, this method needs to store the whole signal, not to research the characteristics of the signal. Data compression cannot be achieved, which results in analyzing the large amount of data.

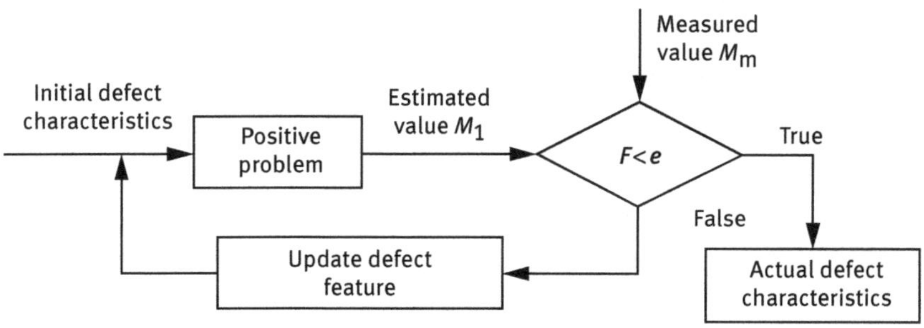

Figure 3.4: Iterative method for solving inverse problem.

Feature-based classification method is to extract the main features (including time domain and frequency domain features), and the feature vector is established. When the most representative features are found, the dimension of feature vector is reduced, and then clustering method or neural network method can be used to classify defects.

Feature extraction has two functions: data compression and variable compression. Signal feature vectors are usually obtained from a large number of testing data. Effective feature extraction method can not only eliminate redundant data and reduce the dimension of feature vectors associated with the signal vector, but also reduce the computation of the inverse of the signal and improve the accuracy of classification. From the current research results, feature extraction has been the difficulty of several signal classification methods. Especially for complex shape defects, it is needed to extract the shape parameters of defects according to the complete and effective feature extraction algorithm.

Classification algorithm includes clustering algorithm and neural network method. In the clustering algorithm, the feature vector of N-dimension can be regarded as a point in an N-dimensional feature space. A set of feature vectors of a class of signals constitutes a point cluster in the feature space. Clustering algorithm can identify point cluster in the regulatory or automatic mode. In neural network, multilayer perceptron neural network is the most widely used classification neural network. A multilayer perceptron network usually includes an input layer, one or more hidden layers and an output layer. The nodes in the same layer are not interconnected, but the nodes are completely connected to the nodes of the next layer, and all nodes in one layer process the data in parallel. Error backpropagation (BP) algorithm is the most commonly used training algorithm. The training mode is entered into the network in order. The algorithm is an iterative modification of interconnection weights, using gradient descent techniques to reduce the error between the actual output and the expected output greatly.

In summary, for the quantitative analysis of MFL signal of the corrosion, the direct method cannot get a unique and stable solution due to the constraint of the inverse problem of the electromagnetic field. While in the indirect method, the mapping method is widely used. Because it is based on open-loop structure, the sample is highly dependent, which needs to obtain a large number of sample data through simulation or experiment. The existing methods mainly map the defect parameters to single characteristics of the magnetic leakage field, without taking into account the real and complex multivariable function, so the error is big, and it cannot meet the requirements of high testing precision.

As for the iterative method in the indirect method, although it has a good approximation ability, however, it cannot achieve a balance between computational efficiency and accuracy. Although signal classification is widely used in defect classification, it can only achieve the classification of the signal, but cannot achieve the accurate parameter quantization, which cannot meet the requirements of high-precision detection.

3.2 Defect Quantification Method based on Statistical Identification

Statistical recognition is a typical pattern matching method, which is based on the standard signal library to establish the mapping relationship between the shape parameters and the characteristics of the MFL signal. According to the characteristics of the actual magnetic leakage testing signals, the size parameter of the defect is determined. At present, the statistical identification has been widely used in the quantitative analysis of the corrosion defect. Therefore, the statistical pattern recognition method based on principal component analysis (PCA), multivariate nonlinear regression and subsection nonlinear discriminant analysis is conducted to identify the defect parameters based on the characteristics of multiple MFL signals.

3.2.1 Pretreatment of Magnetic Flux Leakage Signal

Before the statistical identification of the defect magnetic leakage signals, it is necessary to preprocess the MFL signal, which includes the recognition of foreign body signals and the smooth interpolation of the waveforms.

3.2.1.1 Identification of Foreign Body Signals

In the MFL testing of pipelines, it is often interfered by the presence of some foreign body signals. It is necessary to distinguish the defect signal from the interfering signal in the quantitative analysis of the corrosion defect. Common foreign bodies include welding, casing pipe, flange, three pass and all kinds of valves and so on. By analyzing the waveform characteristics of the signal, we can effectively distinguish these foreign bodies from the corrosion defects.

The casing pipe, the welding and the flange are distributed in the entire circumference of the pipe, and their magnetic leakage signals are more obvious. Their magnetic leakage signal waveforms are also distributed throughout the entire circumference, that is, the circumferential direction of the pipe can be detected. The axial span of the detection signal waveform of the welding and flange is smaller than that of the casing pipe. The waveform is relatively steep and the distribution range is small. Three pass and the valve is fixed in the pipe circumferential part, so only part of the probes can detect the corresponding magnetic leakage signal.

Underground buried pipeline is made of a lot of long welded steel pipe, and the length of each section of the steel pipe is not more than tens of meters. The location of triplets and valves in the pipeline can be known in the construction plans, and the mileage wheel signal from the detector can further determine the existence and precise location of these foreign bodies.

3.2.1.2 Interpolation Smoothing of Magnetic Flux Leakage Waveform

Because of the low sampling frequency of the defect magnetic leakage signal, the waveform obtained from the measurement of the defect leakage field can be a straight line connecting with the actual testing data. The result will be a broken line. There is a large difference between the actual signal and the testing waveform. It is likely to give a large error to the feature extraction of signal waveform. If the sampling frequency is increased, it will lead to the hardware design and software programming of detection system complex. Interpolation smoothing technique can be used to obtain smooth curve and waveform, thus providing effective means to signal sampling waveform replication of time domain. There are many interpolation methods, such as polynomial interpolation, Newton interpolation, piecewise linear interpolation and spline interpolation. This book is based on the excellent properties of the three-spline function, and the three-spline interpolation method is adopted.

The three-spline interpolation method is to use the three-spline function to connect adjacent points to obtain the three-spline interpolation curve. Because the three-spline function has a high degree of smoothness (second order) and approximation order (fourth order), the resulting curve is very smooth.

The interval time of each test is the same, and the test time is set as $t_1 < t_2 < \cdots < t_N$, the corresponding three-spline function $s(t)$ should meet the following conditions:

1. $s(t_i) = x_i, \quad i = 1, 2, \ldots, N;$
2. In each subrange (t_i, t_{i+1}), $s(t_i)$ is a three-degree polynomial, $i = 1, 2, \ldots, N-1;$
3. $s(t)$ has the first- and second-order-continuous derivative in $[t_1, t_N]$, and at the ends, there has $s'(t_1) = s''(t_N) = 0$.

Then the interpolation computation is carried out. According to the Newton interpolation formula, there are

$$s(t) = s(t_i) + s(t_i, t_{i+1}) + (t - t_i)(t - t_{i+1})s(t, t_i, t_{i+1}), \quad t \in (t_i, t_{i+1}). \tag{3.2}$$

In eq. (3.1), the second-order difference is $s(t, t_i, t_{i+1}) = \frac{1}{6}[s'(t)+s'(t_i)+s'(t_{i+1})]$. If Formula (3.1) is processed for first-order derivative, then we can obtain

$$s'(t) = s(t_i, t_{i+1}) + (2t - t_i - t_{i+1})s(t, t_i, t_{i+1}) + \frac{1}{6}(t - t_i)(t - t_{i+1})s'(t_i, t_{i+1}). \tag{3.3}$$

The second-order derivative is

$$s'(t) = s''(t_i) + (t - t_i)s'(t_i, t_{i+1}). \tag{3.4}$$

With the formula, $s''(t)(i = 2, 3, \ldots, N-1)$ meets the three diagonal linear equations:

$$\begin{aligned}(t_i - t_{i+1})s''(t_{i-1}) + 2(t_{i+1} - t_{i-1})s'(t_i) + (t_{i+1} - t_i)s''(t_{i+1})\\ = 6[s(t_i, t_{i+1}) - s(t_{i-1}, t_i)];\ i = 2, 3, \ldots, N-1.\end{aligned} \tag{3.5}$$

With them, $s''(t_1) = s'(t_N) = 0$.

3.2.2 Definition and Extraction of the Waveform Features

The key to the qualitative and quantitative analysis of the MFL signal is the extraction of the signal characteristic quantity or the vector. The corrosion defect signal is an abnormal signal in the local space or time axis, as shown in Figure 3.5.

For ease of analysis, the characteristics of the magnetic flux signal waveform are defined as follows:

1. Peak and valley values of signal waveforms Y_{p-p}. Many kinds of factors affect the magnitude of the magnetic leakage signal, so peak and valley values Y_{p-p} is used as the feature, which can eliminate the influence of the DC component of the magnetic leakage signal and improve the reliability of the detection, as shown in Figure 3.6. In order to calculate Y_{p-p}, the maximum and minimum value of the defect signal should be found out, and the absolute value of the difference between the adjacent pair of extreme value can be obtained.
2. The signal's axial length L_X. The detector is sampled along the pipeline, and the threshold of detection signal is L_X, which can effectively reflect the axial distribution of the defect. Combined with the peak and valley values, the shape of the defect can be roughly determined, which helps to quantitatively detect the defect length and the depth of the defect. For the signal interception threshold G, a fixed value can be set, or the threshold is set dynamically according to the signal characteristics.

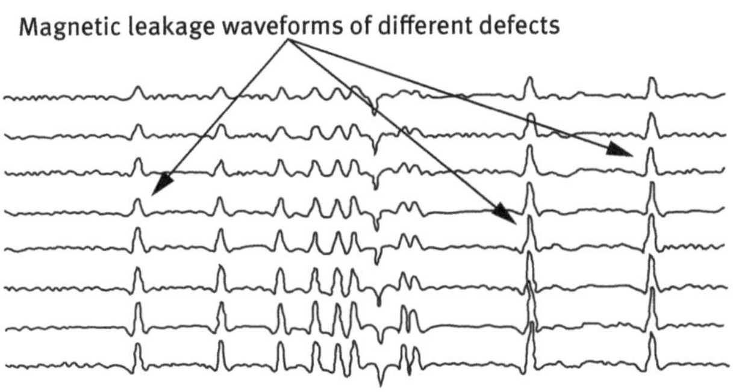

Figure 3.5: Magnetic leakage waveforms of different defects.

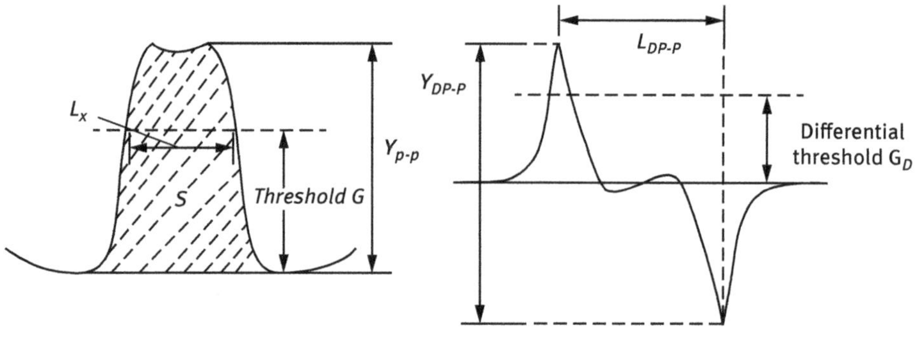

Figure 3.6: The main characteristics of the axial component waveform of the defect magnetic leakage field.

3. The area of the signal waveform S. The area of the waveform is the area of the curve of a wave. The MFL signal of a single defect is composed of a peak and a valley, which is reflected by the short first-order central moment of the signal:

$$S = \sum_{n}^{N} \{x(t) - \min[x(t)]\}. \tag{3.6}$$

4. The energy of the signal waveform E. The wave energy is equivalent to the second-order central moment of the signal, which reflects the discrete degree of the energy or signal of the magnetic signal fluctuations in a certain space. Using a formula for

$$E = \sum_{n}^{N} \{x(t) - \min[x(t)]\}^2. \tag{3.7}$$

5. The peak and valley length of differential signal L_{DP-P}. After differential, the high-frequency part of the magnetic signal can be enhanced, and the variation rate of the magnetic leakage field is reflected by L_{DP-P}: the mechanism and finite element analysis. By the mechanism and finite element analysis results of the magnetic leakage field, we can see that the magnetic leakage field changes seriously at the length boundary of the defect, which can be used to measure the length of the defect.
6. The peak and valley value of first-order differential signal Y_{DP-P}. The feature reflects the change of the highest point and the lowest point of the magnetic leakage field. Since it reflects the depth of the defect, it can be used as an indicator of the depth of the defect.
7. The magnetic leakage field distribution of the defect has a space vector characteristic. Although the detection system is magnetized along the axial direction of the pipeline, the circumferential waveform of the magnetic field is composed of a

plurality of signals. The width of the defect can be analyzed by the circumferential characteristic of the signal waveform. In addition, the evaluation of the width should be based on the number of probes that are covered by the circumferential magnetic leakage field distribution. So it is required to effectively intercept the magnetic leakage field size, in order to determine the number of covered probes. After determining the maximum amplitude of the detected signal, the threshold value G_W is set to determine the number W_Z of covered probes. The signal length is L_Z with the threshold G_W. At the same time, the differential value of the circumferential signal and the corresponding threshold value G_{DW} of the probe can determine the number of probes covered by the differential signal. As a characteristic value of the width, it is denoted as W_{DZ}.

After defining the above characteristics, we can make use of the standard defect library, in which one dimension (length, width and depth) of the defects changes and other dimensions do not change. To screen out the characteristics of the magnetic leakage field signal which is most closely related to the defect size, carry out the statistical identification.

3.2.3 Statistical Identification of Defect Length

There is an approximate linear relationship between the length of the defect MFL and the defect length, and there is a certain relationship between the length of the defect MFL and the defect depth. Because the span of the magnetic leakage field is related to the depth of defects, if a unified threshold value is used to quantify the length of the defect, there will be an obvious error.

As shown in Figure 3.7, for the two defects whose depths are 20 % and 40 % of the wall thickness, whose lengths of the defects are the same, if a fixed threshold G is adopted, then although the two defects have the same length, the spans of their magnetic leakage field are not the same. It is indicated that the depth factor must be corrected in the calculation of the length of the defect, and the span difference of the leakage fields of the defects with the same length and different depth can be reduced effectively.

Figure 3.8 shows the MFL waveforms of the defects whose length and width are $1P$ (P is the thickness of the pipewall), and the depth of the defect is $0.2P$, $0.4P$, $0.6P$, $0.8P$ and $1P$, respectively. Using the dynamic setting of the length threshold, we can see that the axial length of the different waveform is very close, that is, the dynamic threshold can eliminate the influence of depth.

Table 3.1 shows the comparison of the quantization errors of fixed threshold and dynamic threshold. It can be seen that if the quantitative formula is obtained from a series of depth of $0.2P$, when using a fixed threshold, the error will change rapidly with increase of the depth of defects, but a dynamic threshold method will remain about the same error.

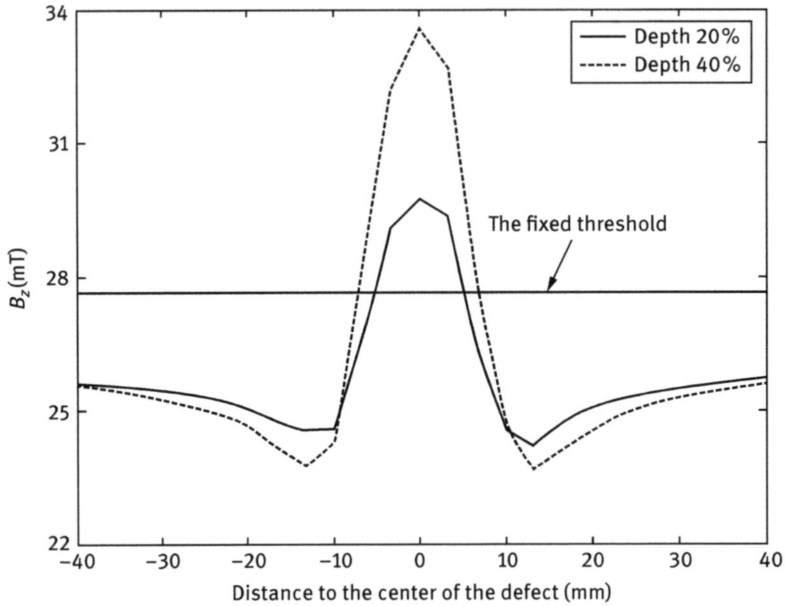

Figure 3.7: Comparison of magnetic leakage fields with different depth of defects.

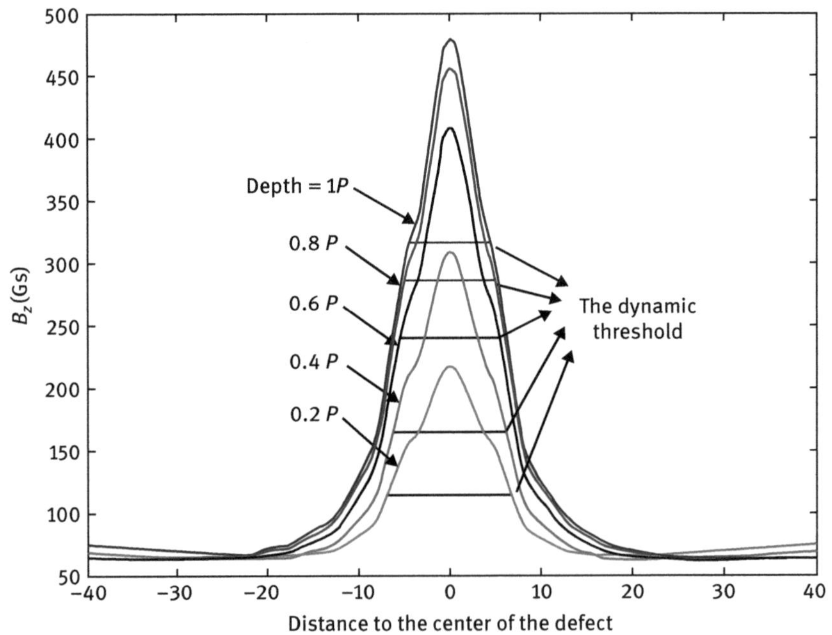

Figure 3.8: The effect of dynamic threshold on the depth of elimination.

Table 3.1: Error comparison of quantization between fixed threshold and dynamic threshold.

Defect depth	Actual length of defect (mm)	Error of fixed threshold (%)	Error of dynamic threshold (%)
0.2*P*	7.3	5.37	5.37
0.4*P*	7.3	36.99	5.37
0.6*P*	7.3	41.2	5.37
0.8*P*	7.3	44.36	5.37
1*P*	7.3	105.48	10.64

The error of the dynamic threshold quantization method is tested. Specific approach is to establish a standard signal library containing 100 defects. Every 20 defects are set as a group. Defects in each group has the same length and different depth from 5 % to 100 % of wall thickness (step is 5 % wall thickness). The lengths of the defects of the five groups are 1, 2, 3, 4 and 5 times of the wall thickness. Although the depth of the defect is different, it can be seen that the average error of the quantization length and the actual length is only 2.19 % (shown in Figure 3.9), which is sufficient to meet the needs of high-precision detection.

3.2.4 Multivariate Statistical Analysis Method

The width and depth of the defect are related to several characteristics of MFL signals, and there is a nonlinear relationship between these characteristics, which constitutes

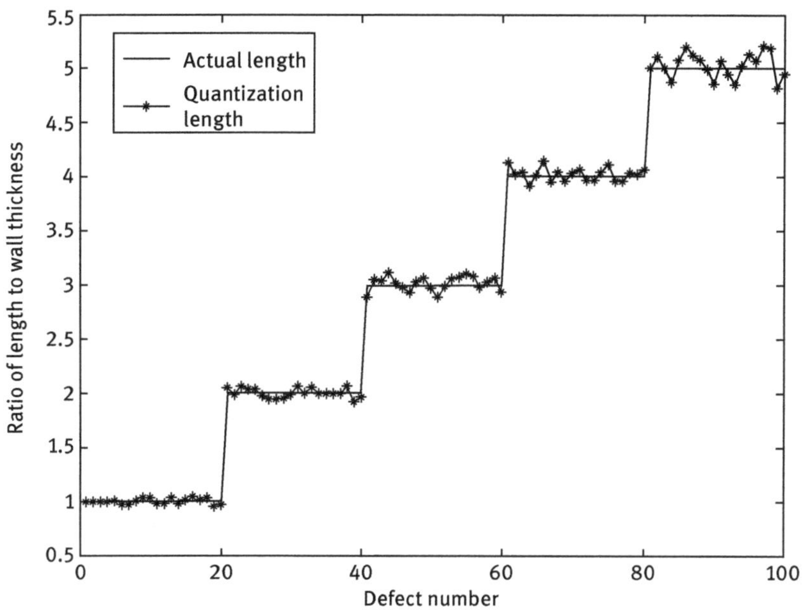

Figure 3.9: Length quantization error curve.

a typical multivariate statistical analysis problem. To solve the problem, we can use multiple nonlinear regression method, PCA method, linear pattern classifier, nonlinear discriminant function and so on.

3.2.4.1 Multivariate Nonlinear Regression

Multiple nonlinear regression method is used to quantify the defect of MFL detection, which can be obtained from the following two aspects:

1. By variable substitution, the nonlinear relationship is transformed into a linear relationship. The scatter plot based on the theory, experience and experimental data is used to decide the function type, and then the function is transformed into linear regression problem to be solved.
2. If the functional type of the actual problem is not easy to judge, a polynomial can be used to approximate it.

Assume that there is a nonlinear relationship between the variable y (length, width or depth of the defect) and $x_1, x_2, \ldots, x_p$ (the characteristics, which can determine the size of the defect, of the magnetic leakage field signal). Test sample size is n. The shape parameter data obtained by the ith test is y_i. The corresponding variables are $x_{i1}, x_{i2}, \ldots, x_{ip}$. And there is a nonlinear relationship between the y_i and the following polynomials:

$$x_{i1} + x_{i1}^2 + \cdots + x_{i1}^m, x_{i2} + x_{i2}^2 + \cdots + x_{i2}^n, \cdots, x_{ip} + x_{ip}^2 + \cdots + x_{ip}^l, \tag{3.8}$$

where y_i can be expressed as

$$\begin{aligned} y_i = &\beta_0 + \beta_1 x_{i1} + \beta_2 x_{i2} + \cdots + \beta_p x_{ip} + \beta_{21} x_{i1}^2 + \beta_{22} x_{i2}^2 \cdots + \beta_{2p} x_{ip}^2 + \cdots + \\ &\beta_{m1} x_{i1}^m + \beta_{n2} x_{i2}^n + \cdots + \beta_{lp} x_{ip}^l \end{aligned} \tag{3.9}$$

Substituting the high order, let $x_{i1} = z_{i1}, x_{i2} = z_{i2}, \ldots, x_{i1}^2 = z_{ip+1}, x_{i2}^2 = z_{ip+2}$. Multivariate polynomial regression can be used to solve the problem of multiple linear regression. The transformation form is

$$y = a_0 + a_1 z_{i1} + a_2 z_{i2} + a_q z_{iq} + \varepsilon_i, \quad i = 1, 2, 3, n, \tag{3.10}$$

In eq. (3.10), a_0, a_1, $a_2, \ldots, a_q$ are $q+1$ parameters to be estimated and ε_i are random variables that are independent of each other and obey the same normal distribution. $z_1, z_2, \ldots, z_q$ are values of the q-independent transformed variables.

The next question is to find the parameter a_0, a_1, $a_2, \ldots, a_q$ values of the variable z. Then reverse substitute to $\beta_0, \beta_1, \beta_2, \ldots, \beta_{2p}, \ldots, \beta_{m1}, \ldots, \beta_{lp}$ so as to obtain the original form of the polynomial.

Use the least squares estimation to calculate the value of the parameter a_i.

From the test sample data, the estimated values of a_0, a_1, $a_2, \ldots, a_q$ can be obtained, and the estimated equation of the regression equation is derived:

$$\hat{y} = b_0 + b_1 z_{i1} + b_2 z_{i2} + \cdots + b_q z_{iq} \tag{3.11}$$

According to the principle of least squares method, $b_0, b_1, b_2, \ldots, b_q$ should make the sum of the deviation squares of all the observed values y_i and the regression values $\hat{y}_i$ the minimum, that is to let the Q in eq. (3.12) the minimum:

$$Q = \sum_{i=1}^{n} (y_i - \hat{y}_i)^2 = \sum \left[y_i - (b_0 + b_1 z_{i1} + b_2 z_{i2} + \cdots + b_q z_{iq})\right]^2. \tag{3.12}$$

With known sample data, Q is a nonnegative quadric expression about b_0, b_1, $b_2, \ldots, b_q$, so its minimum value must exist. According to the extremum principle, $b_0, b_1, b_2, \ldots, b_q$ should be the solution of the following equation:

$$\begin{cases} \dfrac{\partial Q}{\partial b_0} = -2\sum_{i=1}^{n} (y_i - \hat{y}_i) = 0 \\ \dfrac{\partial Q}{\partial b_j} = -2\sum_{i=1}^{n} (y_i - \hat{y}_i)\, z_{ij} = 0 \end{cases} \quad (j = 1, 2, \ldots, q). \tag{3.13}$$

3.2.4.2 PCA Method

PCA simplifies characteristic variables related to the defect shape parameters as a few uncorrelated variables. It simplifies the statistical data and reveals the relationship between the characteristics of internal variables. Each principal component is a linear combination of the initial variables. All principal components are orthogonal, so there is no redundant information, which constitute an orthonormal basis for a space. The basic idea is dimensionality reduction.

The concrete steps of the PCA are as follows:

There are number n of corrosion defects with different shapes. The number of evaluating characteristics of a defect size (such as length, width and depth) is p, so the testing data matrix Z is

$$Z = \begin{bmatrix} z_{11} & z_{12} & \cdots & z_{1p} \\ z_{21} & z_{22} & \cdots & z_{2p} \\ \vdots & \vdots & \ddots & \vdots \\ z_{n1} & z_{n2} & \cdots & z_{np} \end{bmatrix} \tag{3.14}$$

Each characteristic quantity (column) is transformed as follows:

$$\begin{aligned} x_{kj} &= \frac{z_{kj} - \bar{z}_j}{S_j} \\ \bar{z}_j &= \frac{1}{n}\sum_{k=1}^{n} z_{kj} \\ S_j^2 &= \frac{1}{n-1}\sum_{k=1}^{n} (z_{kj} - \bar{z}_j)^2 \quad (k = 1, 2, \ldots, n; j = 1, 2, \ldots, p). \end{aligned} \tag{3.15}$$

Suppose the transformed data matrix as X, then

$$X = \begin{bmatrix} x_{11} & x_{12} & \cdots & x_{1p} \\ x_{21} & x_{22} & \cdots & x_{2p} \\ \vdots & \vdots & \ddots & \vdots \\ x_{n1} & x_{n2} & \cdots & x_{np} \end{bmatrix} = (x_1\, x_2 \cdots x_p) = \begin{bmatrix} x'_{(1)} \\ x'_{(2)} \\ \vdots \\ x'_{(n)} \end{bmatrix}. \tag{3.16}$$

Among them

$$\begin{cases} x_j = \begin{bmatrix} x_{1j} \\ x_{2j} \\ \vdots \\ x_{nj} \end{bmatrix} & (j = 1, 2, \ldots, p), \\ x(k) = \begin{bmatrix} x_{k1} \\ x_{k2} \\ \vdots \\ x_{kp} \end{bmatrix} & (k = 1, 2, \ldots, n). \end{cases} \tag{3.17}$$

Then the correlation coefficient matrix between the original variable (feature) $z_1, z_2, \ldots, z_q$ is

$$R = \frac{1}{n-1} \bullet X' \bullet X, \tag{3.18}$$

where p characteristic value λ_i and the corresponding characteristic component are obtained by the Jacobi method; thus, p principal component equations are established. The principal components in the principal component equations have the same form:

$$Y = a_1 x_1 + a_2 x_2 + \cdots + a_p x_p, \tag{3.19}$$

where a_1, $a_2, \ldots, a_q$ are the corresponding characteristic components, which are called weight coefficients. The equation contribution rate is set as m, then

$$m = \lambda_i / \sum_{i=1}^{p} \lambda_i. \tag{3.20}$$

The largest variance contribution rate is known as the first principal component, followed by the second principal component and so on. From big to small, the cumulative value of the cumulative variance contribution rate reaches at least 80 % as the number k of principal component selection. The standardization values of the original variables are substituted into k corresponding principal component equations. The contents of main components can be calculated.

The linear classifier method includes the minimum distance classifier, linear discriminant function, perceptron criterion function and error criterion function and so on. Nonlinear discriminant classifier consists of quadratic discriminant function, piecewise linear discriminant function and other nonlinear function. The commonly used methods of nearest neighbor and optimization algorithm are nonlinear classification and recognition method.

3.2.5 Statistical Identification of Defect Width

Width of defect quantitative analysis is mainly based on the pipe circumferential span of defect magnetic leakage field, and quantization precision depends mainly on the circumferential distance of the probes, that is the pipe circumferential detection channel spacing. For the high-resolution detector, the distance between the probes in the circumferential direction is 8–12 mm, which can achieve a high precision.

Similar to the length of corrosion defects, the quantitative discrimination of defect width is mainly based on the signal waveform formed by the circumferential distribution of the MFL field. By means of PCA and a large number of statistical experience, the MFL signal features of defect width can be selected as: peak–valley value of circumferential MFL waveform, intercept length threshold of circumferential MFL waveform, probe number covered by circumferential MFL field and probe number covered by circumferential MFL differential waveform.

If the detector has better consistency of the probe, the circumferential signal waveform of the defect should be able to accurately reflect the change of the defect width, and the threshold value of the L_z can effectively evaluate the width of the defect. In the actual testing, the probe liftoff and various kinds of interference cause that the testing values of different probes are not consistent. Even the calibration is processed before the testing, and the random disturbance factors will lead to inconsistent output of various probes, which is the cause of poor linearity between L_z characteristics and the width of defect, as shown in Figure 3.10.

Other dimensions of the defect, especially the influence of the defect depth, should be considered in the evaluation of the defect width, that is, the number of the probes that cover the circumferential magnetic leakage field can be used to evaluate width, but the peak–valley value of the circumferential MFL signals also impacts on the width, and the peak–valley values of the axial and circumferential MFL signals mainly depend on the defect depth. For the defects of the same length and width, there is a big difference between the peak–valley value and the covering probes of the circumferential magnetic leakage field. Therefore, quantification of the width still needs to select a dynamic threshold, and the peak–valley value and peak value are taken as the circumferential wave characteristics. By setting different thresholds for defects of different depths, linear approximation and compensation to the number of the covering probes affected by the nonlinear relationship between defect depth and wave characteristics can be processed.

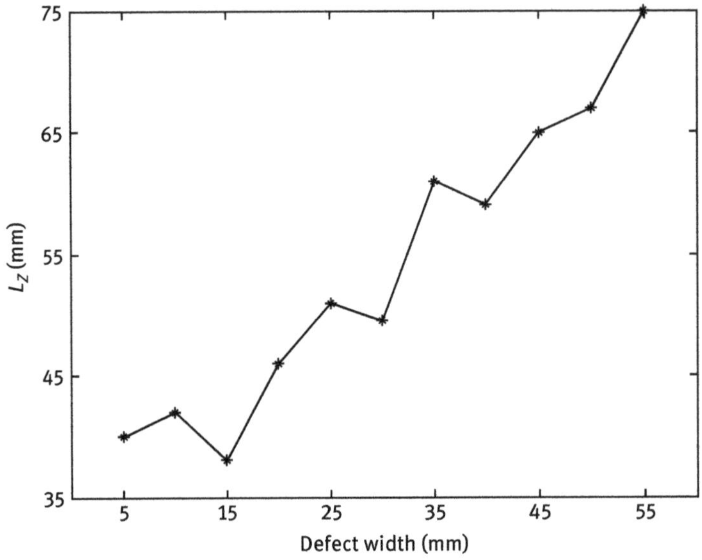

Figure 3.10: The relationship between the defect width and L_Z under other dimensions unchanged.

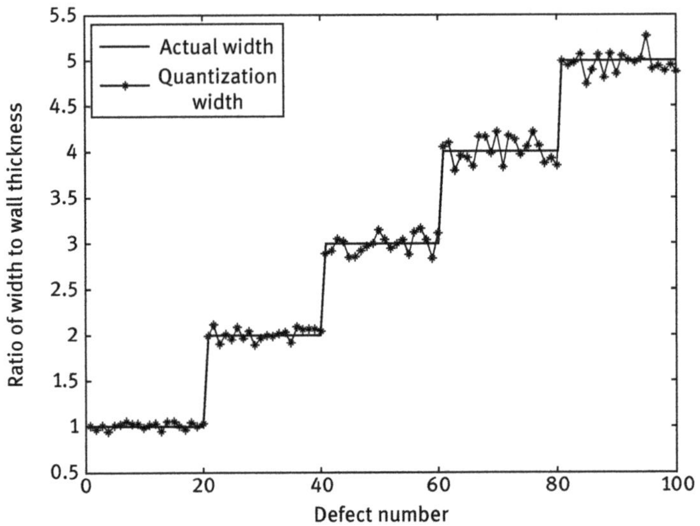

Figure 3.11: The error of the width quantization.

Figure 3.11 shows the verifying results of the width error of the quantization method. A standard signal library containing 100 defects is established. Every 20 defects are classified as a group, which has the same width and length, but different depth from 5 % to 100 % of the wall thickness (5 % wall thickness as a step). The widths of the five groups of defects are 1, 2, 3, 4 and 5 times of the wall thickness, respectively. Figure 3.11 shows that the error between the actual width and that of the quantization is small. The average error is only 4.37 %.

3.2.6 Statistical Identification of Defect Depth

It is difficult to evaluate defect depth in detecting corrosion defects. It can be seen from the above analysis, without considering other factors, only considering the defect shape parameters, the influence of depth on the magnetic leakage field waveform is mainly reflected in the peak–valley value in the axial and circumferential directions. Therefore, axial MFL peak–valley value is the most effective characteristics to quantify defects depth, which is given the maximum weight. Many scholars only use this feature to evaluate defect depth. But the peak–valley value is also influenced by the length and width of the defect. The axial magnetization of pipeline MFL testing determines that the peak–valley value of magnetic leakage field of the defect, which has large circumferential width and small axial length, is greater, and vice versa. Therefore, the depth evaluation should include the width and length characteristics of the MFL of defects.

Through the statistics and experimental analysis by the PCA method to a large number of defect signals, the MFL characteristic of defect depth is finally determined as follows: the peak–valley value of axial MFL waveform, the area of axial MFL waveform, the peak–valley length of axial differential signal and the covering probe number in circumferential direction.

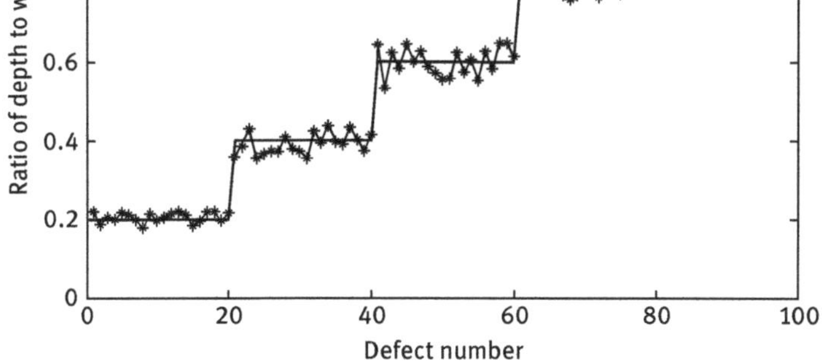

Figure 3.12: The error of the depth quantization.

Figure 3.12 shows the verifying results of the depth error of the quantization method. A standard signal library containing 100 defects is established. Every 20 defects are classified as a group, which has the same width and depth, but different length from 1 to 5 times of the wall thickness (20 % wall thickness as a step). The depths of the five groups of defects are 0.2, 0.4, 0.6, 0.8 and 1 times of the wall thickness, respectively. The widths of the five groups of defects are 1, 2, 3, 4 and 5 times of the wall thickness, respectively. The average error between the actual depth and that of the quantization is 6.75 %.

You can see, from Figure 3.12, that the evaluation error of defect depth is larger than that of the length and width of the defect. The main reason is that the characteristics related to the depth are more than those of the length and width used.

To sum up, evaluating the parameters of the defect shapes (length, width and depth) using statistical methods is a complex process, in which the length of the evaluation is the easiest; the evaluations of width and depth are more complex.

3.3 Defect Quantification Method Based on Radial Basis Function Neural Network

Although the statistical identification method of defects quantitative identification has better performance in practice, it is necessary to realize that the statistical method does not have the ability of adaptive processing, such as not having the ability to self-organizing, self-learning and the ability to identify noises. And this method requires a large amount of test or simulation data. The statistical process is more complex. There is dependence on the sample.

In order to further improve the accuracy of detection, many scholars are committed to the introduction of a variety of new technologies to the quantitative analysis of defects. The neural network method is successfully applied to the processing of MFL signal and the reconstruction of magnetic field. However, the method of quantitative analysis of neural network, on the one hand, has the disadvantage of low computation speed and low recognition accuracy. On the other hand, it focuses on the inductive analysis of the existing detection results, so the adaptive ability of any defect is still insufficient.

Therefore, an iterative method based on radial basis function neural network (RBFNN) is used to solve the problem of adaptive ability. Compared with ordinary neural network method, this method can improve the adaptability of different types of defects by using closed-loop structure. At the same time, it has the advantages of fast convergence speed, good antinoise performance and so on.

3.3.1 The Iterative Method Based on Neural Network

Although the calculation process of the analytical method is simple and the calculation efficiency is high, the complexity of the defects, the arbitrary and the electromagnetic characteristics of the pipe wall material determine that the numerical method is more effective and accurate. At present, the common numerical method is the finite element method. However, the computational efficiency of the finite element method is not high, especially in the calculation of small defects. So the workload of the iterative calculation based on the finite element method is very large, especially in the use of the 3D model, which limits its application in practice.

In order to overcome this difficulty, an RBFNN is used as the solution to the problem. The method combines the advantages of neural network and the iterative method, which can, on the one hand, improve the accuracy and adaptability by using the closed-loop iteration. On the other hand, it can improve the computational efficiency by using neural network.

As shown in Figure 3.13, RBFNN is usually composed of three layers of input layer, hidden layer and output layer. There is no connection among the nerve cells of each layer, and the information is transmitted between layers and layers.

In Figure 3.13, $X = (x_1, x_2, \ldots, x_n)^T$ is the input vector and $Y = (y_1, y_2, \ldots, y_n)^T$ is the output vector. The input to output nonlinear mapping functions are usually the Gauss function, multi-quadric function, Duchon thin-plate spline function and so on. Here, the most commonly used Gauss function is chosen as the radial basis function. At this time, RBFNN's input/output relationship is expressed as

$$Y_l = F(X) = \sum_{j=1}^{H} w_{lj} \exp\left(\frac{\|X - c_j\|^2}{2\sigma_j^2}\right), \quad l = 1, 2, \ldots, M, \tag{3.21}$$

where X is the input, $Y = (y_1, y_2, \ldots, y_n)^T$ is the output, H is the number of basis function used, $\|\bullet\|$ means Euclid norm, c_j represents the center of the hidden layer of the j basis function unit, σ_j is the radius of the basis function, w_{lj} is the connection weights between the j basis functions and the output point l.

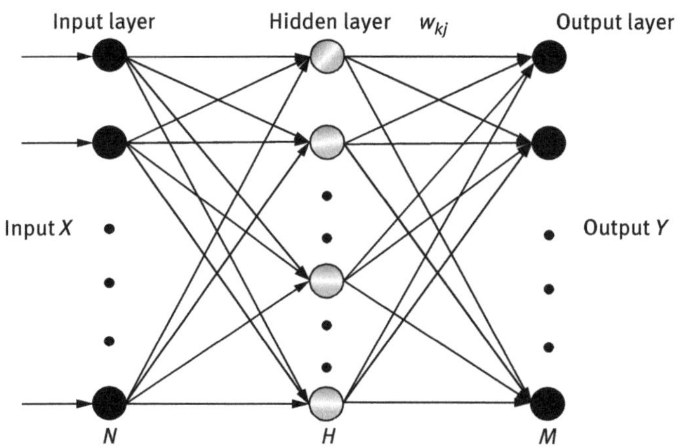

Figure 3.13: The structure of radial basis function neural network.

RBFNN training process can be seen as a set of given input and output data, in the multidimensional space to find a curve fitting or function approximation problem. The objective is to find the best fitting of the functional mapping relationship between the input and output data in the statistical view. This process can be divided into three stages: first, a set of radial basis functions is defined, and the data of the input network is converted to the hidden layer. Second, conduct the linear transformation between the hidden layer and the output layer. Finally, the unknown data are performed interpolation process through the network to obtain the corresponding solution.

Given the data is $X = (x_1, x_2, \ldots, x_n)^T$, its corresponding output is $Y = (y_1, y_2, \ldots, y_n)^T$. The objective is to obtain a function $f : X \to Y$, which satisfies the interpolation condition

$$f(x_i) = y_i,\ i = 1, 2, \ldots, N, \tag{3.22}$$

$f(\bullet)$ applies to all training data for RBFNN. Using the radial basis function method, $f(\bullet)$ can be expressed as

$$f(x_i) = \sum_{j=1}^{N} w_j \phi_j \left(\|\overline{x} - c_j\|\right). \tag{3.23}$$

In the formula, $\{\varphi_j(x)\}$, $j = 1, 2, \ldots, N$, is a set of radial basis functions, $\{c_j\}$ is the center of radial basis functions and $\{w_j\}$ constitute a group of weights.

Formula (3.23) is expressed by matrix as follows:

$$\begin{bmatrix} \phi_{11} & \phi_{12} & \cdots & \phi_{1N} \\ \phi_{21} & \phi_{22} & \cdots & \phi_{2N} \\ \vdots & \vdots & \ddots & \vdots \\ \phi_{N1} & \phi_{N2} & \cdots & \phi_{NN} \end{bmatrix} \begin{bmatrix} w_1 \\ w_2 \\ \vdots \\ w_N \end{bmatrix} = \begin{bmatrix} y_1 \\ y_2 \\ \vdots \\ y_N \end{bmatrix}. \tag{3.24}$$

In eq. (3.24),

$$\phi_{ij} = \phi\left(\|x_i - c_j\|\right), \quad i = 1, 2, \ldots, N, j = 1, 2, \ldots, N. \tag{3.25}$$

So formula (3.25) is equivalent to

$$\Phi W = Y. \tag{3.26}$$

If the matrix Φ is nonsingular, the inverse matrix Φ^{-1} exists:

$$W = \Phi^{-1} Y. \tag{3.27}$$

The number of radial basis functions is limited by the number of training samples. If the number of training samples is too much, the degree of freedom of the physical process is more than the degree of freedom of the physical process. It will lead to deviation and appear over fitting, which lowers the performance of the neural network.

Using some previous information to reflect the mapping relationship between the input and output and adjusting the sample space, the ill-conditioned sample space is converted into a benign sample space, which can solve the problem of RBFNN over fitting. By adding the smoothness constraint, some prior information is embedded into the auxiliary nonnegative functional, and the only and stable solution is obtained by limiting the scope of the solution.

The establishment of RBFNN is mainly divided into the following two steps:

Step 1: The choice of basis functions. Usually choose basis functions based on experience or clustering algorithm. In this book, we use the K-order algorithm. Its principle is to minimize the clustering performance, that is, let the sum of the squares of the distance between the sample points to the center of the class minimum.

The steps of the K-order mean algorithm here are as follows:

1. Randomly select K data as clustering centers.
2. According to the similarity distance formula, each data in the data set is assigned to its nearest cluster.
3. Calculate the new clustering center.
4. If the clustering center does not change, the algorithm ends. Otherwise jump to step 2.

After clustering the sample, we can calculate the radius σ_j of the basis function:

$$\sigma_j = \frac{1}{N_j}(X^P - c_j)^T(X^P - c_j), \tag{3.28}$$

where N_j is the number of samples in the cluster and X^P is the training sample.

Step 2: Training, that is determine the connection weights of the hidden layer to the output layer. The book will have an adaptive mechanism for the gradient descent method applied to the RBFNN parameter adjustment. The error function is defined as

$$J(k) = \frac{1}{2}(Y(k) - y(W, k))^2, \tag{3.29}$$

where $J(k)$ is the error function, $Y(k)$ is the expected output and $Y(W, k)$ represents the actual output of the network.

Therefore, an adaptive adjustment algorithm for the weighted matrix of the hidden layer to output layer is obtained:

$$W(k+1) = W(k) + \mu(k)\left(-\frac{\partial J(W)}{\partial W}\right)\Big|_{W=W(k)} + \alpha(k)(W(k) - W(k-1)). \tag{3.30}$$

The adjustment algorithm for the matrix of the hidden layer centers is

$$C(k+1) = C(k) + \mu(k)\left(-\frac{\partial J(C)}{\partial C}\right)\Big|_{C=C(k)} + \alpha(k)(C(k) - C(k-1)). \tag{3.31}$$

The adjustment algorithm for the hidden layer radius matrix is

$$\sigma(k+1) = \sigma(k) + \mu(k)\left(-\frac{\partial J(\sigma)}{\partial \sigma}\right)\Big|_{\sigma=\sigma(k)} + \alpha(k)(\sigma(k) - \sigma(k-1)), \tag{3.32}$$

where $\mu(k)$ is the learning rate and $a(k)$ is the momentum term, which uses the results of previous adjustment to influence the amount of adjustment, thus speeding up the process of correction. When the previous adjustment was overshot, the momentum term will be opposite to the error of the adjustment. The actual adjustment amount will reduce and the oscillation will be avoided. While the previous adjustment amount is owed, the actual adjustment amount will increase. The role of $a(k)$ is to make the network changes tend to be stable, in order to facilitate the convergence of the network as soon as possible.

Calculation method for learning rate $\mu(k)$ is

$$\text{When } \Delta J < 0, \mu(k) = \varphi\mu(k-1), \quad \varphi > 1, \tag{3.33}$$

$$\text{When } \Delta J > 0, \mu(k) = \beta\mu(k-1), \quad \beta < 1, \tag{3.34}$$

where φ and β are constants, $\Delta J = J(k) - J(k-1)$.

In the BP network, if the learning rate is too small, the network parameters will change slowly, and the convergence is also slow. If it is too large, it can accelerate the learning speed; however, it may lead to the continuous oscillation near the point of stability. So the learning rate is usually chosen a fixed value based on experience. When the change of total error is reduced ($\Delta J < 0$), the training error curve is flat. Then the learning rate should be multiplied by a constant that is more than 1. So the step size increases, the number of iterations reduces, and avoid falling into local minimum point. When the change of total error is increased ($\Delta J > 0$), which means that the training of the error curve changes sharply in the region, the learning rate should be multiplied by a number of less than 1. It is conducive to shorten the step size so that the error decreases.

After the establishment of the RBFNN, the sample data is trained, which makes the error of the network forecast lower than the predetermined threshold.

Combined with the training of the RBFNN, and then according to the iterative method shown in Figure 3.4, we can get the algorithm flow of quantitative analysis of defects:

Randomly set a defect feature, which is called $x(0)$;

During the kth cycle, the input $x(k)$ is trained by the RBFNN, to get a prediction of the magnetic leakage field signal $y(k) = F(x(k))$;

Calculating the sum of error square:

$$E(k) = \frac{1}{2}\left\|d - y(k)\right\|^2. \tag{3.35}$$

Calculating the gradient value of $E(k)$ to $x(k)$, that is $\frac{\partial E(k)}{\partial x(k)}$;

Update x:

$$x(k+1) = x(k) + \eta\left(-\frac{\partial E(k)}{\partial x(k)}\right), \tag{3.36}$$

where η is the learning rate of the iterative process.

If $E(k) \geq$ e, turn back to the second step;

Until $E(k) <$ e, exit cycle.

Here, the defect is characterized as x_i, the corresponding neural network calculation results is y_i, the expected output is d_i and the sum of the error square is

$$E = \frac{1}{2}\sum_{l=1}^{M}(d_l - y_l)^2. \tag{3.37}$$

Here

$$y_i = F(x_i) = \sum_{j=1}^{H} w_{lj}\exp\left(\frac{\left\|x_i - c_j\right\|^2}{2\sigma_j^2}\right). \tag{3.38}$$

So the gradient of E to x_i is

$$\frac{\partial E(x)}{\partial x_i} = -\sum_{l=1}^{M}\left[(d_l - y_l)\left(-\frac{1}{\sigma_j^2}\sum_{j=1}^{H} w_{lj}(x_i - c_{ji}) \times \exp\left(-\frac{\left\|x - c_j\right\|^2}{2\sigma_j^2}\right)\right)\right], \quad i = 1, 2, \ldots, N. \tag{3.39}$$

Take the quantitative analysis of the circumferential width of the defect to verify the validity of the proposed method. In a 3D finite element model of pipeline inspection, the depth of 30 %, 50 %, 80 % pipe wall thickness, the width of 1, 1.5, 2, 3, 3.5, 4, 4.5, 5 and 5.5 times the wall thickness and the length of 1, 2, 3, 4 and 5 times the wall thickness of a total of 150 defects were calculated. Take 120 defect characteristics and simulation results data as a training sample to the RBFNN, the remaining 30 defects data as a test set, that is, there is no duplication between the training samples and the test set. Other parameters required are: the diameter of the pipe is 1,016 mm, the wall thickness of the pipe is 14.6 mm and the liftoff of the probe is 4 mm. And the defect width W is closely related to the circumferential width L_Z and the wave peak and valley value of the defect leakage field circumferential waveform Y_{ZP-P}. As a result, W is used as the input of the RBFNN, and the L_Z and Y_{ZP-P} are used as the output of the network to train the RBFNN. In Figure 3.14, the change of the learning error of the BP based on the sigmoid function and RBFNN is compared, and the target error is 0.001.

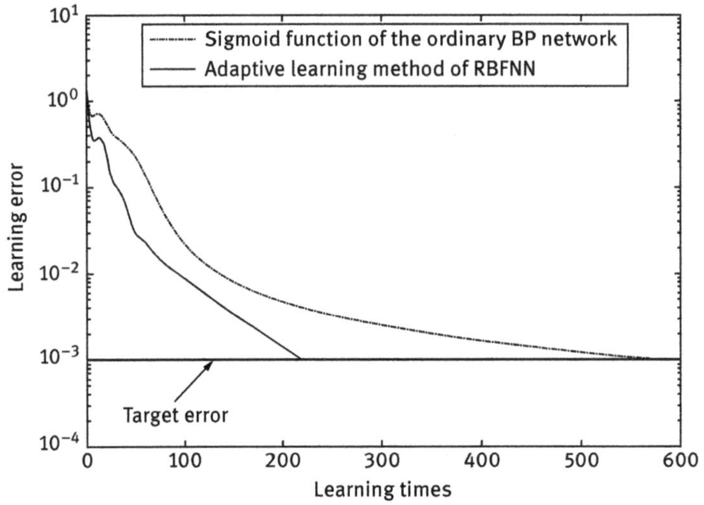

Figure 3.14: Convergence of learning error of two neural networks.

In Figure 3.14, we can know that adaptive learning method of RBFNN achieves the target error after the 221st training and sigmoid function of the ordinary BP network need to go through 572nd training to achieve the target error. The latter is significantly slower than the former.

In order to confirm the accuracy of this method, 30 data in the testing set were chosen to verify the method. Due to the limited space, Table 3.2 shows the comparison between the actual value and the calculated L_Z, Y_{ZP-P} according to W, through the training of the RBFNN to solve the forward problem. The defect length is 43.8 mm (3 times of wall thickness) and the depth is 4.38 mm (30 % wall thickness).

From Table 3.2, we can see the calculating results of the forward problem of RB-FNN after learning are very close to the actual value, and the maximum error is 2.39 % and 2.87 %, respectively.

Table 3.2: Comparison between calculated value and actual value.

Defect no.	W (mm)	L_Z (mm)		Y_{ZP-P} (mT)	
		Calculated value	Actual value	Calculated value	Actual value
1	14.6	111.1	108.5	1.39	1.43
2	29.2	117.8	119.2	2.35	2.31
3	43.8	122.5	123.1	3.04	3.01
4	58.4	131.5	130.7	3.81	3.86
5	73	139.4	138.4	4.41	4.47

Table 3.3: Comparison between calculated value and actual value of the inverse problem.

L_Z (mm)	Y_{ZP-P} (mT)	W (mm)	
		Calculated value	Actual value
108.5	1.43	14.6	14.6
119.2	2.31	29.1	29.2
123.1	3.01	43.9	43.8
130.7	3.86	58.6	58.4
138.4	4.47	72.7	73

Table 3.3 shows the comparison between the calculated *W* and the actual value, using RBFNN as a positive solution, iteratively solving the inverse problem.

From Table 3.3, the maximum error between the actual value and the calculated defect width using the inverse problem-solving method is only 0.26 %. The high quantification accuracy shows that this method is suitable for different defects and has good application prospect. Figure 3.15 shows the convergence of iterative algorithm for inverse problem-solving method, of which, the horizontal line shows the actual wall thickness of 14.6 mm; the up trapezoid line shows the iterative calculation results from an initial value of 30 mm; the low trapezoid line shows the iterative calculation results from an initial value of 5 mm. It can be seen that they both converge fastly

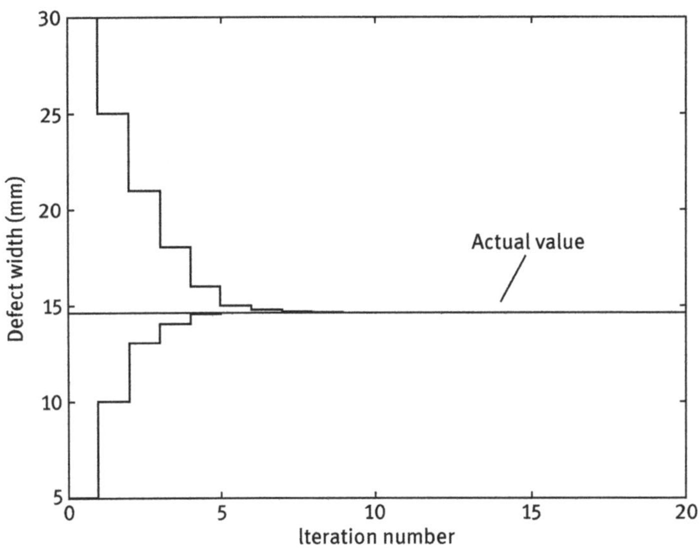

Figure 3.15: The convergence of iterative algorithm for the inverse problem.

to 14.6 mm. This means that, although the initial value is selected randomly, it can converge to the actual value quickly.

3.4 Defect Quantification Method Based on 3D Finite Element Neural Network

The iterative method based on RBFNN used closed-loop structure to reduce the dependence of the statistical sample. However, it is still required to use uniform and standardized defect samples in the training process, so the dependence is not eliminated.

To solve this problem, a new iterative method based on 3D finite element method is proposed, which is embedding the finite element computation into the neural network structure, and the neural network is used as the normal problem-solving model in the iterative method.

3.4.1 Discretization Principle of Finite Element Method

The typical boundary value problem can be defined using the differential equation of the available area Ω and the boundary conditions of the surrounding area. The differential equations can be expressed as

$$L\vec{\phi} = \vec{f}, \tag{3.40}$$

where L represents a differential operator, $\vec{f}$ represents an incentive or an imposed function, and $\vec{\phi}$ is the variable to be found.

Although the above boundary value problem can be solved by analytical method, only a few cases can be resolved. In order to overcome this difficulty, we can use the numerical method to solve the boundary value problem, and the most common method is the finite element method. The finite element method is a variational method, which means that the boundary value problem is represented by a variational expression (i.e. a functional). The minimum value of the functional is corresponding to the control differential equation under the given boundary conditions. The approximate solution can be obtained by solving the minimum value of the function. Functional expressions corresponding to formula (6.31):

$$F(\tilde{\phi}) = \frac{1}{2}\left\langle L\tilde{\phi}, \tilde{\phi}\right\rangle - \frac{1}{2}\left\langle \tilde{\phi}, f\right\rangle - \frac{1}{2}\left\langle f, \tilde{\phi}\right\rangle, \tag{3.41}$$

where $\tilde{\phi}$ represents the testing function. The solution of differential equation is obtained by solving the minimum value of the function to $\tilde{\phi}$. It is a very important step in finite element analysis to find out the testing function that can express or approximate the real solution in the whole solution domain. However, for many problems, this step is very difficult, especially for 2D and 3D problems. To overcome this difficulty, the entire region can be divided into several small subdomains (units), and the testing functions are defined on each subdomain. In the case of unit e, the following expressions $\tilde{\phi}^e$ can be obtained:

$$\tilde{\phi}^e = \sum_{j=1}^{n} N_j^e \phi_j^e = \{N^e\}^T \{\phi^e\} = \{\phi^e\}^T \{N^e\}, \tag{3.42}$$

where $\tilde{\phi}^e$ represents the pending variables in unit e, n represents the number of nodes in unit e, ϕ_j^e represents value ϕ of node j in the unit, N_j^e is a basis function (also known as interpolation function or expansion function) that is associated with node j in unit e. The important feature of N_j^e is that it is not zero in unit e, and it is zero outside the unit. Thus, formula (3.41) can be expressed as

$$F(\tilde{\phi}) = \sum_{e=1}^{M} F^e(\tilde{\phi}^e), \tag{3.43}$$

where $F^e(\tilde{\phi}^e)$ represents the functional value in unit e, M is the number of all the units in the solving region. According to formula (3.41), we can obtain

$$F^e(\tilde{\phi}^e) = \frac{1}{2}\int_{\Omega^e} \tilde{\phi}^e L\tilde{\phi}^e d\Omega - \int_{\Omega^e} f\tilde{\phi}^e d\Omega. \tag{3.44}$$

Substitute eq. (3.42) into eq. (3.44), we can obtain

$$F^e(\tilde{\phi}^e) = \frac{1}{2}\{\phi^e\}^T \int_{\Omega^e} \{N^e\} L \{N^e\}^T d\Omega \{\phi^e\} - \{\phi^e\}^T \int_{\Omega^e} f\{N^e\} d\Omega, \tag{3.45}$$

where $\{\cdots\}^T$ represents the transpose matrix.

Equation (3.45) can be written in matrix as

$$F^e(\tilde{\phi}^e) = \frac{1}{2}\{\phi^e\}^T [K^e] \{\phi^e\} - \{\phi^e\}^T \{b^e\}, \tag{3.46}$$

where $[K^e]$ is an $n \times n$ matrix, $\{b^e\}$ is a $n \times 1$ column vector. Their elements are

$$K_{ij}^e = \int_{\Omega^e} N_i^e L N_j^e d\Omega,$$
$$b_i^e = \int_{\Omega^e} f N_i^e d\Omega. \tag{3.47}$$

Since L is self-adjoint, the element matrix $[K^e]$ is symmetric. Substituting eq. (3.47) into eq. (3.43), we get

$$F(\tilde{\phi}) = \sum_{e=1}^{M} \left(\frac{1}{2} \{\phi^e\}^T [K^e] \{\phi^e\} - \{\phi^e\}^T \{b^e\} \right). \tag{3.48}$$

By summing and computing, and then using the global encoding, eq. (3.48) can be written as

$$F = \frac{1}{2}\{\phi\}^T[K]\{\phi\} - \{\phi\}^T\{b\}, \tag{3.49}$$

where $[K]$ is an $N \times N$ symmetric matrix, N is the number of unknown variables or nodes, $\{\phi\}$ is $N \times 1$ unknown vector whose elements are expansion coefficients of unknown variables, {b} is $N \times 1$ known vector. Let the partial derivative of F to ϕ_i be zero:

$$\frac{\partial F}{\partial \phi_i} = \frac{1}{2} \sum_{j=1}^{N} (K_{ij} + K_{ji})\phi_j - b_i = 0,\ i = 1, 2, 3, \ldots, N, \tag{3.50}$$

because $[K]$ is symmetrical, the formula is changed to

$$\frac{\partial F}{\partial \phi_i} = \sum_{j=1}^{N} K_{ij}\phi_j - b_i = 0, \quad i = 1, 2, 3, \ldots, N, \tag{3.51}$$

or written in the matrix form:

$$[K]\{\phi\} = \{b\}. \tag{3.52}$$

The boundary conditions for the problem are usually of two forms: One is the Dirichlet boundary conditions, that is, given boundary value ϕ; the other is homogeneous Norman boundary condition, also called natural boundary condition, that is, the normal derivative of the boundary is zero. The first boundary conditions are necessary and must be imposed in the calculation. Second kinds of boundary conditions are usually met in the process of solving. In the finite element calculation, the boundary conditions can be satisfied by deleting rows and columns corresponding to the boundary nodes in the K matrix and the b vector.

3.4.2 Finite Element Neural Network

The discretization of the finite element method shows that the $[K]$ matrix and the $\{b\}$ vector are the key to the solution, while the $[K]$ matrix can be decomposed into the neural network. This indicates the direction of establishing the finite element neural network (FENN).

Take a simple 2D problem as an example:

$$-\nabla \bullet (\alpha \nabla \phi) + \beta \phi = f. \tag{3.53}$$

Its boundary condition is: on the boundary Γ_1, $\phi = p$; on the boundary Γ_2, $\alpha \nabla \phi \cdot \hat{n} + r\phi = q$, where α, β are coefficients related to the material properties; f is an external source; $\hat{n}$ is the external normal vector; r, q, p are parameters related to the boundary.

Assuming that the solution region is divided into two units, a total of four nodes, as shown in Figure 3.16, the element matrix K^e and b^e can be expressed as

$$K_{ij}^e = \int_{\Omega^e} (\alpha \nabla N_i^e \bullet \nabla N_i^e + \beta N_i^e N_j^e) d\Omega,$$
$$b_i^e = \int_{\Omega^e} f N_i^e d\Omega. \tag{3.54}$$

The global matrix can be obtained by merging the two element matrices. So we need a set of integers for the unit encoding. Also, in order to identify the nodes at the vertex of the unit, the other group of integers is required to give the node encoding. Because each element is associated with several nodes, a node, in addition to its position in the model, is also in the position of the corresponding element. This position can be used as integer encoding or part encoding. Global encoding shows the location of the node in the entire model.

In order to link these three kinds of encoding, an integer array or an array of contacts can be introduced. It is expressed as $n(i, e)$, where $i = 1, 2, 3$, $e = 1, 2, 3, \ldots, M$. M represents the number of units. In an array of contacts, i means the local encoding of the node, e is the unit encoding, $n(i, e)$ is the global encoding of the node. In this way, the array contains all the information about the unit and the node encoding.

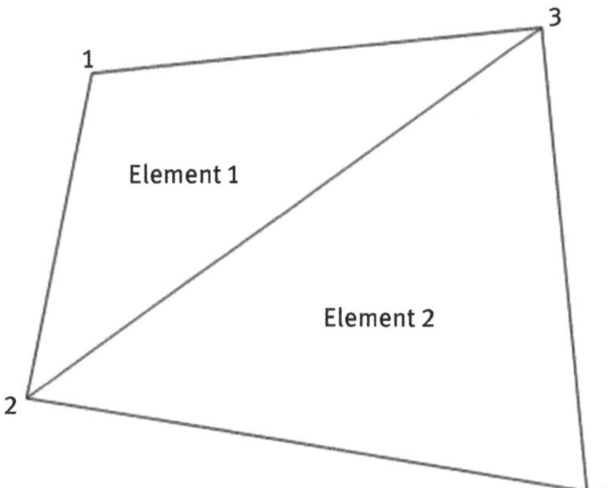

Figure 3.16: Grid partition for the solution area.

Table 3.4: Local and global encoding in the unit nodes.

e	*n*(1, *e*)	*n*(2, *e*)	*n*(3, *e*)
1	1	2	3
2	4	3	2

As shown in Figure 3.16, the array $n(i, e)$ can be set up in the solution area as shown in Table 3.4.

Using the connection information in Table 3.4, we can combine the unit matrix to establish the global matrix K. Take the node 2 as an example, because the node 2 is the second node of unit 1 and the third node of unit 2, respectively, so elements K_{22}^1 and K_{33}^2 in the unit matrix can be merged to obtain $K_{22} = K_{22}^1 + K_{33}^2$. For all four nodes to perform this operation, you can get

$$K = \begin{pmatrix} K_{11}^1 & K_{12}^1 & K_{13}^1 & 0 \\ K_{21}^1 & K_{22}^1 + K_{33}^2 & K_{23}^1 + K_{32}^2 & K_{31}^2 \\ K_{31}^1 & K_{32}^1 + K_{23}^2 & K_{33}^1 + K_{22}^2 & K_{21}^2 \\ 0 & K_{13}^2 & K_{12}^2 & K_{11}^2 \end{pmatrix}. \tag{3.55}$$

The vector b can be obtained by the combination of the elements:

$$b = \begin{pmatrix} b_1^1 \\ b_2^1 + b_3^2 \\ b_3^1 + b_2^2 \\ b_1^2 \end{pmatrix}. \tag{3.56}$$

In order to transform the finite element model into neural networks, it is necessary to decompose K_{ij}^e into two components: one component depends on the material properties α, β, the other is independent of the material. According to this idea, eq. (3.54) can be transformed into

$$K_{ij}^e = \int_{\Omega^e} (\alpha \nabla N_i^e \bullet \nabla N_i^e + \beta N_i^e N_j^e) d\Omega = \alpha S_{ij}^e + \beta T_{ij}^e \tag{3.57}$$

thereinto

$$S_{ij}^e = \int_{\Omega^e} \nabla N_i^e \bullet \nabla N_i^e d\Omega,$$

$$T_{ij}^e = \int_{\Omega^e} N_i^e N_j^e d\Omega. \tag{3.58}$$

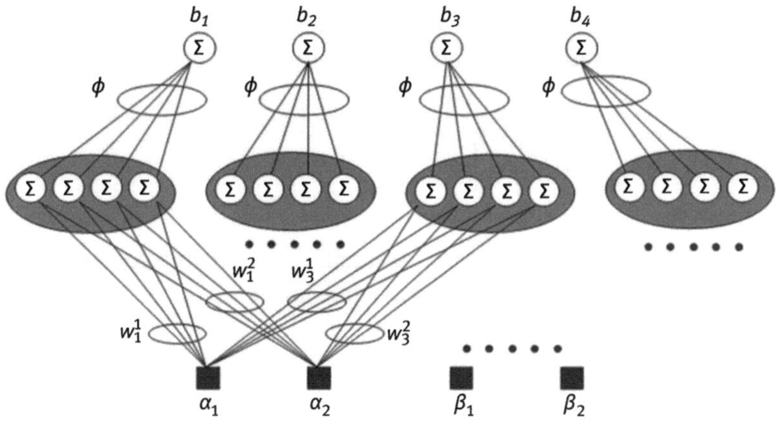

Figure 3.17: The structure of the finite element neural network.

On this basis, the finite element model can be transformed into a neural network, that is, FENN. Its structure is shown in Figure 3.17.

In Figure 3.17, the FENN contains three layers: input layer, output layer and hidden layer. The input layer consists of two sets of neurons. One group of each unit uses α as the input, and the other group uses β as input. The hidden layer contains 16 neurons, and four of them constitute a group. Each neuron in the hidden layer acts as a collection point, and the output is the element K_{ij} of the global matrix K. Its expression is

$$K_{ij} = \sum_{e=1}^{2} (\alpha_e w_{ij}^e + \beta_e g_{ij}^e). \tag{3.59}$$

The output of the hidden layer neurons in each group is a row vector corresponding to a node in the K matrix. The weights between the input layer and the hidden layer are set to the corresponding values of S and T, and the relationship between w_{ij}^e, g_{ij}^e and T, S is: If i and j belong to one node, then $w_{ij}^e = S_{ij}^e$, otherwise, $w_{ij}^e = 0$. If i and j belong to one node, then $g_{ij}^e = T_{ij}^e$, otherwise, $g_{ij}^e = 0$.

Each hidden neurons are connected to an output neuron through a set of weights φ. Each element of φ is corresponding to the node value ϕ_j. Each output neuron is also a collection point, the output value is b_i:

$$b_i = \sum_{j=1}^{4} K_{ij}\phi_j = \sum_{j=1}^{4} \phi_j(\alpha_e w_{ij}^e + \beta_e g_{ij}^e). \tag{3.60}$$

By the above transformation, the finite element model of the solution can be transformed into the form of neural network. To get a solution to the unknown quantity, the boundary conditions must also be considered.

According to the variational principle of finite element method, the natural boundary conditions can be satisfied by adding an additional term $F_b(\phi)$ to the functional. It can be expressed as

$$F_{\mathrm{b}}(\phi) = \int\limits_{\Gamma_2} \left(\frac{r}{2}\phi^2 - q\phi\right) d\Gamma. \tag{3.61}$$

Here, the subscript b indicates the boundary. Assuming that M_s line segments are included in Γ_2, then $F_{\mathrm{b}}(\phi)$ can be expressed as a discrete form:

$$F_{\mathrm{b}}(\phi) = \sum_{s=1}^{M_s} F_b^e(\phi^s), \tag{3.62}$$

where $F_{\mathrm{b}}^e(\phi^s)$ represents the integral on the line s. The unknown function ϕ within each segment can be approximated as

$$\phi^s = \sum_{j=1}^{2} N_j^s \phi_j^s, \tag{3.63}$$

where N is the basis function defined in the unit s, and the expression is

$$N_1^s = 1 - \xi, N_2^s = \xi, \tag{3.64}$$

where ξ is the normalized distance between nodes 1 and 2. In particular, in node 1, $\xi = 0$. In node 2, $\xi = 1$. Between the two nodes, it changes linearly. Then you can get

$$\left\{ \frac{\partial F_{\mathrm{b}}^s}{\partial \phi_j^s} \right\} = [K^s]\{\phi^s\} - \{b^s\}, \tag{3.65}$$

thereby

$$\begin{aligned} K_{ij}^s &= \int_0^1 r N_i^s N_j^s l^s d\xi, \quad i, j = 1, 2, \\ b_i^s &= \int_0^1 q N_i^s l^s d\xi, \quad i = 1, 2, \\ \phi^s &= [\phi_1^s, \phi_2^s]^T. \end{aligned} \tag{3.66}$$

If r and q are constants in each line segment, and they are expressed as r^s and q^s, respectively. So we can find the solution by using the analytic method:

$$\begin{aligned} K_{ij}^s &= r^s \frac{l^s}{6}(1 + \delta_{ij}), \\ b_i^s &= q^s \frac{l^s}{2}. \end{aligned} \tag{3.67}$$

In order to take F_{b} into account the global equation, we obtain

$$\begin{aligned}\left\{\frac{\partial F}{\partial \phi}\right\} &= \sum_{e=1}^{M}\left\{\frac{\partial F^{e}}{\partial \phi^{e}}\right\} + \sum_{s=1}^{M_s}\left\{\frac{\partial F_{\mathrm{b}}^{s}}{\partial \phi^{s}}\right\} \\ &= \sum_{e=1}^{M}([K^{e}]\{\phi^{e}\} - \{b^{e}\}) + \sum_{s=1}^{M_s}([K^{s}]\{\phi^{s}\} - \{b^{s}\}) \\ &= \{0\}.\end{aligned} \tag{3.68}$$

The formula can be simplified to

$$\frac{\partial F}{\partial \phi} = (K\phi - b) + (K^{s}\phi^{s} - b^{s}) = 0. \tag{3.69}$$

In order to expand the vector and matrix, it is clear that an array, linked to the global encoding of the line segment and the associated node, is needed. This array can be used to express $ns(i,s)$ $(i = 1,2,3,\ s = 1,2,3,\ldots,M_s)$. It has a similar effect on the contact array $n(i,e)$ discussed earlier. Specifically, the value of $ns(i,s)$ represents the global encoding of the i node on the s line segment. Then, add every K_{ij}^{s} to $K_{ns(i,s),ns(j,s)}$ and add b_i^s to $b_{ns(i,s)}$. From the above derivation, we can see that $[K^s]$ and $\{b^s\}$ do not rely on the characteristics of the material, that is unrelated to α and β. This indicates that the natural boundary conditions can be used as the boundary of the hidden layer nodes and the corresponding output nodes of the bias input.

3.4.3 From One Dimension to Three Dimensions

The method of constructing the finite element network and satisfying the boundary conditions is analyzed. However, the example is a 2D problem. The actual quantifying problem of the MFL testing need to use 3D model. Therefore, it is necessary to build up a 3D FENN to study the FENN. For the finite element network, the key lies in the solution of w and g matrices, that is, the calculation of the weights between the input layer and the hidden layer. To obtain the elements of the two matrices, the value and the analytical expression of the interpolation function are needed. Taking one-dimensional problem as an example, the usual practice is to set the variable ϕ to make a linear change in each subdomain (i.e. each line), and thus get the corresponding interpolation function

$$N_1^e(x) = \frac{x_2^e - x}{l^e},\quad N_1^e(x) = \frac{x - x_1^e}{l^e},\quad l^e = x_2^e - x_1^e. \tag{3.70}$$

And there are

$$N_j^e(x_i^e) = \delta_{ij} = \begin{cases}1, & i = j\\ 0, & i \neq j\end{cases}. \tag{3.71}$$

For 2D problems, the solution region is divided into n linear triangular elements, then the unknown function ϕ can be expressed as the form of $a+bx+cy$. The corresponding interpolation function can be obtained by solving the equations set up to the three vertices:

$$\begin{aligned}
&N_j^e(x,y) = \frac{1}{2\Delta^e}(a_j^e + b_j^e x + c_j^e y),\ j = 1,2,3,\\
&a_1^e = x_2^e y_3^e - y_2^e x_3^e;\quad b_1^e = y_2^e - y_3^e;\quad c_1^e = x_3^e - x_2^e;\\
&a_2^e = x_3^e y_1^e - y_3^e x_1^e;\quad b_2^e = y_3^e - y_1^e;\quad c_2^e = x_1^e - x_3^e;\\
&a_3^e = x_1^e y_2^e - y_1^e x_2^e;\quad b_3^e = y_1^e - y_2^e;\quad c_3^e = x_2^e - x_1^e.
\end{aligned} \tag{3.72}$$

In eq. (3.72),

$$\Delta^e = \frac{1}{2}\begin{vmatrix} 1 & x_1^e & y_1^e \\ 1 & x_2^e & y_2^e \\ 1 & x_3^e & y_3^e \end{vmatrix} = \frac{1}{2}(b_1^e c_2^e - b_2^e c_1^e), \tag{3.73}$$

so to get the value of the interpolation function:

$$N_j^e(x_i^e, y_j^e) = \delta_{ij} = \begin{cases} 1, & i = j \\ 0, & i \neq j \end{cases}, \tag{3.74}$$

where x_j^e and y_j^e $(j = 1,2,3)$ represent the coordinates of the vertices.

For 3D problems, the solution region is divided into n linear tetrahedral elements, as shown in Figure 3.18.

The unknown function ϕ within each region can be approximated as

$$\phi^e(x,y,z) = a^e + b^e x + c^e y + d^e z. \tag{3.75}$$

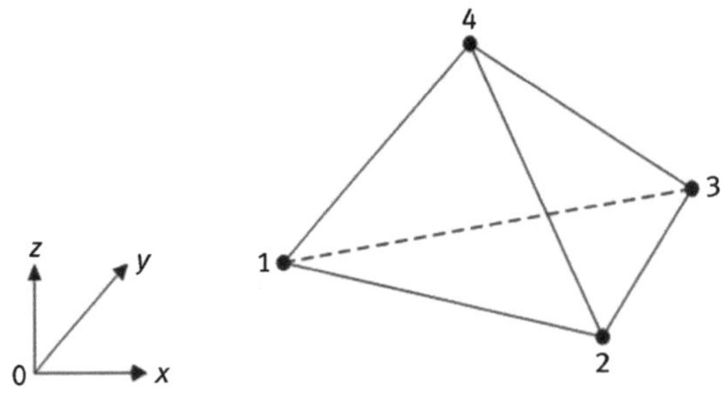

Figure 3.18: The unit of linear tetrahedral element.

The four vertices of the unit are imposed on eq. (3.75), and the equations are solved simultaneously. a^e, b^e, c^e, d^e can be obtained accordingly. The value of ϕ on the j node is signed as ϕ_j^e

$$\begin{aligned}
\phi_1^e(x,y,z) &= a^e + b^e x_1^e + c^e y_1^e + d^e z_1^e,\\
\phi_2^e(x,y,z) &= a^e + b^e x_2^e + c^e y_2^e + d^e z_2^e,\\
\phi_3^e(x,y,z) &= a^e + b^e x_3^e + c^e y_3^e + d^e z_3^e,\\
\phi_4^e(x,y,z) &= a^e + b^e x_4^e + c^e y_4^e + d^e z_4^e.
\end{aligned} \tag{3.76}$$

From eq. (3.76), we can obtain

$$\begin{aligned}
a^e &= \frac{1}{6V^e}\begin{vmatrix} \phi_1^e & \phi_2^e & \phi_3^e & \phi_4^e \\ x_1^e & x_2^e & x_3^e & x_4^e \\ y_1^e & y_2^e & y_3^e & y_4^e \\ z_1^e & z_2^e & z_3^e & z_4^e \end{vmatrix} = \frac{1}{6V^e}(a_1^e\phi_1^e + a_2^e\phi_2^e + a_3^e\phi_3^e + a_4^e\phi_4^e),\\
b^e &= \frac{1}{6V^e}\begin{vmatrix} 1 & 1 & 1 & 1 \\ \phi_1^e & \phi_2^e & \phi_3^e & \phi_4^e \\ y_1^e & y_2^e & y_3^e & y_4^e \\ z_1^e & z_2^e & z_3^e & z_4^e \end{vmatrix} = \frac{1}{6V^e}(b_1^e\phi_1^e + b_2^e\phi_2^e + b_3^e\phi_3^e + b_4^e\phi_4^e),\\
c^e &= \frac{1}{6V^e}\begin{vmatrix} 1 & 1 & 1 & 1 \\ x_1^e & x_2^e & x_3^e & x_4^e \\ \phi_1^e & \phi_2^e & \phi_3^e & \phi_4^e \\ z_1^e & z_2^e & z_3^e & z_4^e \end{vmatrix} = \frac{1}{6V^e}(c_1^e\phi_1^e + c_2^e\phi_2^e + c_3^e\phi_3^e + c_4^e\phi_4^e),\\
d^e &= \frac{1}{6V^e}\begin{vmatrix} 1 & 1 & 1 & 1 \\ x_1^e & x_2^e & x_3^e & x_4^e \\ y_1^e & y_2^e & y_3^e & y_4^e \\ \phi_1^e & \phi_2^e & \phi_3^e & \phi_4^e \end{vmatrix} = \frac{1}{6V^e}(d_1^e\phi_1^e + d_2^e\phi_2^e + d_3^e\phi_3^e + d_4^e\phi_4^e),
\end{aligned} \tag{3.77}$$

thereby

$$V^e = \frac{1}{6}\begin{vmatrix} 1 & 1 & 1 & 1 \\ x_1^e & x_2^e & x_3^e & x_4^e \\ y_1^e & y_2^e & y_3^e & y_4^e \\ z_1^e & z_2^e & z_3^e & z_4^e \end{vmatrix} = \text{Elementary volume}. \tag{3.78}$$

The expressions of a^e, b^e, c^e, d^e are replaced by eq. (3.48), then we get

$$\begin{aligned} \phi^e(x,y,z) &= \sum_{j=1}^{4} N_j^e(x,y,z)\phi_j^e, \\ N_j^e(x,y,z) &= \frac{1}{6V^e}(a_j^e + b_j^e x + c_j^e y + d_j^e z) = \delta_{ij}\begin{cases} 1, & i=j \\ 0, & i \neq j \end{cases} . \end{aligned} \tag{3.79}$$

By the expression and value of the above-mentioned interpolation function, the elements of w and g matrices of FENN can be obtained, and the construction of 3D FENN is realized.

A 2D example of the FENN construction method is introduced in the front section. For a general case, the FENN construction can be set up according to the way shown in Figure 3.19. It is assumed that the finite element mesh structure built in advance has M elements, N nodes, and the characteristic parameters of each element are α and β. Then the input layer of the neural network structure contains $2M$ neurons, namely M number of α and M number of β. They are then multiplied by the weight matrix element w_{ij}^e, g_{ij}^e as the input of the hidden layer. The hidden layer contains N^2 neurons, N neurons consist a group, the output of each group is a row vector of K matrix. The output b_i of the hidden layer is then multiplied by the value ϕ of the corresponding node. The output layer has N neurons, representing the value b_i of the N nodes.

3.4.4 Solving Positive and Inverse Problems Using FENN

The essence of the FENN solution of the direct problem is seeking the weights ϕ of the FENN hidden layer to the output layer at the condition of material parameters and excitation source. Steps are as follows:

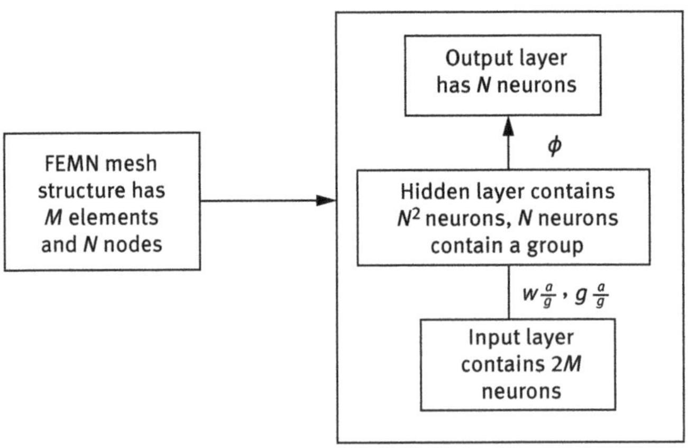

Figure 3.19: The FENN construction method in general.

1. A finite element mesh is built, which includes M elements and N nodes. The material parameters α_i and β_i of each unit are added into the input neurons to calculate the output of each neuron. Any natural boundary conditions are used as the input of the hidden layer neuron. According to the Dirichlet boundary condition, some weights can be fixed. For all free weights, the initial values are randomly selected. Set the weight of t cycle as $\phi(t)$.
2. The output of the neural network in the t cycle is

$$\hat{b}_i(t) = \sum_{j=1}^{N} K_{ij}\phi_j(t), \quad t = 1, 2, \ldots, N. \tag{3.80}$$

3. To calculate the error of the output of the neural network:

$$E = \frac{1}{2}\left\|b - \hat{b}(t)\right\| = \frac{1}{2}\sum_{i=1}^{N}(b - \hat{b}(t))^2 = \frac{1}{2}\sum_{i=1}^{N}E_i(t)^2. \tag{3.81}$$

4. The gradient value of the error relative to the free hidden layer weight is calculated:

$$\frac{\partial E(t)}{\partial \phi_j} = -\sum_{i=1}^{N}E_i(t)K_{ij}. \tag{3.82}$$

5. Use gradient descent method to update the free weights:

$$\phi_j(t+1) = \phi_j(t) + \eta\left(-\frac{\partial E(t)}{\partial \phi_j}\right), \tag{3.83}$$

   where η is the learning rate, which can be chosen as a suitable value according to the experience.
6. Repeat steps 2–5 until the output error is less than a predetermined threshold.

FENN can also be used to solve the inverse problem, that is, the known measured value ϕ and the external excitation source b to solve the material parameters α and β. Specific algorithms are as follows:

1. Randomly set the initial value of the material parameters α and β. $\alpha(t)$ and $\beta(t)$ represent material parameter values in the t cycle. The weights between the hidden layer and the output layer are fixed as ϕ.
2. In the tth cycle, the output of the network will be calculated by the input of $\alpha(t)$ and $\beta(t)$:

$$\hat{b}_i(t) = \sum_{j=1}^{N} K_{ij}\phi_j(t), \quad t = 1, 2, \ldots, N. \tag{3.84}$$

3. To calculate the error of the output of the neural network:

$$E = \frac{1}{2}\left\| b - \hat{b}(t) \right\| = \frac{1}{2}\sum_{i=1}^{N}(b - \hat{b}(t))^2 = \frac{1}{2}\sum_{i=1}^{N}E_i(t)^2. \tag{3.85}$$

4. To calculate the gradient value of the error relative to $\alpha(t)$ and $\beta(t)$:

$$\begin{aligned}\frac{\partial E(t)}{\partial \alpha_e} &= -\sum_{i=1}^{N}E_i\left(\sum_{j=1}^{N}\phi_j w_{ij}^e\right),\\ \frac{\partial E(t)}{\partial \beta_e} &= -\sum_{i=1}^{N}E_i\left(\sum_{j=1}^{N}\phi_j g_{ij}^e\right).\end{aligned} \tag{3.86}$$

5. Use gradient descent method to update $\alpha(t)$ and $\beta(t)$:

$$\begin{aligned}\alpha_e(t+1) &= \alpha_e(t) + \eta\left(-\frac{\partial E(t)}{\partial \alpha_e}\right),\\ \beta_e(t+1) &= \beta_e(t) + \eta\left(-\frac{\partial E(t)}{\partial \beta_e}\right).\end{aligned} \tag{3.87}$$

6. Repeat steps 2–5 until convergence is reached, that is, the output error is less than a predetermined threshold.

The method of solving the positive and inverse problems with FENN is given above. In the problem of the quantitative detection of defects, the problem is that the shape parameters (i.e. α and β) and the external magnetic field strength are known, and the distribution of the MFL (i.e. the weight ϕ) is solved. The inverse problem is that the external excitation source and the distribution of the magnetic leakage induction intensity are known, and the shape parameters of the defect is solved. Combined with the iterative method, the 3D FENN is used as the positive problem model, and the iterative method based on 3D FENN can be obtained.

This method has the characteristics of finite element calculation that is based on the Maxwell equation, considering the material nonlinearity, so it has high accuracy. And it has the characteristics of neural network, based on parallel computing, so the speed is fast. At the same time, it can also enhance the approximation ability by the feedback structure with iterative method.

3.4.5 Analysis on the Advantages of the FENN

The main advantages of 3D FENN is the parallel implementation of the finite element calculation, so it is easy to achieve the parallel computing in hardware and software, which can greatly accelerate the speed of finite element calculation. And error gradient calculation is very convenient in FENN, which can also use a parallel calculating mode, so the gradient value can be obtained rapidly using a gradient descent method to solve the forward problem and inverse problem. In addition, when FENN solves the inverse problem, inverse matrix does not need to solve the global matrix in each cycle. So it can reduce the difficulty of the inverse problem calculation, saving computing time.

Another important advantage of FENN is that it does not need any training because most of the weights have been computed and stored in advance. Because the weights depend on the differential equations and boundary conditions, if the two conditions do not change, it does not need to change the weights. This means that as a quantitative method of corrosion defects, FENN does not depend on the training sample, which can improve the ability to identify defects with irregular shapes in practice.

From the construction of FENN, one of its main disadvantage is the need for a large number of neurons and weights, which makes the system memory capacity requirements higher. However, because the most weights between the input layer and hidden layer is 0 (determined by w and g matrices), these weights and the corresponding connection will be discarded in the calculation, so the demand of memory will be greatly reduced. Also, because the matrix K is a symmetric banded matrix, in which the most elements is 0, so neurons of the hidden layer and related links can also be discarded in the calculation. Furthermore, each group of weights between hidden layer neuron and output layer is the same (value ϕ for the corresponding node).

3.4.6 The Optimization of FENN

For the optimization of FENN, you can start from the optimization problem of finite element mesh node number. Because the matrix K is symmetric, an up triangular (or lower triangular) matrix need only to be stored when programming. However, because of the sparsity of the matrix, the memory space is still large. Therefore, according to the zonal distribution characteristics of the nonzero elements in matrix K, namely, the nonzero elements in the upper triangular matrix are all contained in an equivalent bandwidth on one side of the diagonal, the 2D equivalent bandwidth storage is often used, whose memory space is proportional to the bandwidth. For n-order matrix, usually the maximum bandwidth is selected as D, which is the maximum number among all rows of the matrix from the main diagonal to the last nonzero elements. The detailed method is to seek the maximum half-bandwidth D_e of any row, and then to calculate the maximum value of all the rows. D_e can be expressed as

$$D_e = (n_{\max} - n_{\min})p, \tag{3.88}$$

where $n_{\max}$ and $n_{\min}$ are the minimum and maximum values of global encoding in node e, p is the degree of freedom for the node. Therefore, the bandwidth of matrix

K is associated with the maximum difference between node numbers of all elements. Bandwidth for the same finite element model may not be the same because of different node encoding methods corresponding to matrix K. In order to get the optimal bandwidth value, the node encoding should be appropriate to arrange nonzero elements in matrix K as much as possible on both sides close to the diagonal, thus saving computer memory space and computing time, and improving the calculation speed and efficiency.

The optimization method of grid node numbering is an inductive and deductive method, which is based on an improved 2D grid node numbering (the most typical grid consists of eight units). According to the characteristics of the best node numbering (minimum bandwidth), some rules have been summed up. And then the nodes without optimization are adjusted by the rules repeatedly until the semibandwidth value remains unchanged.

Figure 3.20 shows a grid with eight units whose node numbers are in the optimum condition, and the maximum interpolation (i.e. bandwidth value) in the same unit is 4. It found the following three rules:

1. Node ratio (i.e. the sum of node numbers in related unit divided by the sum of relevant units) is in ascending order from small to large.
2. The sum of the minimum node number and the maximum node number also grows up in ascending order.
3. The maximum node is basically in ascending order, but with some repeated numbers.

Thus, a new optimization algorithm of node encoding can be summed up. The specific steps are as follows:

1. Read the node number, unit number and unit topology array from the grid data file.
2. According to the sequence of nodes, node number around each node is searched repeatedly. The number of related node is established.
3. Calculate the sum of the related node numbers of each node.
4. Calculate the maximum node number and the minimum node number in the related units with that node.

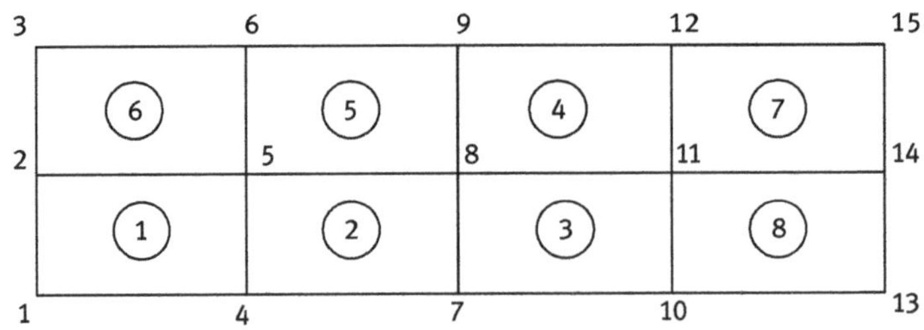

Figure 3.20: The grid with better node number.

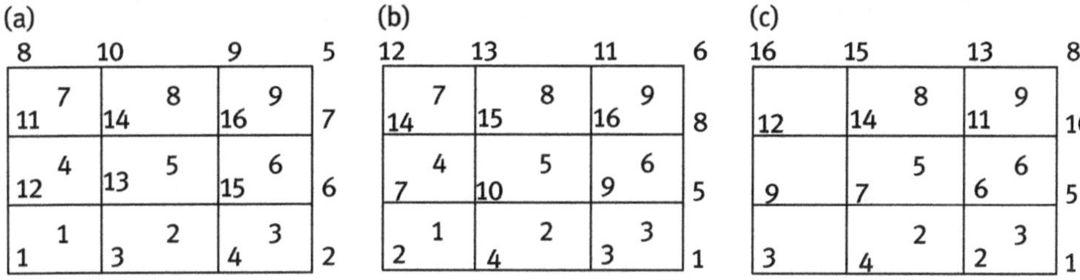

Figure 3.21: Mesh optimization example. (a) Initial grid and node number, half bandwidth is 28 (b) The grid after once renumbering, half bandwidth is 22 (c) The grid many times renumbering, half bandwidth is 16

5. Calculate the sum of the minimum node number and the maximum node number.
6. Calculate the node ratio, which is the ratio of the sum of node numbers in related unit to the sum of relevant units.
7. Renumber the nodes according to the node ratio. Small node ratio has small node number, and large node ratio has large node number; if node ratio is equal, the sum of the minimum node number and maximum node number will be the renumbering rule.
8. Repeat steps 3–7, until the bandwidth does not decrease.

Using this algorithm, the node number of a six-unit grid is optimized, and the results are shown in Figure 3.21. As you can see, by optimizing the number, the bandwidth is reduced from 28 to 16, which decreases to 42.86 %.

(a) Initial grid and node number, half bandwidth is 28 (b) The grid after once renumbering, half bandwidth is 22 (c) The grid many times renumbering, half bandwidth is 16

This new node numbering optimization method can be used in a variety of finite element analysis and calculation, including the plane triangle unit, quadrilateral element, 3D tetrahedron, hexahedron elements and shell unit. It can reduce the difference between the node numbers in the same element, thereby reducing the bandwidth of finite element calculation. The numbers of grid nodes after optimization is input to the FENN; it can improve the calculation efficiency of FENN and reduce the demand of memory.

4 Defect Profile Inversion of Three-Dimensional MFL Detection

4.1 The Characteristics of Three-Dimensional MFL Signal

Figure 4.1 shows the schematic diagram of oil and gas pipeline three-dimensional (3D) magnetic flux leakage (MFL) detection, including the principal structure of pipeline MFL detector's measurement part and the corresponding pipeline MFL detection's circuit structure [108]. The magnetic circuit of 3D MFL testing is mainly composed of steel brush, permanent magnet, back iron and pipe wall. In the detector, there are arranged measuring probes evenly along the circumferential direction of the pipe. In each probe, there are several coil sensors or Hall sensors for the measurement of magnetic leakage signals, which include axial signal *B*a, radial signal *B*r and circumferential signal *B*c, caused by wall defects.

In order to obtain the optimal detection effect, permanent magnets are used along the pipe axis to magnetize the wall to saturation or near-saturation state. At the same time, the magnetic sensors are arranged in the center of the magnetic circuit, to make them in a uniform magnetic field, so as not to be affected by the irregular magnetic field near the steel brushes.

In order to verify the detector's performance, a series of standard defects, including pinholes, pits, circumferential grooves and slots, axial groove and slots, were processed on a piece of pipe. According to the geometry division, these defects can be divided into standard rectangular defects, spherical defects and cylindrical defects. Based on this, the three kinds of defects with the rectangle, spherical and cylindrical shapes are chosen for simulation analysis and theoretical research.

4.1.1 The Basic Characteristics of the Signal

Based on the symmetry of oil and gas pipeline 3D MFL detection's circuit structure, pipeline circumferential 90° finite element simulation model is established, using finite element analysis software ANSYS, for the 3D MFL signal simulations of any defects, as shown in Figure 4.2. The 3D MFL detector size and magnetic properties of materials using the simulation model are shown in Table 4.1. Figure 4.3 shows magnetization curves of back iron, steel brush and wall materials in the 3D finite element simulation model.

Figures 4.4–4.6 show the contour plots of three axial MFL signals for a rectangular defect of 28.6 mm × 14.3 mm × 7.2 mm (length × width × depth), a globoid defect of 28.6 mm × 7.2 mm (diameter × depth), a cylindrical defect of 28.6 mm × 7.2 mm (diameter × depth), respectively. Several general properties of the signals could be summarized as follows.

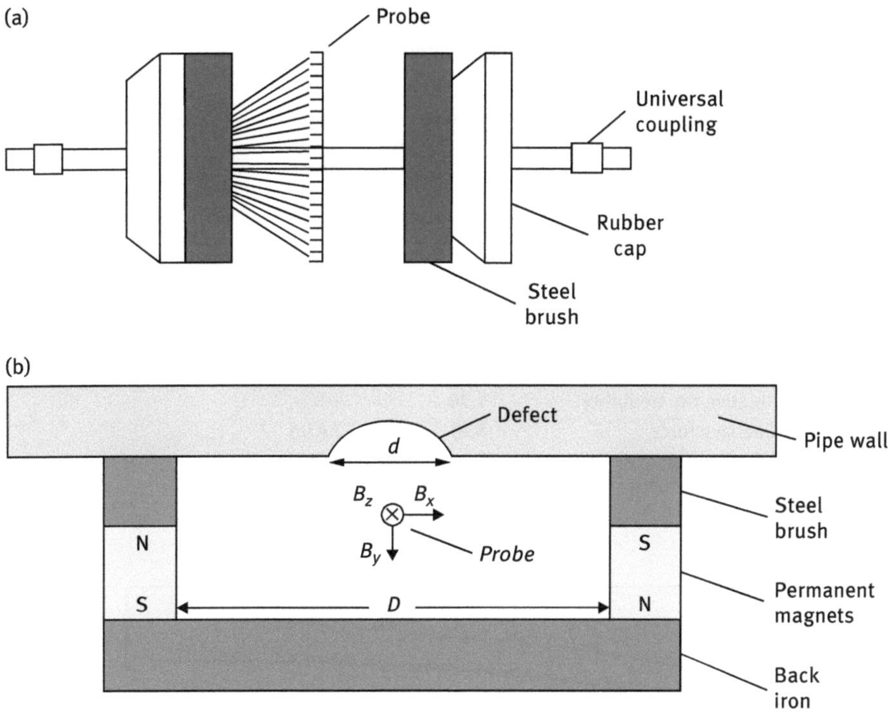

Figure 4.1: The schematic diagram of oil and gas pipeline 3D MFL detection: (a) the principal structure of pipeline MFL detector's measurement part and (b) the magnetic circuit of 3D MFL testing.

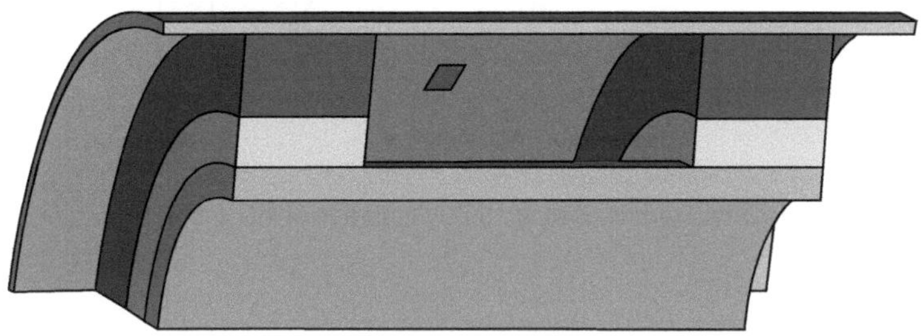

Figure 4.2: A finite element simulation model of oil and gas pipeline 3D MFL detection.

1. The radial component consists of a pair of peaks, one positive and one negative. There are one plane of symmetry and one plane of antisymmetry. To a large extent, the radial component characterizes the opening contour of the defect.
2. The circumferential component consists of four peaks, two positive and two negative. There are two planes of antisymmetry, and the circumferential component is a key reference representing the turning points of the side boundary of the defect.

Table 4.1: The sizes of a three-dimensional magnetic flux leakage detector and characteristic parameters of magnetic materials.

Symbol	Parameter	Value	Unit
OD	Pipe diameter	457	mm
T	Wall thickness	14.3	mm
W_p	Width of magnet	80	mm
H_p	Height of magnet	30	mm
W_s	Width of brush	80	mm
H_s	Height of brush	50	mm
H_b	Height of back iron	20	mm
D	Spacing of poles	1,000	mm
μ_r	Relative permeability	1.26	–
Hcb	Coercive force	836	kA/m
L	Liftoff value	3	mm

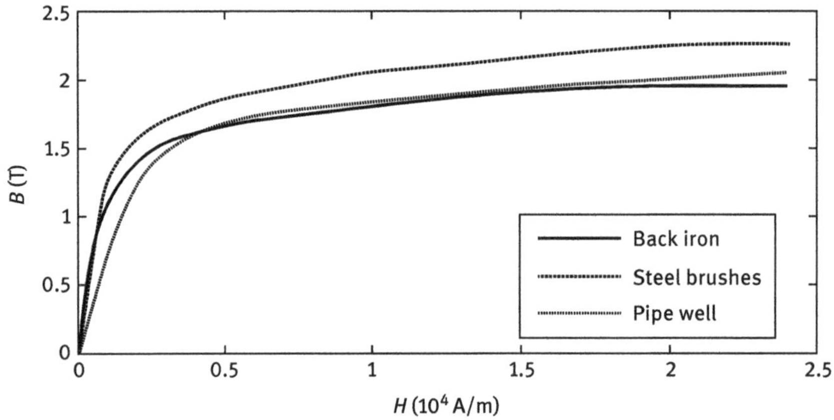

Figure 4.3: Nonlinear characteristics of materials used in 3D FE simulation of MFL inspection.

3. The axial component consists of two small troughs and the peak part between them. There are two planes of symmetry, and the two troughs represent the turning points of the depth profile of the defect.

In Figures 4.4–4.6, the dotted lines identify the actual open contours of defects. Comparison of defect opening contours and 3D MFL signals shows that

1. In the axial direction along which the external magnetizing field is applied, the two valley points of the axial component are located at the left and right edges of defects, the peak is located in the center of the defect area.

2. The peak and valley points of the radial component are located at the left and right edges of defects.
3. The peak values of the circumferential components are located on the edges of upper left and bottom right, and valley values are located on the edges of bottom left and upper right.

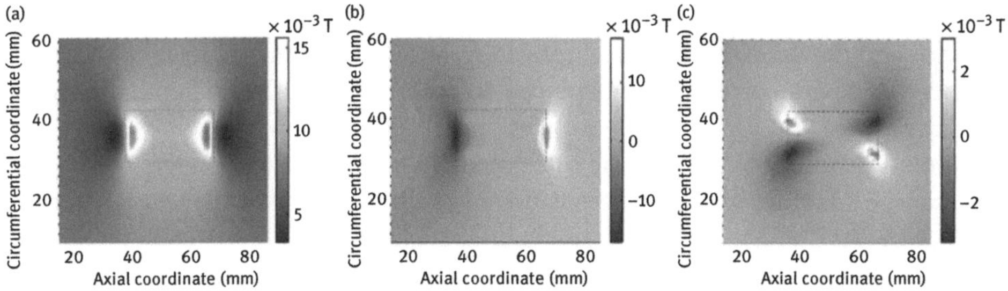

Figure 4.4: Three axial MFL signals for a rectangular defect of 28.6 mm × 14.3 mm × 7.2 mm: (a) axial component; (b) radial component; and (c) circumferential component.

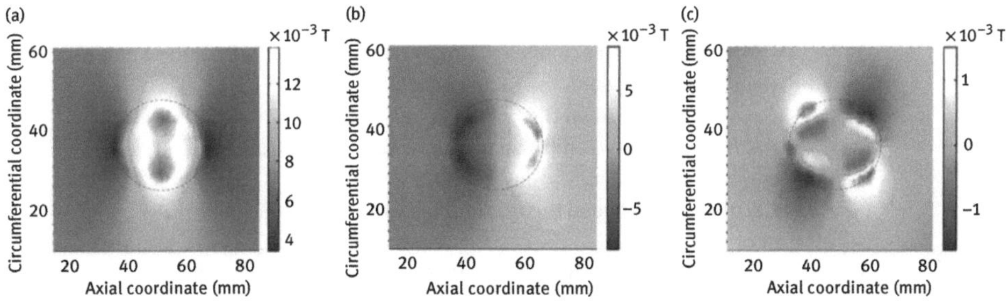

Figure 4.5: Three axial MFL signals for a globoid defect of 28.6 mm × 7.2 mm: (a) axial component; (b) radial component; and (c) circumferential component.

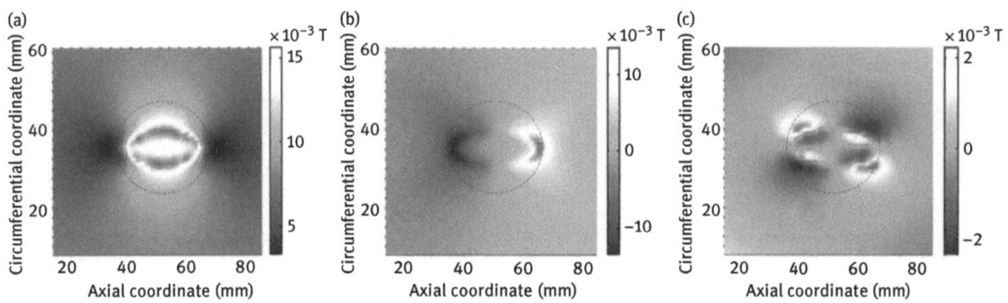

Figure 4.6: Three axial MFL signals for a cylindrical defect of 28.6 mm × 7.2 mm: (a) axial component; (b) radial component; and (c) circumferential component.

4.1.2 The Signal Changes with the Size of Defects

In order to research the change rules of 3D MFL signal varying with the defect parameters, characteristic parameters of 3D MFL signals, based on Figure 4.4, are defined as shown in Table 4.2.

Based on 3D finite element simulations, a series of rectangular defect signal distribution in 3D MFL detection were obtained. Among them, the variation range of the defect length was from 15 to 105 mm, the width of defect ranged from 17.5 mm to 39.5 mm and the depth of the defect ranged from 1.8 to 10 mm. Changes of statistical characteristic parameters with rectangular defect length, width and depth are shown in Figures 4.7–4.9.

From Figures 4.7–4.9, the basic rules of 3D characteristic parameters of MFL signals can be obtained under different defect sizes. The peak–valley differences of

Table 4.2: Definition of characteristic parameters of three axial MFL signals for rectangular defect.

Parameter	Unit	Definition
P_1	Gs	Peak-to-peak difference of axial data
P_2	Gs	Peak-to-peak difference of radial data
P_3	Gs	Peak-to-peak difference of circumferential data
P_4	mm	Axial distance between the peaks of axial data
P_5	mm	Axial distance between the troughs of axial data
P_6	mm	Axial distance between the maximum and minimum of radial data
P_7	mm	Axial distance between the maximum and minimum of circumferential data
P_8	mm	Circumferential distance between the maximum and minimum of circumferential data

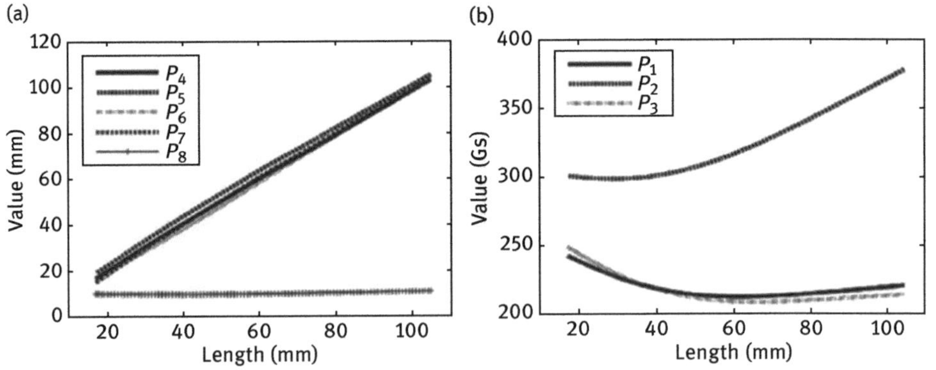

Figure 4.7: Relationship between 3D MFL characteristic parameters and corresponding defect length: (a) P_1–P_3 and (b) P_4–P_8.

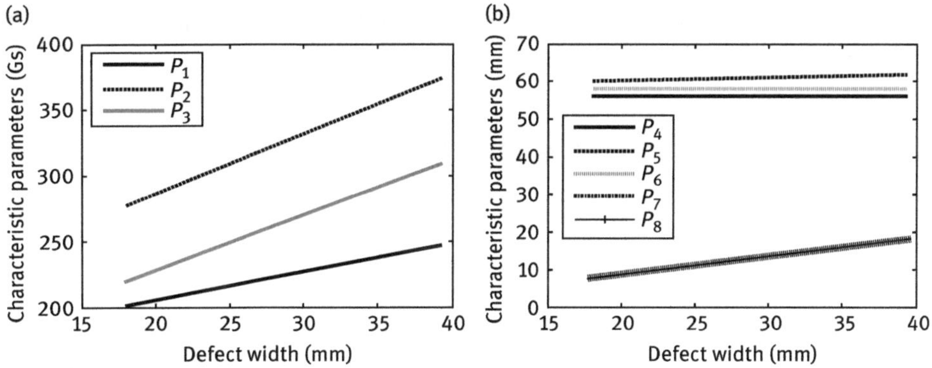

Figure 4.8: Relationship between 3D MFL characteristic parameters and corresponding defect width: (a) P_1–P_3 and (b) P_4–P_8.

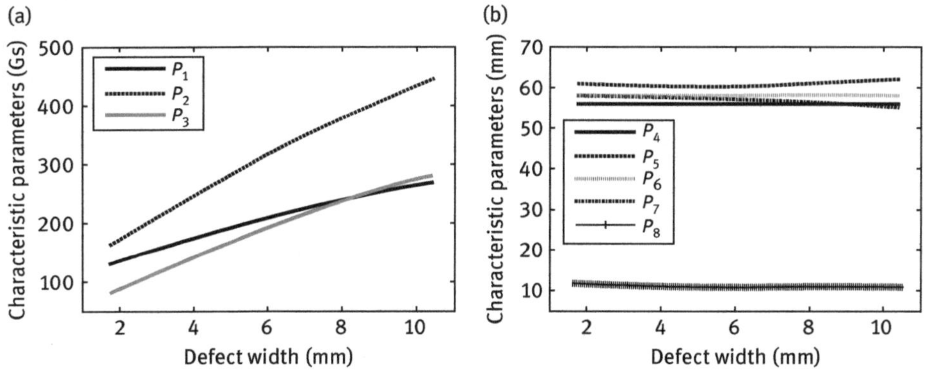

Figure 4.9: Relationship between 3D MFL characteristic parameters and corresponding defect depth: (a) P_1–P_3 and (b) P_4–P_8.

the axial, radial and circumferential components increase with the defect length, width and depth obviously, showing the overall monotonically increasing relationship. When the defect length increases, the circumferential spacing between the peak value and valley value of the circumferential component will remain unchanged; however, the rest of the characteristic parameters associated with axial spacing are increased. When the defect width increases, the circumferential spacing between the peak value and valley value of the circumferential component will increase, and the remaining characteristic parameters associated with the axial spacing are unchanged. When the defect depth increases, all characteristic parameters associated with axial spacing and circumferential spacing show no obvious change.

As shown, characteristic parameters of all the three components might be influenced when the defect profile changes. This further demonstrates the fact that all the three components of MFL signals carry useful information corresponding to the defect profile. Therefore, three axial MFL inspection should be carried out to collect enough information for defect reconstruction.

4.2 Random Search Iterative Inversion Method of Defect 3D Contour

4.2.1 The Region Segmentation and Recognition

In order to realize the 3D profile inversion of the defect, the target area, that is region of interest (ROI), to be solved is first identified based on the 3D MFL detection signal. Because all MFL signals generated by the defects contain the potential information of defect profile, in order to invert the defect profile more accurately, the original defect MFL signal should be reserved. At the same time, in the condition of the complete MFL signal with defects, the identified ROI should be as small as possible, which will reduce the amount of data needed to be processed in the inversion process.

The amplitude of the axial component of the 3D MFL detection signal is the most basic criterion for ROI. In the nondefect area, the amplitude of the axial MFL signal is smaller and corresponding detection curve is relatively flat. However, in the defect region, the amplitude of the axial MFL signal is increased significantly and the corresponding detection curve changes obviously. In view of this, the defect region with its adjacent area is defined as the ROI area to be solved, corresponding to the no defect area is called the non-ROI area.

Based on the above characteristics of the axial MFL detection signal, in order to designate the ROI area and the non-ROI area, the following segmentation method can be used:

1. The MFL testing data obtained from the pipeline is divided according to the circumferential direction, and the measured data of each sensor is detected separately.
2. For the axial MFL data of each channel to be detected, it will be divided into many data segments whose length along the axial direction is L, find the maximum value $\max\{X\}$ and minimum value $\min\{X\}$ within each data segment.
3. Define the threshold TH_1 and TH_2, if $\max\{X\} - \min\{X\} > TH_2$ or $\max\{X\} > TH_1$ is found; it is judged that the detection data is in the ROI, otherwise, the detection data is in non-ROI.

The key of the segment recognition method is the definition of appropriate data length L, and thresholds TH_1 and TH_2, and these parameters need to be chosen reasonably according to the wall magnetization and the detector sampling interval.

Figure 4.10 is the pipe axial MFL detection data of 457 mm diameter, 14.3 mm wall thickness. The signal maximum value is 4,095, corresponding to 500 Gs of the magnetic leakage field. Based on this data, the identification test of the ROI of the pipe

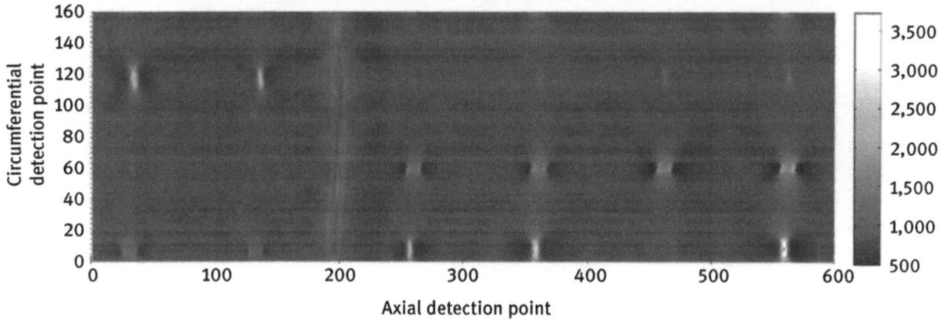

Figure 4.10: Pipe axial MFL test data sample with 457 mm diameter and 14.3 mm wall thickness.

wall is carried out. Analysis of the axial MFL data shows that in the section without defect, the corresponding amplitude of the magnetic induction intensity is about 125 Gs; in the defect area, the corresponding amplitude of the magnetic induction intensity is over 200 Gs. Taking into account the noise, which is generally less than 2 Gs, the thresholds TH_1 and TH_2 are set to 200 and 20 Gs, respectively.

The data segment length is 5, 10, 20 and 40, respectively, and the ROI recognition results, as shown in Figures 4.11–4.14, are obtained from the axial MFL detection data. From the identification results, regardless of the size of ROI is suitable or not, all the artificial defects on pipe wall are recognized. The recognition rate of the defects on the wall is more than 98 %. The high recognition rate of the defects shows that the threshold fully meets the requirements of defect inspection.

Whether ROI region size is appropriate or not will be directly affected by the length L of the data segment. When $L = 5$, compared with the real defect MFL signal range, the identification of ROI shows obvious deficiencies; when $L = 10$, the ROI recognition outcomes have improved, but still there is a partial deletion of irregular region; when $L = 40$, ROI identification results completely contain defects and their surrounding areas, but at the same time, ROI contains a large number of unrelated region, which leads to the solving area of defect contour inversion that is too large; when $L = 20$,

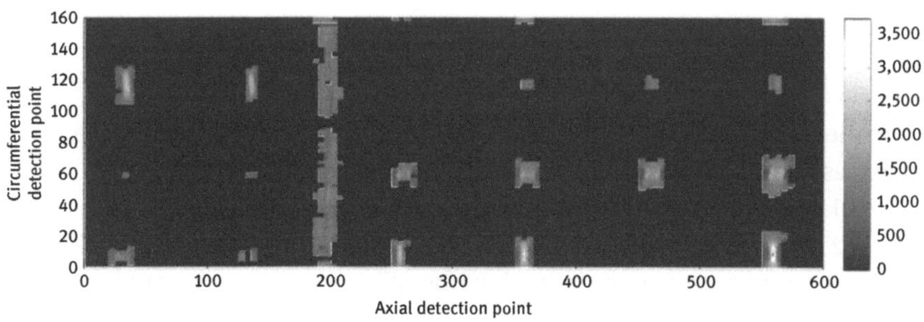

Figure 4.11: The ROI recognition results of the axial MFL detection data ($L = 5$).

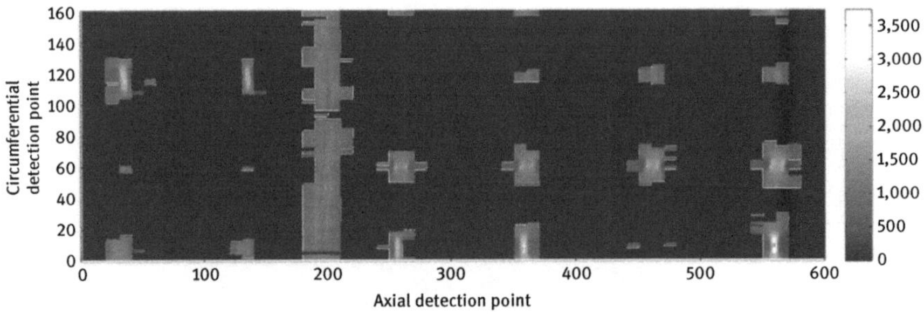

Figure 4.12: The ROI recognition results of the axial MFL detection data (L = 10).

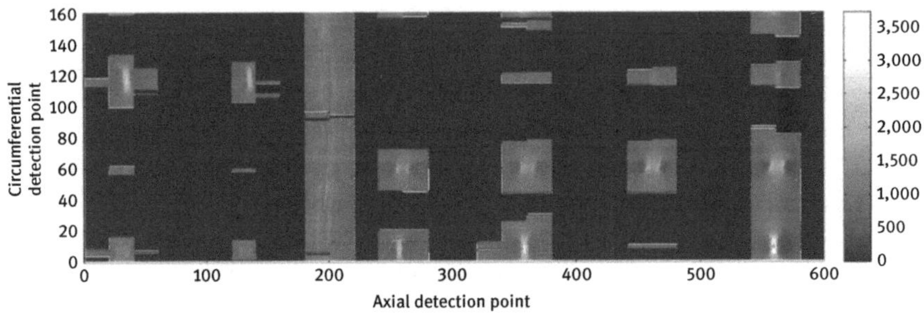

Figure 4.13: The ROI recognition results of the axial MFL detection data (L = 20).

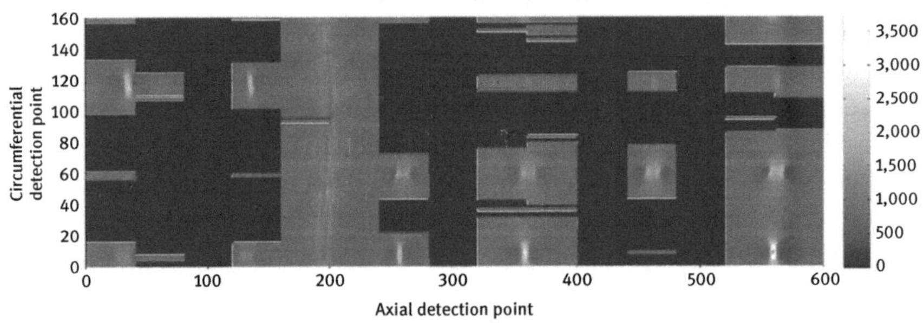

Figure 4.14: The ROI recognition results of the axial MFL detection data (L = 40).

ROI identification results completely contain defects and their surrounding areas, and there is no excess unrelated region, the recognition results are the best.

A rectangular defect of 28.6 mm × 14.3 mm × 7.2 mm and a cambered surface defect of 28.6 mm × 7.2 mm are analyzed in detail; Figures 4.15 and 4.16 show the ROI recognition results in different data segmentation lengths. When the data segment length is 5 or 10, the ROI area is smaller than that of the real defect profile, and there is an obvious lack of detection area. When the data segment length is 40, identification results

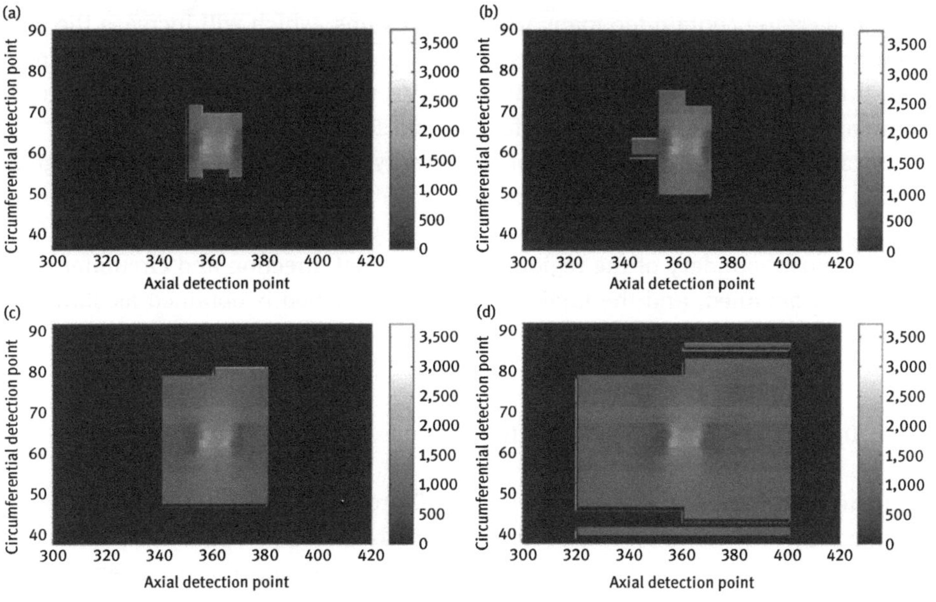

Figure 4.15: The ROI region recognition results of the rectangular defect of 28.6 mm × 14.3 mm × 7.2 mm: (a) L = 5; (b) L = 10; (c) L = 20; and (d) L = 40.

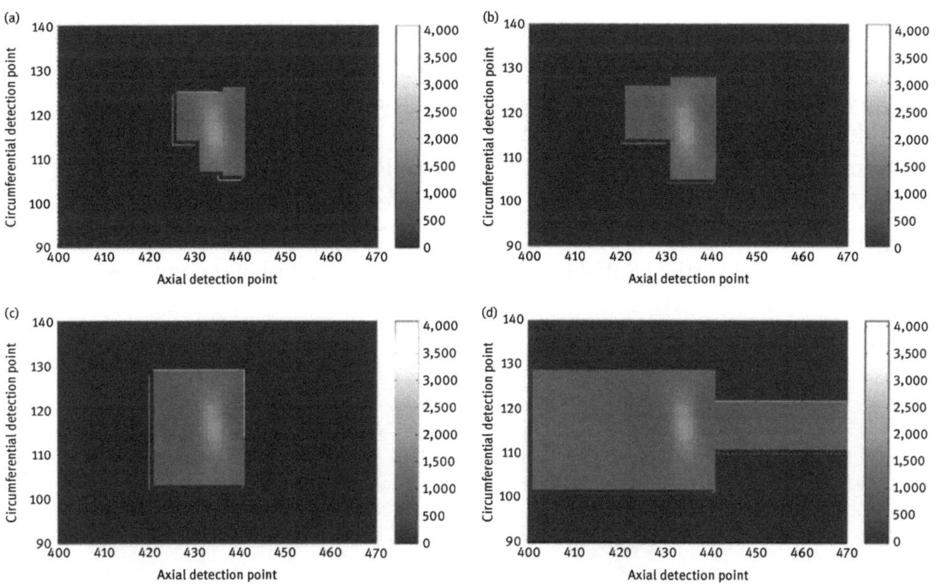

Figure 4.16: The ROI region recognition results of the cambered surface defect of 28.6 mm × 7.2 mm: (a) L = 5; (b) L = 10; (c) L = 20; and (d) L = 40.

of the ROI region contain too many irrelevant regions, which will increase the data quantity of the defect 3D profile inversion operation. When the data segment length is 20, the ROI region is identified, which contains a complete defect profile, and the redundancy is small, so 20 is suitable for the data segment length.

Because the ROI identified is not necessarily for the rules of the rectangular area, in order to facilitate subsequent data processing, the recognition results are rule processing. For the ROI recognition results shown in Figures 4.15(c) and 4.16(c), the outermost boundary of the region along the axial direction and circumferential direction is obtained, and the final rectangle to be solved is obtained as shown in Figure 4.17.

4.2.2 Defect Opening Contour Detection Method

In the iterative inversion process of a defect 3D profile, in order to reduce the defect model complexity and reduce the search space of the defect profile iterative inversion, defect opening contour is detected using the 3D MFL signal, so as to determine the solved region.

According to characteristics of 3D MFL detection signal, the amplitude of the detection signal changes obviously at the edge of the defect. Thus, the opening profile of the defect can be estimated by calculating the gradient value of the corresponding MFL detection signal to each point. For the 3D MFL detection, the defect opening profiles can be detected by each component signal, respectively; then the detection results of three components are weighted synthesis.

Taking the 3D MFL detection signal of a cambered surface defect of 28.6 mm × 7.2 mm shown in Figure 4.18 for example, the gradient values of each point in the signal image of the axial, radial and circumferential components are calculated, respectively, by using the discrete Sobel operator. For any component signal image A, the horizontal and vertical Sobel convolution operators are planar convolved with the

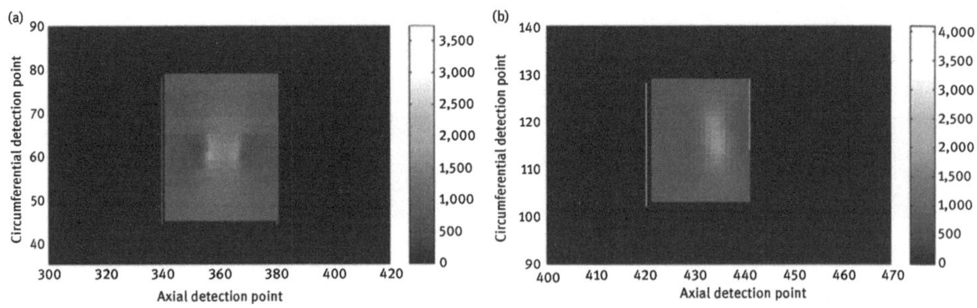

Figure 4.17: The final rectangle region to be solved: (a) the rectangular defect of 28.6 mm × 14.3 mm × 7.2 mm and (b) the cambered surface defect of 28.6 mm × 7.2 mm.

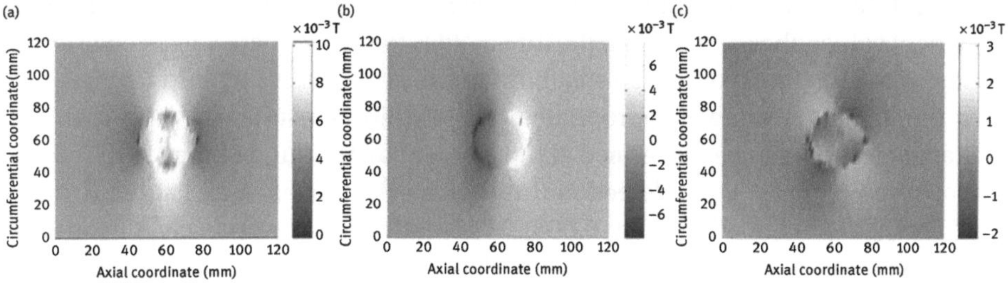

Figure 4.18: The 3D MFL detection signal of a cambered surface defect of 28.6 mm × 7.2 mm: (a) axial component; (b) radial component; and (c) circumferential component.

image A, respectively, which can get the approximate gradients $G(x)$ and $G(y)$ in the transverse and longitudinal directions. That is

$$G(x) = \begin{bmatrix} -1 & 0 & +1 \\ -2 & 0 & +2 \\ -1 & 0 & +1 \end{bmatrix} \times \mathbf{A}, \quad G(y) = \begin{bmatrix} +1 & +2 & +1 \\ 0 & 0 & 0 \\ -1 & -2 & -1 \end{bmatrix} \times \mathbf{A}. \tag{4.1}$$

In eq. (4.1), the units of the magnetic induction intensity of the MFL signal image in each point are Tesla (T).

Then, the gradient value G and the direction angle θ of each point in the image A are calculated by the following formula:

$$G = \sqrt{G^2(x) + G^2(y)}, \tag{4.2}$$

$$\theta = \arctan(G(y)/G(x)). \tag{4.3}$$

Three-dimensional MFL was calculated to get the gradient distribution of each component signal, as shown in Figure 4.19.

In order to detect the defect opening profile, the threshold $G(\delta)$ is set, and all the points whose gradient values are larger than $G(\delta)$ are recognized as the defect opening

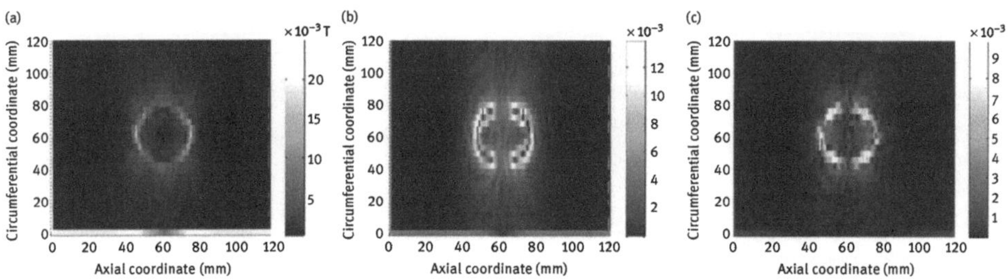

Figure 4.19: The gradient distribution of the 3D MFL detection signal of a cambered surface defect of 28.6 mm × 7.2 mm: (a) axial component; (b) radial component; and (c) circumferential component.

edges, and the remaining points are recognized as non-edge points. The distribution of gradient values shows that when the threshold is too large, there will be edge loss of defects; when the threshold is too small, there will be unnecessary interference; only when the threshold value is appropriate, the recognition results will get a clear and accurate defect opening edge. Aimed at the gradient distribution of the 3D MFL signal as shown in Figure 4.19, the detection thresholds of the axial, radial and circumferential components are adjusted to 0.004, 0.006 and 0.004, respectively, the opening edge recognition results are shown as in Figure 4.20. In order to eliminate the internal point of the direct opening edge recognition, the outermost edge point of the direct recognition result is extracted and the recognition result is obtained as shown in Figure 4.21.

Straight lines are made through each recognized edge point and perpendicular to the gradient direction. The closed region enclosed by these lines, which is the identification result of the defect opening profile, is shown in Figure 4.22. Among them, the gray area is the result of the opening profile recognition, and the black circle is the actual defect profile. Comparison shows that the identified opening profile based on 3D MFL signal component of defects contains the complete defect opening shape. The recognition result of the circumferential component is the closest to

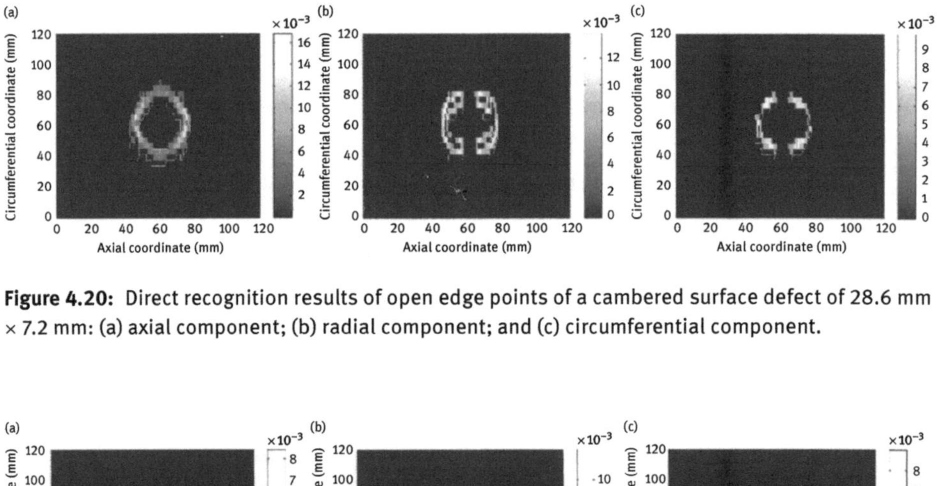

Figure 4.20: Direct recognition results of open edge points of a cambered surface defect of 28.6 mm × 7.2 mm: (a) axial component; (b) radial component; and (c) circumferential component.

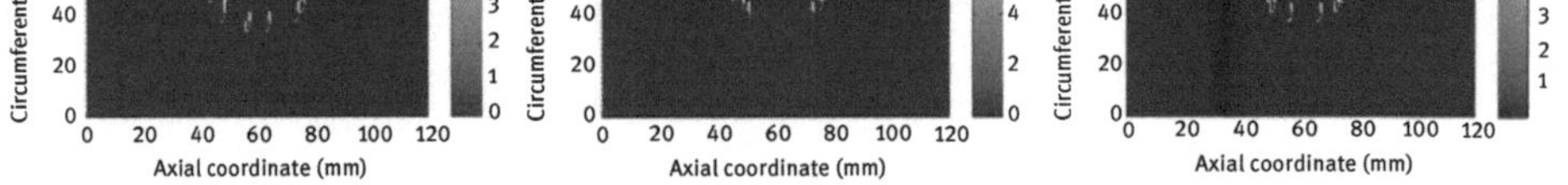

Figure 4.21: Open edge points recognition results of a cambered surface defect of 28.6 mm × 7.2 mm after correction: (a) axial component; (b) radial component; and (c) circumferential component.

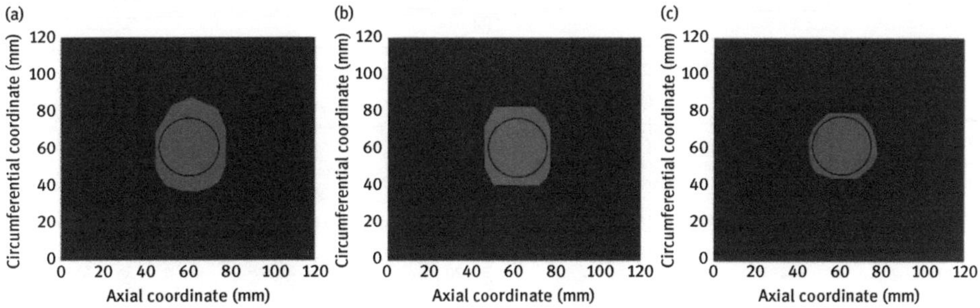

Figure 4.22: The opening profile recognition results of a cambered surface defect of 28.6 mm × 7.2 mm: (a) axial component; (b) radial component; and (c) circumferential component.

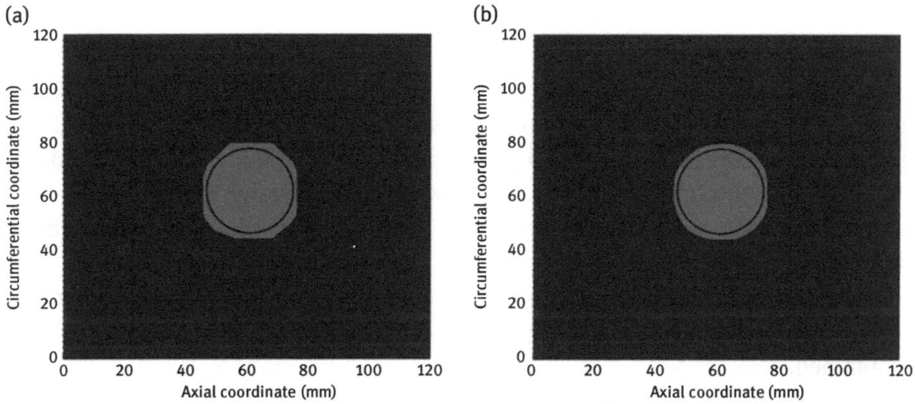

Figure 4.23: Final opening profile recognition result of a cambered surface defect of 28.6 mm × 7.2 mm: (a) before smoothing and (b) after smoothing.

the real contour. The axial component identification result exceeds the real contour the most.

Take the common part of the defect area recognized by the opening profile recognition of axial, radial and circumferential components of 3D MFL defect detection signal, and the edge is smoothed appropriately, the final opening profile of the defect is obtained, as shown in Figure 4.23. The comparison shows that the identified opening is slightly larger than that of the real defect which reflects the actual situation of defect opening profile.

To validate the accuracy of the proposed profile detection method, an irregular defect was simulated by the finite element. The defect shape is shown in Figure 4.24(a), and Figure 4.24(b) shows the contour recognition result based on the 3D MFL signal. Comparison shows that the identified open contour of the irregular defect has little difference with the real defect profile.

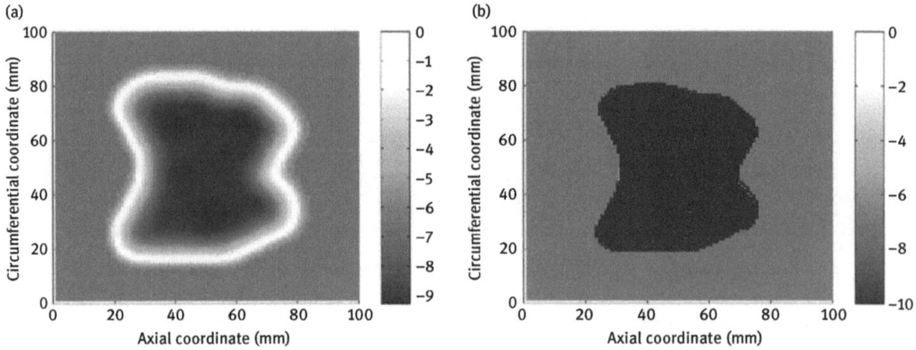

Figure 4.24: The opening contour recognition result of an irregular defect: (a) the real defect and (b) opening contour detection result.

4.2.3 Defect 3D Profile Mesh Model

In order to reduce the search space of the iterative inversion algorithm of the defect profile, a method of establishing a 3D profile mesh model is proposed, which is based on the opening contour of any defect. The flow diagram is shown in Figure 4.25.
The specific steps, by which the proposed 3D contour mesh model of the defect can be established, are as follows:

1. In order to avoid the influence of the detection error of the defect opening contour, the opening profile is enlarged to 1.05 times, and the new contour area S' is obtained.
2. To determine the size of the mesh model in the horizontal plane, the region S' is divided into a $N_1 \times N_2$ grid based on the center of the region.
3. To select all the overlapping grids with S', the new area of defect 3D profile inversion to be solved is obtained.
4. To determine the size of the defect profile mesh model in the depth direction and add depth parameter d to all the meshes to be solved, the final 3D profile mesh model M_n of the defect is established.

Figure 4.26 is an example of the final 3D profile mesh model of a defect after adding depth parameter d, and the unknown parameters are the depth of each grid area. Let $d_{i,j}$ express the depth of the grid of the ith row and the jth column, then the arbitrary defect based on the mesh model can be expressed in the following form:

$$M_n = \left\{ \begin{array}{lll} d_{1,1}, & d_{1,2}, \ldots, d_{1,j}, \ldots, d_{1,N_2-1}, & d_{1,N_2} \\ & \cdots & \\ d_{i,1}, & d_{i,2}, \ldots, d_{i,j}, \ldots, d_{i,N_2-1}, & d_{i,N_2} \\ & \cdots & \\ d_{N_1,1}, & d_{N_1,2}, \ldots, d_{N_1,j}, \ldots, d_{N_1,N_2-1}, & d_{N_1,N_2} \end{array} \right\}. \tag{4.4}$$

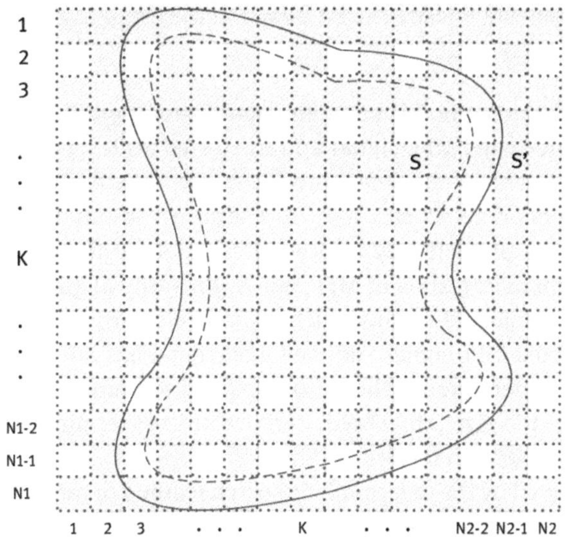

Figure 4.25: The flowchart of the 3D contour mesh model.

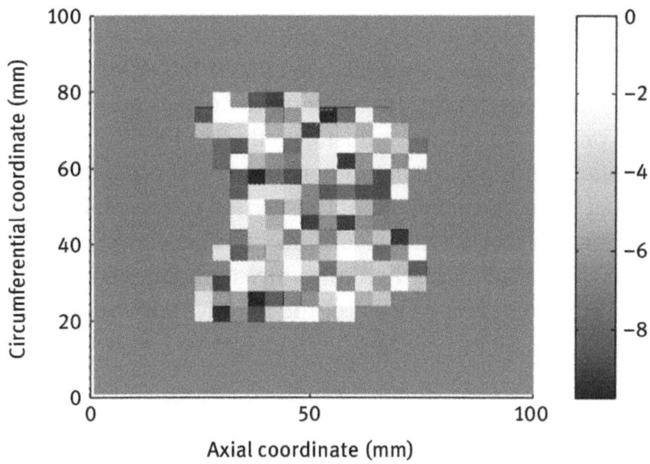

Figure 4.26: Defect 3D profile mesh model sample.

In the defect expression based on the mesh model, the depth of the exterior of the defect opening profile is 0. In the process of iterative inversion of the defect 3D profile, the depth of the interior of the defect opening profile is modified.

In addition, after obtaining the inversion results based on the mesh model, appropriate smoothing is performed in order to obtain a smooth edge which is more consistent with the real defect.

4.2.4 Random Searching Iterative Inversion for the 3D Profile of Defects

4.2.4.1 Random Search Strategy

The basic idea of the iterative inversion method is using the optimization problem of the minimum error between the forecasted MFL signal and the tested MFL signal to replace the magnetic leakage field reverse of the defect profile. As the basic flowchart shown in Figure 4.27, iterative inversion defect can start from arbitrary initial defect contour, and MFL detection signal is predicted by a forward calculation model. In order to obtain more accurate calculation results of MFL signal, forward calculation model is usually calculated by the numerical method such as finite element calculation. If the error between the predicted MFL signal and the target signal has not been limited to the given threshold, the defect profile should be updated to iterative and calculate again; if the error is within the given threshold, iterative process terminates, the defect profile shows the newest inversion results.

The optimization problem is solved by the deterministic search method or random search method, which can be used to update the defect profile in the iterative inversion method. Deterministic search method, including the steepest descent method, conjugate gradient method, quasi-Newton method and so on, uses the information of the current search location to determine the search direction and the step size. Deterministic method is fast, but in essence it is still a local optimization method. In the absence of proper initial guess defect profile, the deterministic method is easy to fall into the local optimal solution of the iterative inversion of the defect profile, so that the real defect profile cannot be found. In addition, the defect profile inversion is ill-conditioned inverse problem, and deterministic method in iterative process usually needs additional regularization processing; otherwise, it will produce larger calculation errors and even lead to iteration that cannot be carried out normally.

Random search methods include tabu search algorithm, simulated annealing algorithm and genetic algorithm. This kind of method introduces random variables or pseudorandom variables into the generation process of the iterative solution so that

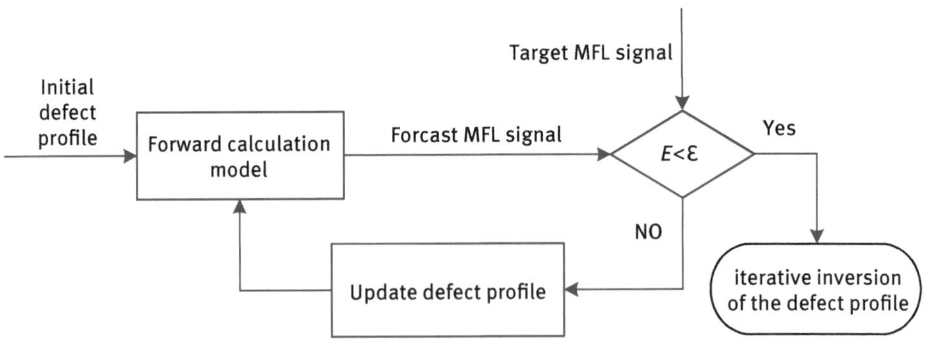

Figure 4.27: Basic flow of iterative inversion method for defect profile.

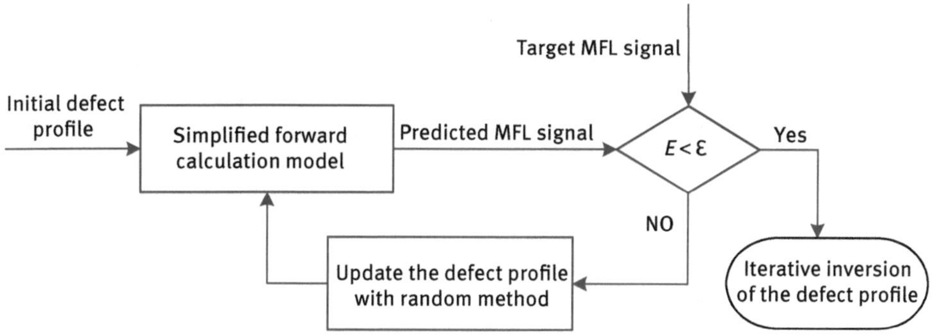

Figure 4.28: The basic flow of the iterative inversion method for the 3D profile of the defects in the MFL testing.

the search direction and the step size are random to avoid the local optimal solution. Compared with conventional methods, the disadvantage of random method is that the search speed is slow, which will lead to relatively longer time of iterative inversion.

Considering the advantages and disadvantages of the above two kinds of methods, in order to obtain more accurate results, the random search method is selected as the optimization strategy of the iterative inversion method of the defect profile. Figure 4.28 shows the basic process of the iterative inversion method for the 3D contour of the defects in the MFL detection based on the random search method. Among them, 3D MFL signals are adopted as the measured and predicted MFL signals, and the defect profile is updated in each iteration based on random search strategy.

In addition, considering the random search method is slow at the search speed, the following will be in the optimization objective function, simplifying the forward calculation model, thus speeding up defect profile inversion speed.

4.2.4.2 Similarity Objective Function

In the iterative inversion algorithm, the common definition of objective function is to predict the absolute error of predicted MFL signal and the measured MFL signal, which can be realized by solving the root mean square (RMS) error of the two MFL signals:

$$E = \left\{ \sum_{j=1}^{n} \sum_{i=1}^{m} \left[(sa_{ij}^{p} - sa_{ij}^{d})^2 + (sr_{ij}^{p} - sr_{ij}^{d})^2 + (sc_{ij}^{p} - sc_{ij}^{d})^2 \right] / 3mn \right\}^{\frac{1}{2}}, \tag{4.5}$$

where sa_{ij}^{p}, sr_{ij}^{p}, sc_{ij}^{p} are the axial, radial and circumferential MFL signals predicted by the forward model, whose unit is T; sa_{ij}^{p}, sr_{ij}^{p}, sc_{ij}^{p} are the axial, radial and circumferential target MFL signals, respectively, and their unit is T; i and j are the number of sampled data points along the axial direction and circumferential direction of the pipe, respectively.

The objective function defined by the absolute error requires that the finite element forward calculation of MFL testing has absolutely accurate results. Although this definition is intuitive and accurate, it is not suitable for the objective function of the defect iterative inversion method of the MFL testing. In fact, in the condition of defect profile iterative inversion, the objective function value of the optimal solution is the minimum value of all possible solutions in the objective function, but not necessarily zero. In the discriminant condition of the iterative process, for any two feasible solutions, as long as the feasible solution of the relatively more close to the real defect has a smaller value of the objective function, the definition of objective function in any form will not affect the result of the iterative process. So if you can find other objective functions that meet the requirements of the definition, it is not necessarily required to calculate the MFL testing with absolute precision, which can be appropriate to reduce the finite element calculation of dividing precision without affecting the identification result of the iterative process, thereby reducing the finite element forward calculation time and memory consumption.

Pearson correlation coefficient is a kind of effective method for measuring the correlation between two variables, which can be used to forecast the degree of similarity between predicted MFL signal and the measured MFL signal, and the objective function of the defect outline iterative inversion algorithm can be defined based on it. For any two data samples x and y, the corresponding Pearson correlation coefficient can be obtained as follows:

$$P(x,y) = \frac{\sum x_i y_i - n\overline{xy}}{(n-1)S_x S_y} = \frac{n\sum x_i y_i - \sum x_i \sum y_i}{\sqrt{n\sum x_i^2 - (\sum x_i)^2}\sqrt{n\sum y_i^2 - (\sum y_i)^2}}. \tag{4.6}$$

In the formula, S_x and S_y are the standard deviations of samples x and y, respectively.

In order to compare the similarity between the predicted MFL signal and the target MFL signal, the objective function of the similarity of one-dimensional MFL signal is defined as follows:

$$\begin{aligned} E_O &= \sum_{i=1}^{N} P(x_p^i, x_d^i) = \sum_{i=1}^{N}\left(\frac{\sum_{j=1}^{K} x_p^{ij} x_d^{ij} - n\overline{x_p^i x_d^i}}{(K-1)S_{x_p^i} S_{x_d^i}}\right) \\ &= \sum_{i=1}^{N}\left(\frac{K\sum_{j=1}^{K} x_p^{ij} x_d^{ij} - \sum_{j=1}^{K} x_p^{ij}\sum_{j=1}^{K} x_d^{ij}}{\sqrt{K\sum_{j=1}^{K} x_p^{ij2} - \left(\sum_{j=1}^{K} x_p^{ij}\right)^2}\sqrt{K\sum_{j=1}^{K} x_d^{ij2} - \left(\sum_{j=1}^{K} x_d^{ij}\right)^2}}\right), \end{aligned} \tag{4.7}$$

where x_p^i is the predicted MFL data along the axial direction of the pipe, whose unit is T; x_d^i is the detection MFL data along the axial direction of the pipe, whose unit is T; N is the number of MFL sensors along the circumferential direction of the pipeline; K is the total sample points along the axial direction of the pipeline.

Accordingly, the objective function of the similarity between the 3D MFL signals is defined as follows:

$$
\begin{aligned}
E_T &= E_{OA} + E_{OR} + E_{OC} \\
&= \sum_{i=1}^{N} \left[P(a_p^i, a_d^i) + P(r_p^i, r_d^i) + P(c_p^i, c_d^i) \right] \\
&= \sum_{i=1}^{N} \left(\frac{\sum_{j=1}^{K} a_p^{ij} a_d^{ij} - n\overline{a_p^i a_d^i}}{(K-1) S_{a_p^i} S_{a_d^i}} + \frac{\sum_{j=1}^{K} r_p^{ij} r_d^{ij} - n\overline{r_p^i r_d^i}}{(K-1) S_{r_p^i} S_{r_d^i}} + \frac{\sum_{j=1}^{K} c_p^{ij} c_d^{ij} - n\overline{c_p^i c_d^i}}{(K-1) S_{c_p^i} S_{c_d^i}} \right),
\end{aligned}
\tag{4.8}
$$

where E_{OA}, E_{OR} and E_{OC} are the objective functions of the similarity of the axial, radial and circumferential MFL signals, respectively.

In order to transform the defect 3D profile inversion problem into solving the minimum value of the objective function, the minimum of the objective function value is set to 0. Finally, the 3D MFL signal similarity objective function is defined as follows:

$$
\begin{aligned}
E_P &= 1 - \frac{1}{3}(E_{OA} + E_{OR} + E_{OC}) \\
&= 1 - \frac{1}{3}\sum_{i=1}^{N} \left[P(a_p^i, a_d^i) + P(r_p^i, r_d^i) + P(c_p^i, c_d^i) \right] \\
&= 1 - \frac{1}{3}\sum_{i=1}^{N} \left(\frac{\sum_{j=1}^{K} a_p^{ij} a_d^{ij} - n\overline{a_p^i a_d^i}}{(K-1) S_{a_p^i} S_{a_d^i}} + \frac{\sum_{j=1}^{K} r_p^{ij} r_d^{ij} - n\overline{r_p^i r_d^i}}{(K-1) S_{r_p^i} S_{r_d^i}} + \frac{\sum_{j=1}^{K} c_p^{ij} c_d^{ij} - n\overline{c_p^i c_d^i}}{(K-1) S_{c_p^i} S_{c_d^i}} \right),
\end{aligned}
\tag{4.9}
$$

where a_p^i, r_p^i and c_p^i are the predicted data sequences of the axial, radial and circumferential MFL signals along the axial direction of the pipeline, respectively, whose unit is T; a_d^i, r_d^i and c_d^i are the detected data sequences of the axial, radial and circumferential MFL signals along the axial direction; S_{ap}, S_{rp} and S_{cp} are the standard deviations for predicted data sequences; S_{ad}, S_{rd} and S_{cd} are the standard deviations for tested data sequences; N is the number of MFL sensors along the circumferential direction of the pipe; K is the number of sample points along the axial direction of the pipe.

In order to verify the feasibility of the similarity objective function, the following experiments are carried out. The 3D MFL signal is defined as the target MFL signal by using the 3 mm × 30 mm × 7.2 mm rectangular defect as the target defect. To change the length and width of the defect, the corresponding 3D MFL signal is calculated by the finite element simulation, and the similarity objective function value between the corresponding defect MFL signal and the target MFL signal is calculated. Finally, the statistical similarity value of the objective function varies with the size of the defect length and width, as shown in Figure 4.29.

From the statistical results, when the inversion defect size is the same as the target defect size, the similarity function value is 0; with the increase of the deviation

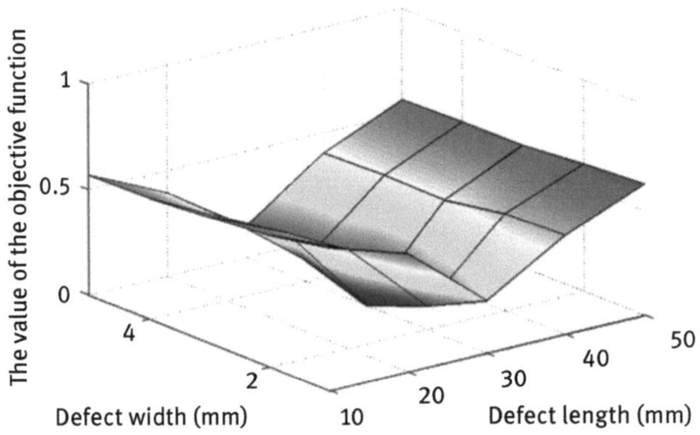

Figure 4.29: The similarity value of the objective function varies with the size of the defect.

between the inversion defect size and the size of the target, the value of the objective function is monotonically increasing.

In order to validate the effectiveness of the similarity objective function that can appropriately reduce the finite element calculation mesh precision and does not affect the discrimination results of the iterative process, experiments are repeated in different finite element computation analyses precision. The target defect is still rectangular defect of 30 mm × 3 mm × 7.2 mm and the target MFL signal is its 3D MFL signal. The statistical objective function includes the RMS error objective function and the similarity objective function. In addition, there are three kinds of finite element mesh size: the first size is the initial size, the second size increases to 5 times the initial size and the third size increases to 10 times its initial size.
In the three mesh sizes, the RMS error objective function and the similarity objective function with the change of the size of the defects are examined, respectively, and the results are shown in Figures 4.30 and 4.31.

The statistical results show that in the finite element calculation with the initial mesh size, with the increase of the deviation between the inversion defect size and the target defect size, the values of the RMS error objective function and similarity function are monotonically increasing, which can be both as the objective function of defect profile iterative inversion process. When the mesh size of finite element calculation increases to 5 times the initial mesh size, the similarity objective function value increases monotonically with the increase of the deviation between the inversion defect size and the target defect size, but RMS error objective function values decrease in local with the increase of defect size deviation, which is unable to meet the requirements of the defect outline iterative inversion due to influencing the discriminant results of the iterative process. When the mesh size of finite element calculation increases to 10 times the initial mesh size, because of the big errors in the calculation of the 3D MFL signal, the two kinds of objective function values both decrease with the increasing defect size deviation, which cannot be used anymore.

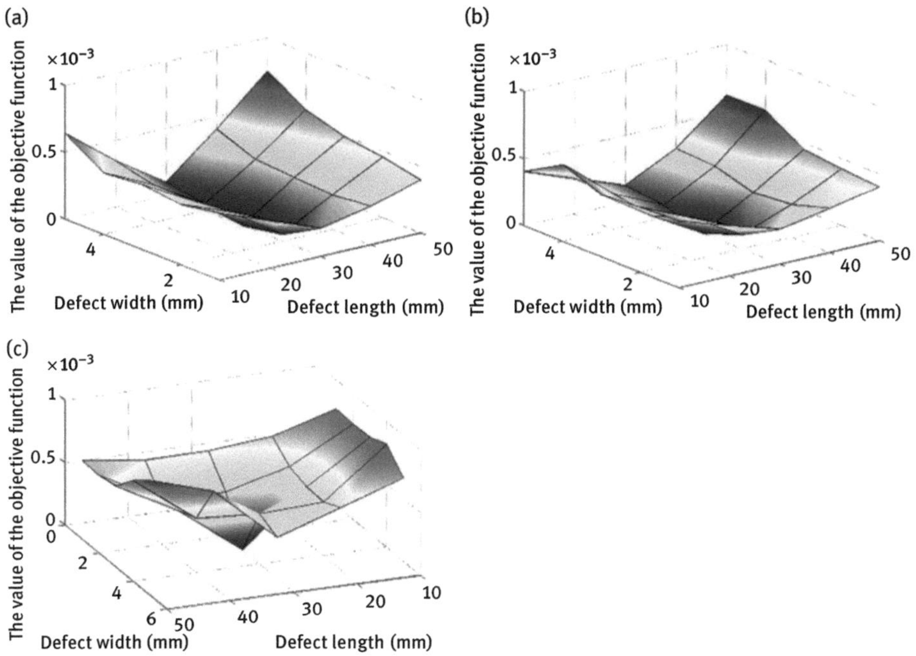

Figure 4.30: When the finite element mesh is changed, the root mean square error of the objective function value varies with the change of defect size: (a) initial mesh size; (b) 5 times the initial size; and (c) 10 times the initial size.

Thus, if the similarity objective function is adopted, under the premise of not affecting the iterative process of the defect contour, the finite element mesh size can be properly increased, so as to reduce the time consumed by finite element forward calculation. Therefore, with respect to the mean square error and other commonly used absolute error objective function, the discriminant effect of similarity objective function is better, and it is more suitable for the iterative inversion of defect 3D profile.

Based on the above conclusions, the similarity objective function applied to the iterative inversion process of contour defects can appropriately reduce the finite element forward calculation mesh precision and does not affect the judgment of the objective function value in the iteration process.

4.2.4.3 Simplified Forward Calculation Model

When the velocity of the detector is not more than a certain value, the velocity effect can be neglected, and the calculation of the magnetic leakage field can be solved as a static magnetic field problem. Based on the magnetic circuit structure of the 3D MFL detector, the basic forward calculation model is established as in Figure 4.32. In this model, the main components of the magnetic circuit include steel brush, permanent

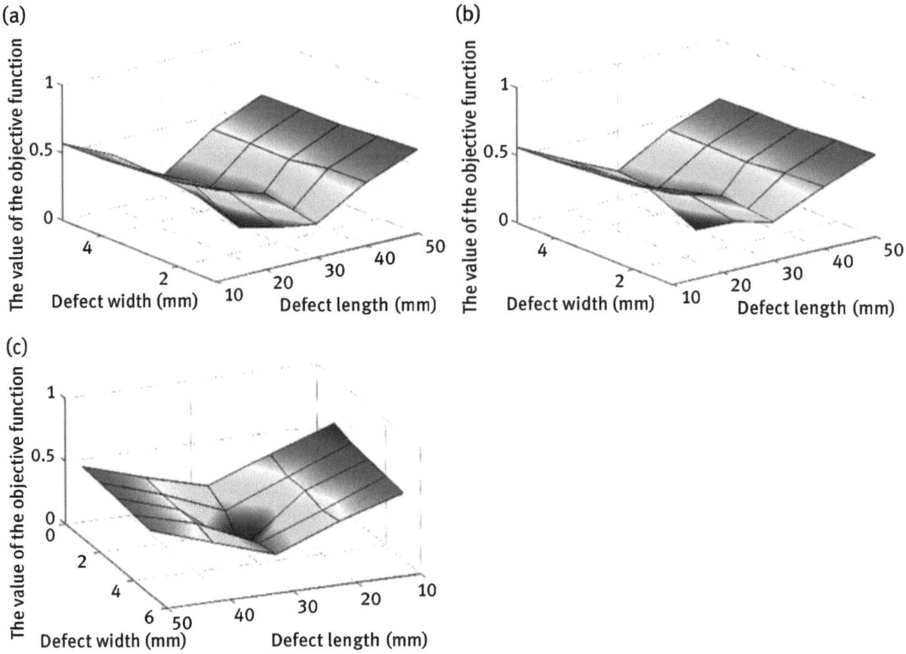

Figure 4.31: When the finite element mesh is changed, the similarity objective function value varies with the change of defect size: (a) initial mesh size; (b) 5 times the initial size; and (c) 10 times the initial size.

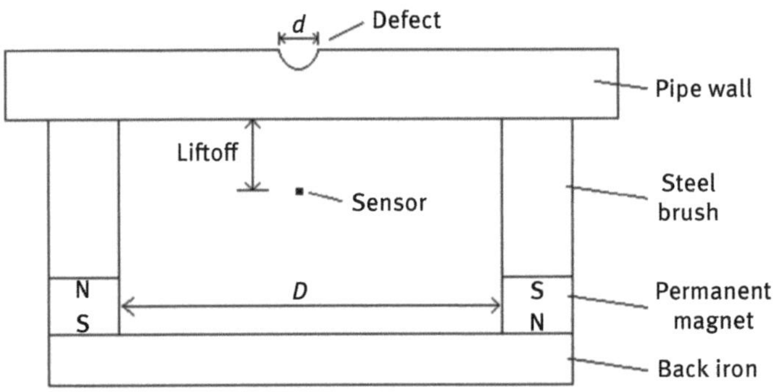

Figure 4.32: Basic forward calculation model for 3D MFL detection of oil and gas pipelines.

magnet, back iron, pipe wall and air. The magnetic properties of permanent magnet and air have fixed value, and the steel brush, the back iron and the pipe wall are nonlinear magnetic materials.

In the numerical calculation with finite element method, the calculation workload is proportional to the size of the space to be solved, and the complexity of the nonlinear magnetic circuit is much higher than that of the linear magnetic circuit. In the iterative inversion method of defect profile, an iterative algorithm is needed to solve the above-mentioned basic forward model to predict the MFL signal. Because the basic forward model has a large space to be solved and more nonlinear materials, the computational load of the solution process is very large, which seriously limits the practicality of the iterative inversion method. Therefore, it is necessary to consider the principles and requirements of the oil and gas pipeline leakage detection and to study the basic forward calculation model, and then optimize the model under the condition of maintaining a certain degree of accuracy.

In order to obtain the best detection result, the pipe wall near the defect should be placed in a uniform near-saturation magnetic field in order to get the best testing result. Therefore, the permanent magnet of the detector usually magnetizes the pipe to the near-saturation state along the axial direction. At the same time, in order to avoid the influence of the irregular MFL near the steel brush on the magnetic sensor, magnetic sensors are arranged at the center of the magnetic circuit, and the axial span D between two steel brushes and the defect length d should satisfy the following relationship:

$$D \geq 10d. \tag{4.10}$$

Based on the actual demand and design concept of the oil and gas pipeline leakage detection, a simplified forward calculation model is proposed for the 3D MFL detection of oil and gas pipelines as shown in Figure 4.33. The simplified model omits the back iron and steel brush in the basic model. At the same time, the permanent magnets embed pipe wall to ensure the pipe wall near the defect in a uniform magnetizing field. In order to ensure the near-saturation magnetization of the pipe wall, the magnetic field intensity in the pipe wall can be adjusted to the appropriate value by modifying coercive force parameter value of the permanent magnet.

In order to verify the validity of the simplified forward calculation model, calculation was conducted on a dual-CPU Intel Xeon E7-4820 2 GHz server. Based on the basic model and the simplified model, the 3D MFL signals of some typical defects are calculated, respectively, and the consumption time of the two models is recorded. At the

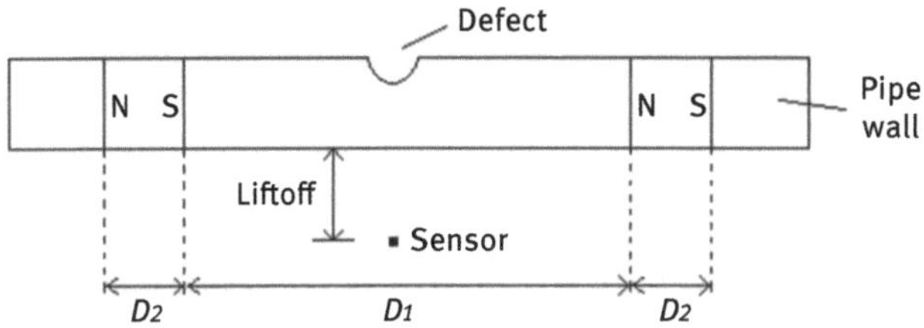

Figure 4.33: Simplified forward calculation model for 3D MFL detection of oil and gas pipelines.

same time, the following formula is defined to calculate the relative error between the two models for solving the MFL signal:

$$E = \left(\frac{\sum_{j=1}^{n} \sum_{i=1}^{m} \left[(sa_{ij}^{\mathrm{r}} - sa_{ij}^{\mathrm{b}})^2 + (sr_{ij}^{\mathrm{r}} - sr_{ij}^{\mathrm{b}})^2 + (sc_{ij}^{\mathrm{r}} - sc_{ij}^{\mathrm{b}})^2 \right]}{\sum_{j=1}^{n} \sum_{i=1}^{m} \left[(sa_{ij}^{\mathrm{b}})^2 + (sr_{ij}^{\mathrm{b}})^2 + (sc_{ij}^{\mathrm{b}})^2 \right]} \right)^{\frac{1}{2}}, \tag{4.11}$$

where sa_{ij}^{b}, sr_{ij}^{b}, sc_{ij}^{b} are the axial, radial, circumferential MFL signals calculated by the basic model, respectively; sa_{ij}^{r}, sr_{ij}^{r}, sc_{ij}^{r} are the axial, radial and circumferential MFL signals calculated by the simplified model, respectively, and their unit is T.

Figure 4.34 shows the 3D MFL signals of a 100 mm × 14.3 mm × 4.3 mm rectangular defect based on the basic calculation model and the simplified calculation model. Through direct comparison, based on the calculation results of the two models, the position and amplitude of the feature points have no obvious difference, the results of the simplified model are basically consistent with that of the basic model, and the relative error between the MFL signals of the two kinds of models is only 4.25 %. This shows that the simplified model can replace the original MFL detection model with higher accuracy.

According to some typical defects, the calculation time of the two models and the relative error between the two MFL signals is calculated. The results are provided in Table 4.3. According to the statistical results, the calculation time of the simplified

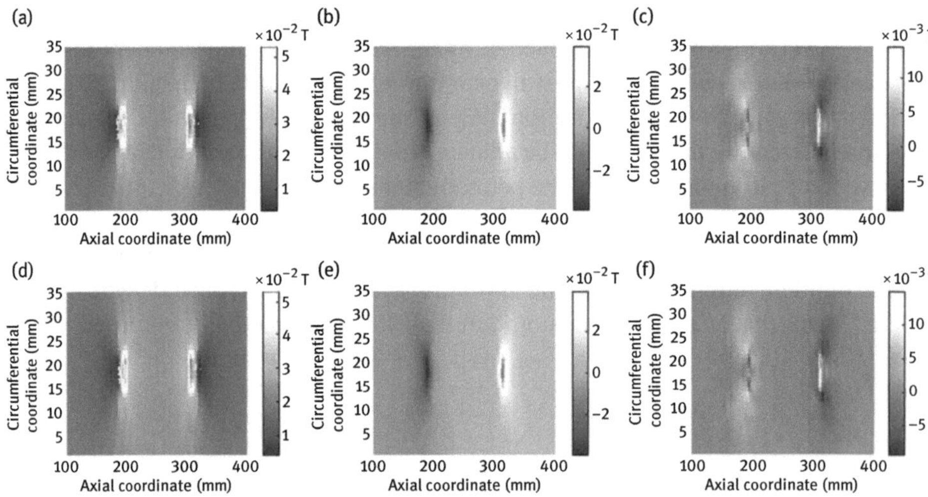

Figure 4.34: The three-dimensional MFL signals of a 100 mm × 14.3 mm × 4.3 mm rectangular defect based on the basic calculation model and the simplified calculation model. (a)–(c) Axial, radial and circumferential MFL signals based on the basic model. (d)–(f) Axial, radial and circumferential MFL signals based on the simplified model.

Table 4.3: The calculation time and the error of MFL signals based on the two models.

Defect (mm)	Calculation time (min)		The relative error between the two models (%)
	Basic model	Simplified model	
Rectangular 100 × 14.3 × 4.3	152.8	11.2	4.25
Rectangular 42.9 × 42.9 × 5.7	189.2	14.1	4.14
Globoid 100 × 7.2	186.4	13.7	4.18
Globoid 42.9 × 7.2	51.4	3.2	4.47
Cylinder 100 × 7.2	215.3	16.0	4.06
Cylinder 42.9 × 7.2	59.7	3.7	4.38

model is less than 1/10 that of the basic model, and the relative error between the MFL signals calculated by the two models is not more than 5 %. Therefore, in the case of maintaining a high degree of accuracy, the simplified model can effectively shorten the calculation time of forward MFL signal so as to speed up the iterative inversion of the 3D contour of the MFL testing.

4.2.4.4 Iterative Inversion Algorithm of Simulated Annealing Tabu Search

Tabu search algorithm is a kind of effective and commonly used random search method [109]. The basic strategy is to generate randomly a series of new feasible solution from the current solution neighborhood and select the optimal solution or the first improved solutions as new solutions. Because the feasible solutions are randomly selected in the neighborhood, tabu search algorithm has the ability to jump out of local minima. However, the structure of the algorithm is likely to fall into the cycle of death in the iteration process, so we need to set up a tabu list, which is used to record and exclude the solution that has been searched for a certain period of time.

Simulated annealing algorithm is another kind of common random search method. The feasible solution of the objective function value is accepted by the algorithm with a certain probability. It has the ability to jump out of local minima. By gradually adjusting the control parameter T, the algorithm can search the optimal solution in the whole region with less computational cost. The theoretical research in the optimization field has proved that the algorithm is able to solve the optimization problem with continuous and bounded objective function for MFL detection.

The above two algorithms have different advantages: the tabu search algorithm has more stable search speed, but it jumps out of local minima slowly; the simulated annealing algorithm has beating property in the initial stage, but it can quickly jump out of the local optima and converge to the optimal solution.

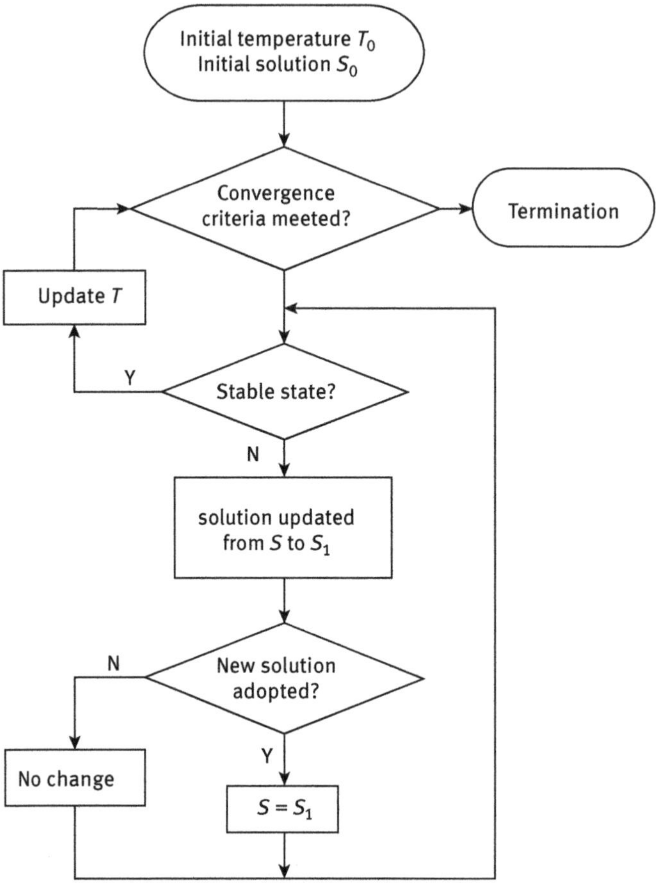

Figure 4.35: Iterative inversion algorithm of simulated annealing tabu search for 3D profile of defects.

For comprehensive utilization of the advantages of the two search algorithms, the simulation annealing tabu search algorithm is put forward in Figure 4.35, which introduces the simulated annealing strategy into tabu search algorithm for iterative inversion. On the basis of this, the iterative inversion of the 3D profile of the defects of MFL testing is carried out.

In the random search iterative inversion method for the 3D profile of the defects in the MFL detection, the defect profile is updated by using the tabu search algorithm, and the simulated annealing strategy is introduced. The similarity of objective function is used as the objective function of the iterative inversion of the defect 3D profile, and the feasible region and neighborhood of the solution are defined based on the mesh model of the 3D model of the defect.

In the 3D mesh model of defects, the pipe wall can be divided into N_1, N_2 and N_3 equal parts along the axial, radial and circumferential directions, respectively. And to make $x_{i,j}$ represent the grid depth of the ith row, jth column, any feasible solution X of the defect profile can be obtained:

$$X = \begin{Bmatrix} x_{1,1}, & x_{1,2}, \ldots, x_{1,j}, \ldots, x_{1,N_2-1}, & x_{1,N_2} \\ & \ldots & \\ x_{i,1}, & x_{i,2}, \ldots, x_{i,j}, \ldots, x_{i,N_2-1}, & x_{i,N_2} \\ & \ldots & \\ x_{N_1,1}, & x_{N_1,2}, \ldots, x_{N_1,j}, \ldots, x_{N_1,N_2-1}, & x_{N_1,N_2} \end{Bmatrix}. \tag{4.12}$$

At the same time, in the process of iterative inversion of the defect profile, the feasible region of any solution is

$$D = \left\{x | 0 \leqslant x_{i,j} \leqslant N_3 (i = 1, 2, \ldots, N_1; j = 1, 2, \ldots, N_2)\right\}. \tag{4.13}$$

Set the search step size of the tabu search algorithm as λ, then an arbitrary solution X' in the neighborhood of current solution X is obtained:

$$X' = \begin{Bmatrix} x_{1,1}, & x_{1,2}, \ldots, x_{1,j}, \ldots, x_{1,N_2-1}, & x_{1,N_2} \\ & \ldots & \\ x_{i,1}, & x_{i,2}, \ldots, x_{i,j} \pm \lambda, \ldots, x_{i,N_2-1}, & x_{i,N_2} \\ & \ldots & \\ x_{N_1,1}, & x_{N_1,2}, \ldots, x_{N_1,j}, \ldots, x_{N_1,N_2-1}, & x_{N_1,N_2} \end{Bmatrix}. \tag{4.14}$$

For the solution of arbitrary neighborhoods, based on the proposed MFL forward simplified calculation model, the 3D MFL detection signal is calculated, and the similarity of the corresponding objective function is calculated, too:

$$E_{\mathrm{P}} = 1 - \frac{1}{3} \sum_{i=1}^{N} \left(\frac{\sum_{j=1}^{K} a_{\mathrm{p}}^{ij} a_{\mathrm{d}}^{ij} - n\overline{a_{\mathrm{p}}^{i} a_{\mathrm{d}}^{i}}}{(K-1) S_{a_{\mathrm{p}}^{i}} S_{a_{\mathrm{d}}^{i}}} + \frac{\sum_{j=1}^{K} r_{\mathrm{p}}^{ij} r_{\mathrm{d}}^{ij} - n\overline{r_{\mathrm{p}}^{i} r_{\mathrm{d}}^{i}}}{(K-1) S_{r_{\mathrm{p}}^{i}} S_{r_{\mathrm{d}}^{i}}} + \frac{\sum_{j=1}^{K} c_{\mathrm{p}}^{ij} c_{\mathrm{d}}^{ij} - n\overline{c_{\mathrm{p}}^{i} c_{\mathrm{d}}^{i}}}{(K-1) S_{c_{\mathrm{p}}^{i}} S_{c_{\mathrm{d}}^{i}}} \right). \tag{4.15}$$

In the formula, a_{p}^{i}, r_{p}^{i} and c_{p}^{i} are the data sequences of the predicted axial, radial and circumferential MFL signals along the pipeline axial detection, respectively, their unit is T; a_{d}^{i}, r_{d}^{i} and c_{d}^{i} are the testing data sequences of the axial, radial and circumferential MFL signals along the pipeline axial detection, respectively, their unit is T; S_{ap}, S_{rp} and S_{cp} are the standard deviations of the predicted data series, respectively; S_{ad}, S_{rd} and S_{cd} are the standard deviations of the tested data series, respectively; N is the number of magnetic sensors along the circumferential direction of the pipeline; K is the number of sampling points along the axial direction of the pipeline.

In order to obtain faster search speed, according to the similarity value of the objective function, the first improved solution in the neighborhood of the current solution is selected as the new solution. If there is no improved solution in the current neighborhood, all similarity objective function values in the neighborhood solutions are compared, and the optimal solution X^* is selected as the new solution. In fact, if the objective function value E_P^* of solution X^* is greater than the objective function value E_P of the current solution X, the current solution X is a local minimum. To choose X^* as a new solution is out of the local minimum point value.

In addition, in order to avoid the tabu search process into a loop, set a tabu list with length N to record and remove the N feasible solutions previously searched.

In order to speed up the jump out speed from the local optimum, simulated annealing algorithm is introduced in tabu search, and any solution selection criteria in the neighborhood are modified as follows. For any solution X' in the neighborhood, if the objective function value E_P' is less than the objective function value E_P of the current solution X, directly set X' as the new solution; otherwise, the new profile is accepted as a new solution only if the following criterion comes into existence:

$$\exp\left(\frac{E_P - E_P'}{T}\right) \geq \text{rand}(0, 1), \tag{4.16}$$

where T is the control parameter; rand(0, 1) represents the random number distributed uniformly in (0, 1); E_P and E_P' are the similarity objective functions corresponding to the defect contours X and X', respectively.

When the number of iterations reaches a certain value k_1, the searching process has reached the equilibrium state under the current control parameter. At this point, the control parameter T is modified as follows:

$$T_{k+1} = \alpha \cdot T_k, \ 0 < \alpha < 1. \tag{4.17}$$

The α in the formula is usually between 0.8 and 0.95.

There are three termination criteria for judging the search process to achieve globally optimum: control parameter is less than a certain threshold δ; the number of iterations has reached a certain value k_2; the objective function value of the optimal solution has not changed in k_3-continuous searches.

When one of the above criteria is established, the search procedure of the iterative inversion is terminated, and the inversion results of the defect profile based on the mesh model are obtained. At this time, in order to obtain the smooth edge which is consistent with the actual defect, the contour of the defect is smoothed:

$$d_{i,j} = \left(d_{i+1,j} + d_{i-1,j} + d_{i,j+1} + d_{i,j-1} + d_{i,j}\right)/5, \tag{4.18}$$

where $d_{i,j}$ is the mesh depth of ith row, jth column in the mesh model.

4.2.4.5 Performance of Iterative Inversion Method on Random Search

Through the defect 3D inversion test, the precision, anti-interference ability and applicability to irregular defects of random search iterative inversion method were verified on a dual-CPU Intel Xeon E7-4820 2 GHz server, and the effect of initial defect profile was studied.

In the tests, the outer diameter and the wall thickness of the pipe were 457 mm and 14.3 mm, respectively; the liftoff value of the MFL sensors was 3 mm; the typical defects were rectangular defect of 30 mm × 20 mm × 5 mm, cambered surface defect of 30 mm × 5 mm and cylindrical defect of 30 mm × 5 mm.

In the mesh model of defect 3D profile, the pipe wall was divided along the axial, radial and circumferential directions at spaces of 2 mm, 1 mm and 1.4 mm, respectively. Because the pipeline is divided into 10 equal parts along the radial direction, feasible solutions for defect contours are as follows:

$$D = \left\{x|0 \le d_{i,j} \le 10 \ (i = 1, 2, \ldots, N_1;\ j = 1, 2, \ldots, N_2)\right\}, \tag{4.19}$$

where $d_{i,j}$ is the defect profile depth of ith row, jth column in the inversion results.

In order to select the appropriate parameters of the iterative inversion algorithm, the parameters are compared and selected within the set range. Finally, the algorithm parameters are provided in Table 4.4.

In the selected parameter settings, the total number of iterations in the iterative inversion algorithm is 2,000, and the iteration time is only dependent on the time of the forward calculation of the finite element. Thus, the iterative inversion time of the same defect is not different, and there is no comparability between inversion time of different defects.

Research on defect profile inversion of MFL detection is mostly based on two-dimensional (2D) profile of defect section. There is no uniform error definition for the 3D profile inversion of the defect. In order to comprehensively and thoroughly evaluate the gap between the defect predicted contour and the real contour, the various error definitions of 2D contour are applied to the defect 3D profile inversion.

For the inversion results of the 3D profile of any defect, the defect opening profile is taken as the base, and the measurement points N_1 and N_2 are evenly distributed along the axial direction and circumferential direction in the region.

On the basis of this, the book defines three kinds of defect 3D profile inversion error, such as RMS error E_{m}, mean positive error E_{p} and mean negative error E_{n}. The schematic diagram of the definitions E_{p} and E_{n} is as shown in Figure 4.36.

Table 4.4: Parameter setting of iterative algorithm.

Parameters	Symbol	Minimum value	Maximum value	Final selection value
Search step size	λ	1	10	2
Initial control parameter	T	1	20	5
Iteration number	k_1	1	50	20
Update rate	α	0.8	0.99	0.92
Threshold of control parameters	δ	0.05	0.5	0.2
Total number of iterations	k_2	1,000	3,000	2,000
Continuous search times	k_3	10	100	50

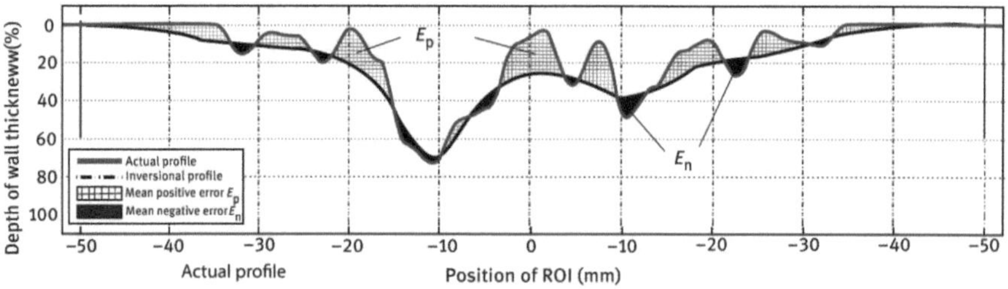

Figure 4.36: Error definition of the inversion results of 3D contour.

1. For the overall deviation between the inversion calculation of defect profile and real defect outlines, the RMS error E_m between the two is defined:

$$E_m = \sqrt{\sum_{i=1}^{N_1}\sum_{j=1}^{N_2}(d_{i,j}^p - d_{i,j}^r)^2 / \sum_{i=1}^{N_1}\sum_{j=1}^{N_2}(d_{i,j}^r)^2}, \tag{4.20}$$

   where $d_{i,j}^p$ and $d_{i,j}^r$ are the depths of the inversion 3D contour and the real contour in the ith measurement points along the axial direction and in the jth measurement points along the circumferential direction, respectively.
2. In order to calculate the average percentage of the depth of the inversion exceeding the real depth, the mean positive error E_p is defined as follows:

$$E_p = \frac{1}{DN_{pos}}\sum_{i=1}^{N_1}\sum_{j=1}^{N_2}\varepsilon_{(d_{i,j}^p - d_{i,j}^r)}(d_{i,j}^p - d_{i,j}^r), \tag{4.21}$$

   where $d_{i,j}^p$ and $d_{i,j}^r$ are the depths of the inversion 3D contour and the real contour in the ith measurement points along the axial direction and in the jth measurement points along the circumferential direction, respectively; ε is the unit step function; D is pipe wall thickness; N_{pos} is the number of measurement points $d_{i,j}^p$ more than $d_{i,j}^r$.
3. In order to calculate the average percentage of the depth of the inversion less than the real depth, the mean negative error E_n is defined as follows:

$$E_n = \frac{1}{DN_{neg}}\sum_{i=1}^{N_1}\sum_{j=1}^{N_2}\varepsilon_{(d_{i,j}^r - d_{i,j}^p)}(d_{i,j}^r - d_{i,j}^p), \tag{4.22}$$

where $d_{i,j}^p$ and $d_{i,j}^r$ are the depths of the inversion 3D contour and the real contour in the ith measurement points along the axial direction and in the jth measurement points along the circumferential direction, respectively; ε is the unit step function; D is pipe wall thickness; N_{neg} is the number of measurement points $d_{i,j}^p$ less than $d_{i,j}^r$.

RMS error E_m is mainly used for evaluating the results of the iterative inversion process and the final inversion results, and the mean positive error E_p and the mean negative error E_n are also used for complementary evaluation of 3D defects contour inversion results. By comparing E_p with E_n, whether the error distribution between the inversion profile and the real profile is uniform or not can be judged, and then whether the iterative inversion process has searched sufficiently in the solution space can be determined.

In addition, due to the random character of the iterative inversion process, the search path of each iteration is different from the final results. Therefore, the contour of each defect is inverted 10 times, and the least RMS error is taken as the final result.

On the basis of the recognition results of the defect openings, the initial defect profile based on the mesh model is set to a flat bottom defect with the depth of the 1/2 pipe wall thickness. And based on random search iterative inversion method, 3D inversion profiles of rectangular defect of 30 mm × 20 mm × 5 mm, cambered surface defect of 30 mm × 5 mm and cylindrical defect of 30 mm × 5 mm were tested.

Figures 4.37–4.39 give 3D iterative inversion process, with the corresponding results of three defects. It can be known that the difference between the identified profiles of the sample defects and those of the real defects is not significant. Based

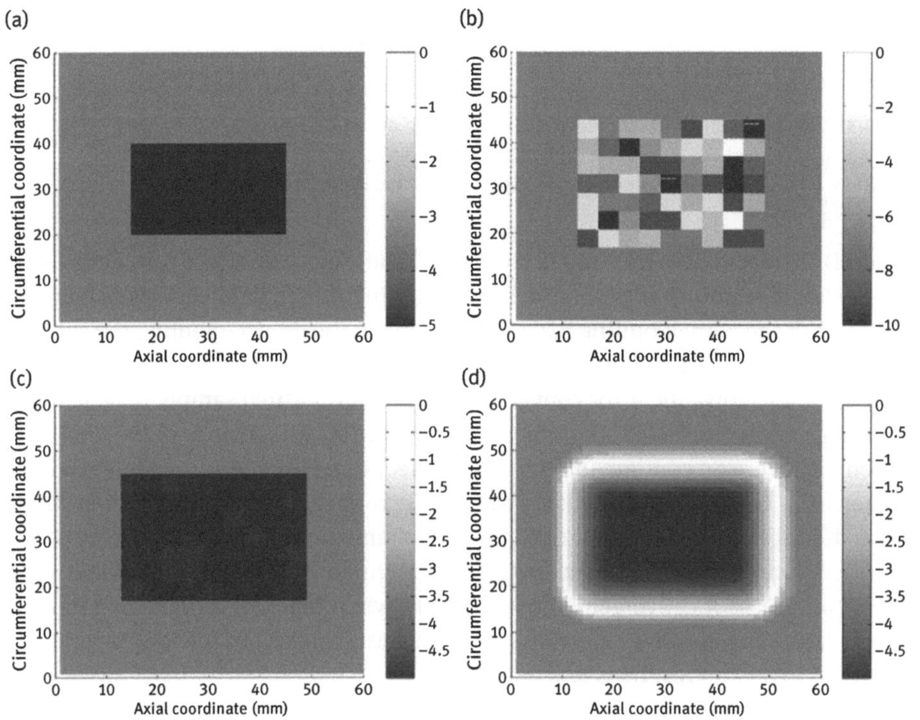

Figure 4.37: Three-dimensional profile iterative inversion of rectangular defect of 30 mm × 20 mm × 5 mm: (a) real defect profile; (b) mesh model based on the results of open profile recognition; (c) inversion results of mesh model; and (d) final defect profile.

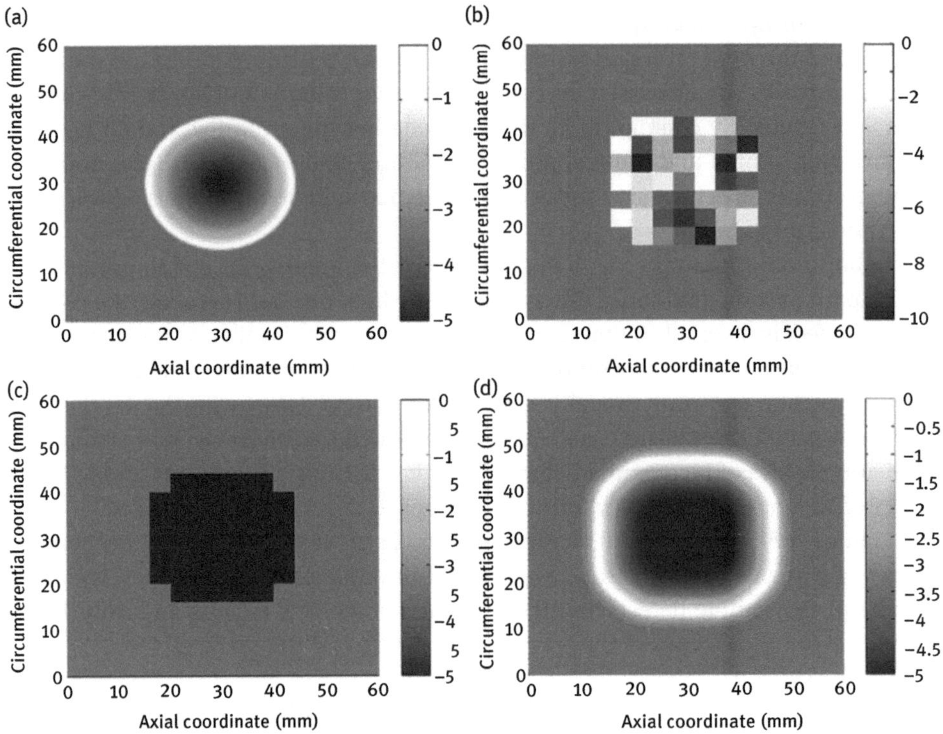

Figure 4.38: Three-dimensional profile iterative inversion of cambered surface defect of 30 mm × 5 mm: (a) real defect profile; (b) mesh model based on the results of open profile recognition; (c) inversion results of mesh model; and (d) final defect profile.

on the 3D mesh model, the initial contour of the defect is effectively delineated. Mesh model inversion results obtained by smoothing the final defect profile coincide basically with the corresponding real defect profile in the defect opening, the extreme point position and depth.

Figure 4.40 shows the error convergence curve of the three sample defects in the iterative inversion process of 3D profile, including the statistical error of the RMS error between the inversion contour and the real profile. From the figure, we can see that the inversion error is gradually decreasing with the increase of the number of iterations. In the search process of the local extreme points, to jump out of the local extreme points the random search algorithm turns to the defect profile with larger error, resulting in the jump on the error convergence curve. Finally, when the total number of iterations achieves 2,000, the iterative inversion process is terminated, and the RMS error of the three defect inversion results is less than 10 %.

The final 3D profile inversion error is shown in Table 4.5. From the statistical results, the RMS error of rectangle, cambered surface and cylindrical defect inversion results were 9.2 %, 7.6 % and 8.5 %. The small deviation between inversion defect profile and real defect outlines proves that the random search iterative inversion

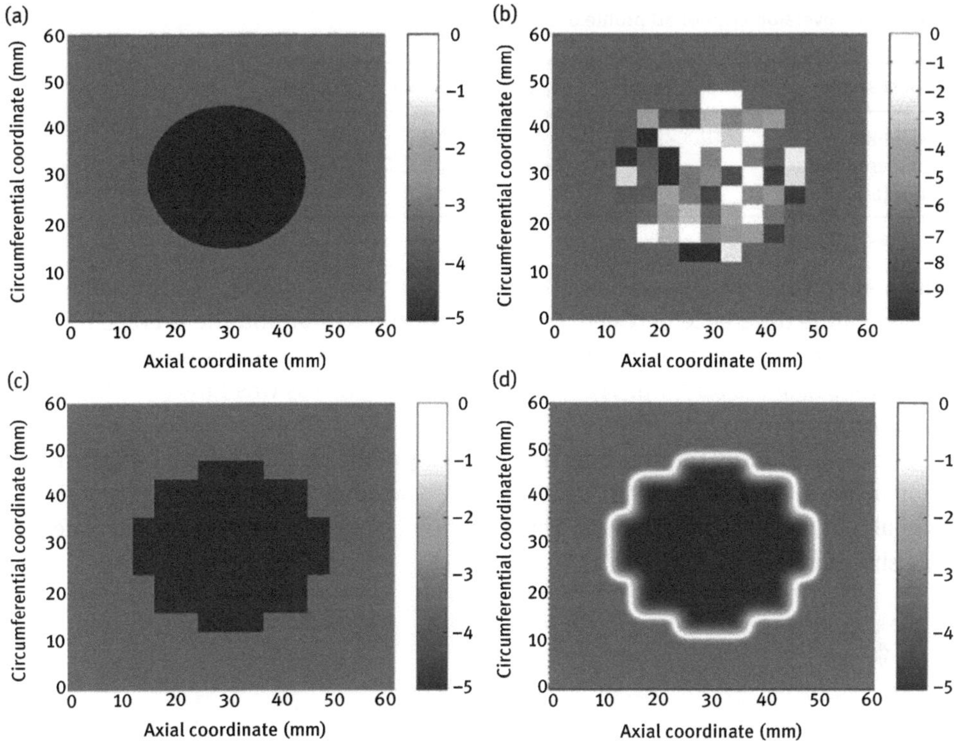

Figure 4.39: Three-dimensional profile iterative inversion of cylindrical defect of 30 mm × 5 mm: (a) teal defect profile; (b) mesh model based on the results of open profile recognition; (c) inversion results of mesh model; and (d) final defect profile.

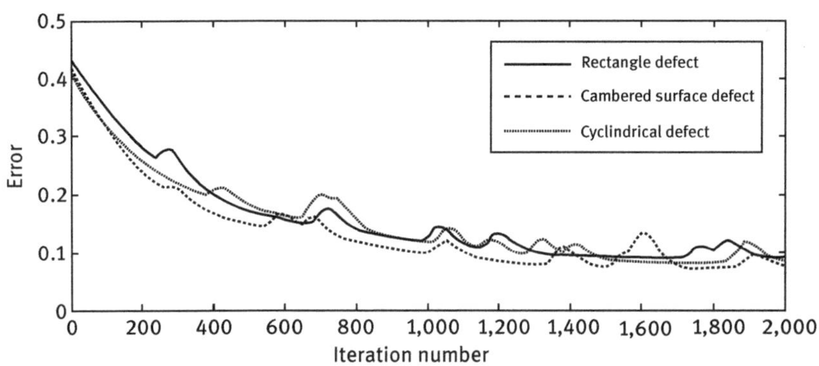

Figure 4.40: Error convergence curve of iterative inversion of defect 3D profile.

Table 4.5: Inversion error of 3D profile of defects.

Defect shape	Size (mm)	E_m (%)	E_p (%)	E_n (%)
Rectangle	30 × 20 × 5	9.2	3.9	3.1
Cambered surface	30 × 5	7.6	3.2	2.6
Cylinder	30 × 5	8.5	3.6	2.8

method achieves high-precision 3D contour inversion of defects. At the same time, there is little difference between the average positive error and the average negative error, which indicates that the error distribution between the inversion profile and the real profile is more uniform. The iterative inversion process was fully searched in the solution space.

In order to verify the anti-interference ability of the random search iterative inversion method, random noises of 1 %, 2 % and 5 % were added in the 3D MFL simulation signals of the above three defects. The results of the contour reinversion test of defects are shown in Figures 4.41–4.43.

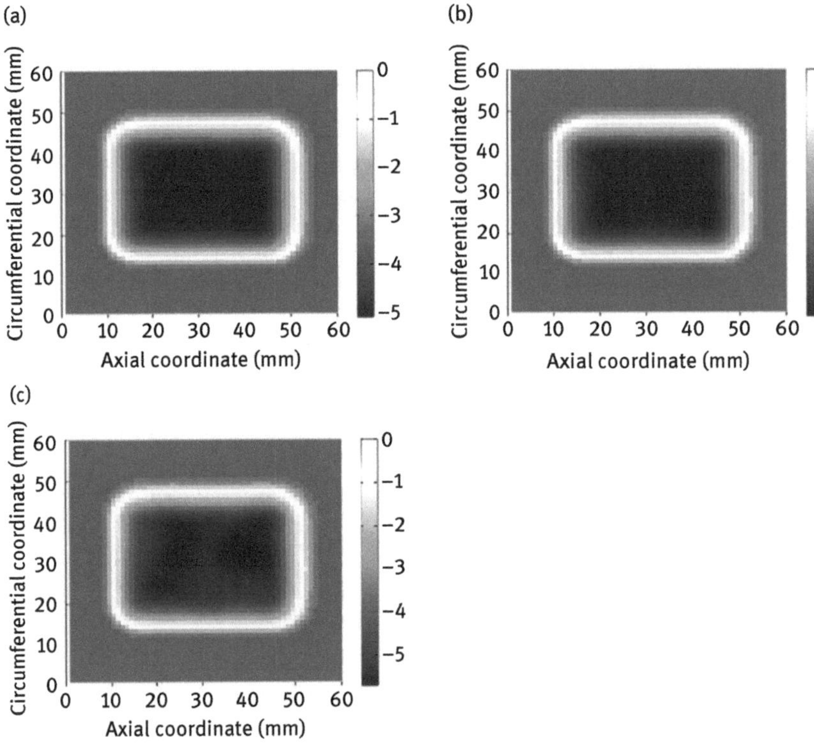

Figure 4.41: Contour inversion results of 30 mm × 20 mm × 5 mm rectangular defect at different noise levels: (a) 1 % noise; (b) 2 % noise; and (c) 5 % noise.

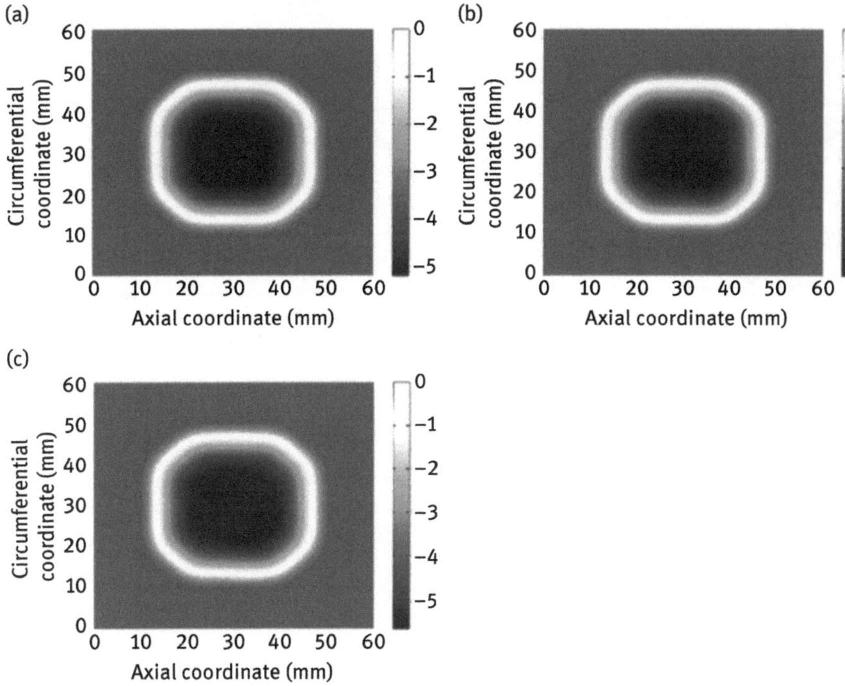

Figure 4.42: Contour inversion results of 30 mm × 5 mm cambered surface defect at different noise levels: (a) 1 % noise; (b) 2 % noise; and (c) 5 % noise.

The inversion error of each defect at different noise levels is shown in Table 4.6. According to the statistical results, with the increase of the noise level of the 3D MFL signal, the 3D profile error of the defect is gradually increased. In 1 % noise level, the RMS error, the average positive error and the average negative error of 3D profile inversion of the defects are no more than 10.5 %, 4.3 % and 3.4 %, respectively. In 5 % noise level, the RMS error, the average positive error and the average negative error of 3D profile inversion of the 30 mm × 20 mm × 5 mm rectangular defect are no more than 15.2 %, 7.5 % and 6.3 %, respectively.

However, in the MFL testing on oil and gas pipeline, the amplitude of MFL defect detection signal will reach 400 Gs or more, while the amplitude of the detection noise in general is less than 2 Gs. Therefore, the noise level of the actual MFL detection is less than 1 %, which will not apparently interfere with the results of the defect 3D shape of the random search iterative inversion.

In order to verify the applicability of the random search iterative inversion method in irregular defects, a defect shown in Figure 4.44 was tested based on 3D profile inversion. Among them, the initial defect profile is set to a flat bottom defect with the depth of the 1/2 pipe wall thickness, and the total number of termination iteration is set to 4,000.

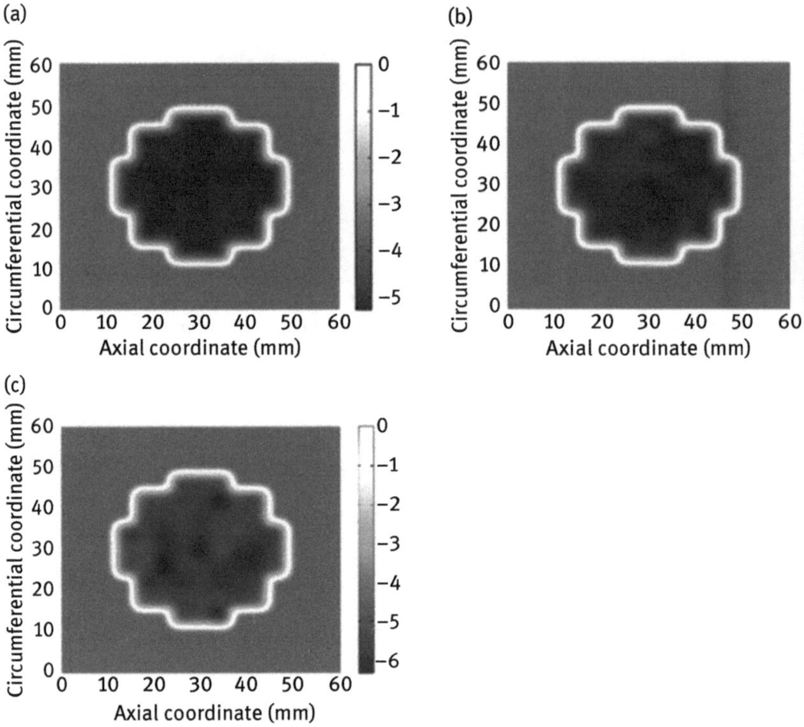

Figure 4.43: Contour inversion results of 30 mm × 5 mm cylindrical defect at different noise levels: (a) 1 % noise; (b) 2 % noise; and (c) 5 % noise.

Table 4.6: Inversion error of defect 3D profile under different noise levels.

Defect (mm)	1 % noise (%)			2 % noise (%)			5 % noise (%)		
	E_m	E_p	E_n	E_m	E_p	E_n	E_m	E_p	E_n
Rectangle 30 × 20 × 5	10.5	4.3	3.4	11.6	5.4	4.4	15.2	7.5	6.3
Cambered surface 30 × 5	8.4	3.6	2.9	9.1	4.7	4.0	13.1	6.7	6.0
Cylinder 30 × 5	9.7	3.9	3.1	10.3	4.9	4.2	14.4	7.2	6.2

As shown in Figure 4.44, the recognition result of the open contour of the irregular defect is basically in accordance with the actual opening profile. Finally, the inversion result of the defect profile is obtained by appropriate smoothing of the iterative inversion result of the defect 3D mesh model. Compared with the real contour, the position and the value of the deepest point have high similarity.

By calculation, the RMS error of the inversion result of the irregular 3D contour defect is 12.4 %, which is acceptable within the actual scope. This shows that the random search iterative inversion method has a good applicability for the irregular defects in the actual pipeline testing.

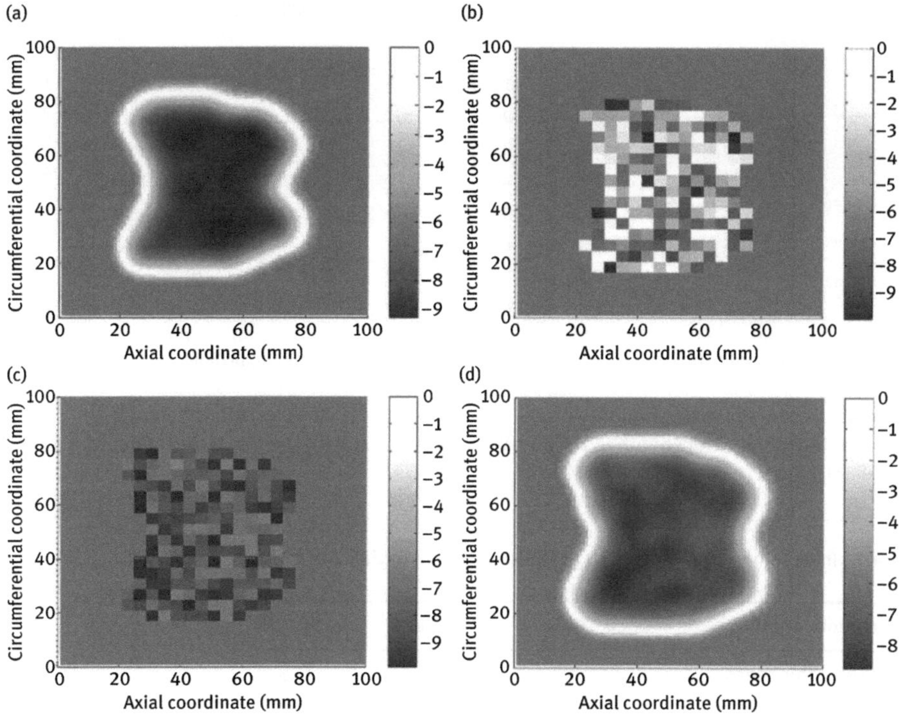

Figure 4.44: Three-dimensional profile iterative inversion of irregular defect: (a) real defect profile; (b) mesh model based on the results of open profile recognition; (c) inversion results of mesh model; and (d) final defect profile.

In the iterative inversion process of the above irregular defect, the initial defect profile was set to flat defects with depths of 1/4 and 3/4 of the pipe wall thickness, respectively, and the total number of iterations in the termination condition was still set to 4,000. Further, an iterative inversion test of the defect 3D shape was carried out, and the results obtained were shown in Figure 4.45.

Table 4.7 shows the inversion errors of the irregular defect in different initial defect profiles. From the table, we can know that when initial defect depth is 1/4 and 3/4 of the pipe wall thickness, RMS errors of the inversion defect profile and the real contours are 28.1 % and 9.7 %, respectively, and these have great changes with the inversion error of initial defect depth of 1/2 pipe wall thickness. At the same time, the mean positive error and the mean negative error of the inversion results are significantly changed after the initial defect profile is changed. It shows that the searching state of the iterative inversion process in the solution space has changed obviously. It can be known that the choice of the initial defect profile will have a significant impact on the results of the random search iterative inversion.

Figure 4.46 shows the error convergence curves of the iterative inversion process for different initial defect profiles, in which the error is the RMS error of the overall

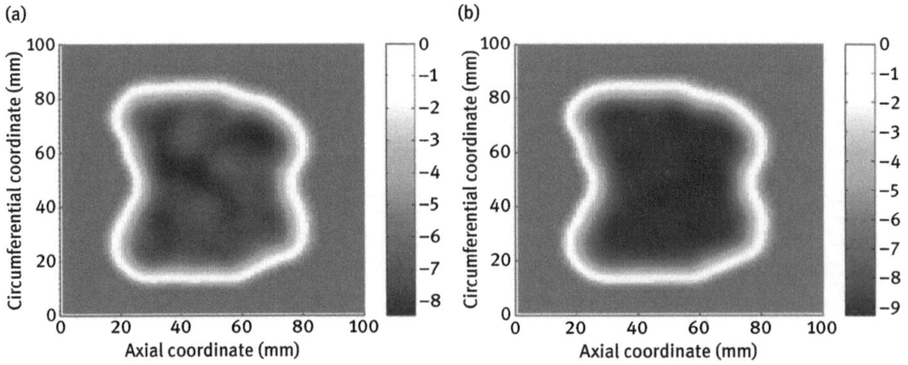

Figure 4.45: Three-dimensional profile inversion results of irregular defects in different initial profiles: (a) the initial defect depth is 1/4 pipe wall thickness and (b) the initial defect depth is 3/4 pipe wall thickness.

Table 4.7: Inversion error of defect 3D profile in different initial contours.

Initial contour depth	E_m (%)	E_p (%)	E_n (%)
1/4 pipe wall thickness	28.1	3.5	11.3
1/2 pipe wall thickness	12.4	6.8	5.6
3/4 pipe wall thickness	9.7	5.4	4.3

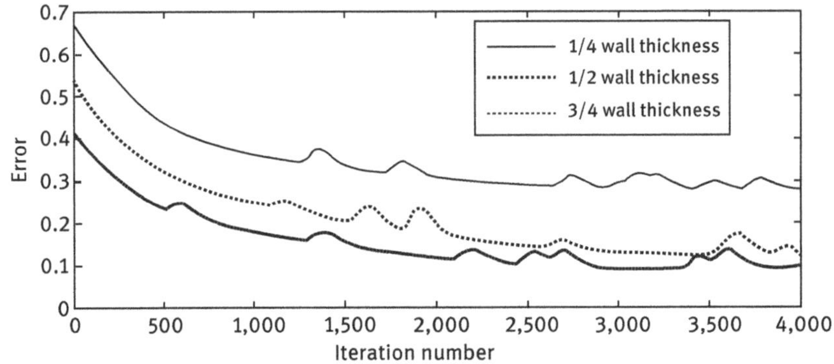

Figure 4.46: Error convergence curves of irregular defect iterative inversions in different initial defect profiles.

deviation of the inversion results. When the depth of initial defect is set to 1/4, 1/2 and 3/4 pipe wall thickness, the initial error of the iterative inversion process is 0.67, 0.54 and 0.41, respectively. Under the same total number of iterations, the different initial profiles directly lead to different final inversion errors. This shows that the initial defect profile will have a direct impact on the results of random search for the 3D profile of the defect.

4.3 Iterative Inversion of Neural Networks for Defect 3D Profiles

Conventional defect quantitative inversion method based on direct neural network uses the trained neural network to forecast the defects contour parameters, which can be called neural network direct inverse method. Neural network direct inverse method has many problems, which make it difficult to be used in the inversion of defect 3D profiles. First, the direct inversion method is not necessarily convergent to the global optimal solution because of the ill-conditioned quantitative inversion problem. Second, this kind of method is highly dependent on the training samples, and the generalization ability of the unknown shape defects is insufficient. Finally, the neural network method has the problem of dimension disaster. When the neural network is used for the defect 3D contour inversion directly, the complex parameters of 3D MFL signal and the defect profile will cause the dramatic increase in the number of neurons, resulting in an increase in network training time dramatically, and even leading to network training algorithms simply failing to converge.

To make use of the quickness of neural network, by adopting the idea of iterative method to solve the low accuracy and generalization ability of the neural network inversion, the 3D contour iteration inversion method of neural network was researched. In order to reduce the number of neurons in neural network, the principal component analysis was applied to extract the main features of the MFL signal defect profile parameters, and a defect 3D contour strip model applicable for the neural network method was established. And then, the equivalent strip models and radial basis function (RBF) neural networks for 3D defect MFL signal forward prediction of a rectangular, cylindrical, cambered surface defect samples were established. Then, the forward predicting neural network was embedded into the iteration and a neural network iterative inversion method of MFL defect 3D profile was proposed. Based on the gradient descent algorithm, the parameters of the strip model were modified.

4.3.1 Main Feature Extraction of 3D MFL Detection Signal

The number of neurons in the neural network is positively correlated with the size of the input data, so there is a curse of dimensionality. When the 3D MFL data is directly used for 3D inversion defects, the large data will cause a dramatic increase in the number of neurons, which will easily lead to a sharp increase in network training time.

In order to reduce the number of neurons in neural network so as to reduce the training complexity, the 3D MFL signal redundancy is analyzed, then the characteristics of the 3D MFL signal are defined, and the main features of the MFL signal defect profile parameters are extracted effectively, so as to achieve the goal of eliminating data redundancy, reducing the size of the input data and reducing the number of neural network neurons.

Based on cambered surface defect of 28.6 mm × 7.2 mm, cylindrical defect of 28.6 mm × 7.2 mm and rectangular defect of 28.6 mm × 14.3 mm × 7.2 mm, the 3D MFL signal data redundancy was analyzed. The axial and circumferential sampling intervals of the 3D MFL detection signal are 1 mm, and the 3D signal components are represented by 90 × 90 matrix.

Figure 4.47 shows the 3D MFL signal of 28.6 mm × 7.2 mm cambered surface defect. The respective feature vectors are extracted from the 3D signal components, the obtained feature vectors are arranged according to the proportion from high to low and the cumulative results of the characteristics proportion are shown in Figure 4.48. According to the statistical results, the proportion of the three main feature vectors of the axial and radial components has reached 95 %, and the proportion of the six main feature vectors of the circumferential component is 95 %. In order to reduce data redundancy and maintaining the highest possible accuracy, the threshold proportion of feature vectors is set to 95 %, so the axial component and the radial component of the sample 3D defect MFL signal can be represented by three principal feature vectors, respectively, and the circumferential component can be represented by six principal feature vectors.

Therefore, in maintaining the accuracy of the original 3D MFL detection signal 95 % characteristics, the axial, radial and circumferential MFL detection signals can be used 3 × 90, 90 × 3 and 90 × 6 matrix for simplified representation. Compared to the original 90 × 90 matrix, the simplified representation based on main feature vectors reduces the amount of data to 96.7 %, 96.7 % and 93.3 %, respectively. It can be known that there is a large amount of data redundancy in the original 3D MFL detection signal.

In the same way, 3D MFL signals of 28.6 mm × 7.2 mm cylindrical defect and 28.6 mm × 14.3 mm × 7.2 mm rectangular defect are analyzed. The cumulative results and the feature vector proportions are shown in Figures 4.49 and 4.50. The threshold

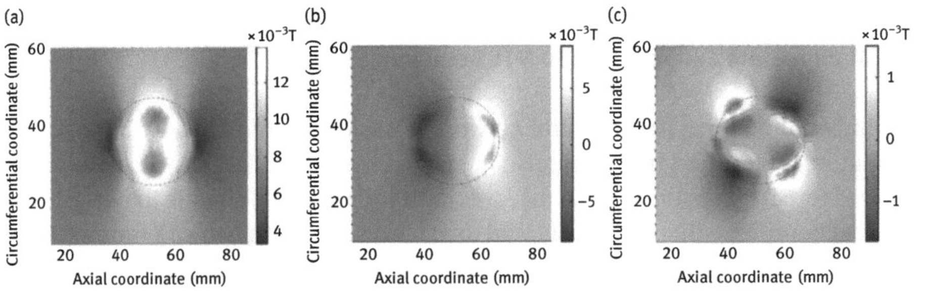

Figure 4.47: Three-dimensional MFL signal of 28.6 mm × 7.2 mm cambered surface defect: (a) axial component; (b) radial component; and (c) circumferential component.

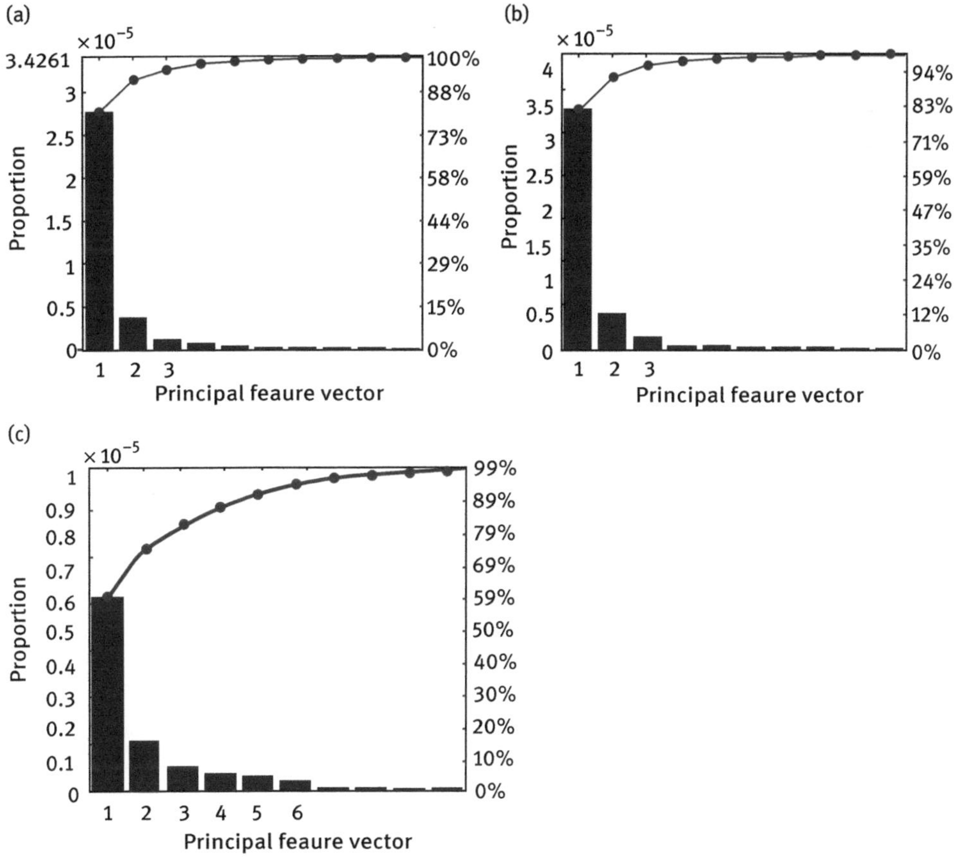

Figure 4.48: The cumulative results and the feature vector proportion of the 3D MFL signal of 28.6 mm × 7.2 mm cambered surface defect: (a) axial component; (b) radial component; and (c) circumferential component.

proportion of feature vectors is set to 95 %, so the axial, radial and circumferential components of the 3D MFL signal of cylindrical defect can be represented by 3, 3 and 6 main feature vectors, respectively, and the axial, radial and circumferential components of the 3D MFL signal of rectangular defect can be represented by 3, 3 and 5 main feature vectors, respectively. The main feature vector is used to simplify the 3D component of the MFL signal of the cylinder defect and the rectangular defect, which can reduce the amount of data to be processed by more than 90 %. It shows that there is a lot of data redundancy in the original 3D MFL signal.

In order to accurately reflect the contour parameters of the defect, the characteristic value of the 3D MFL detection signal is defined. Because of the difference of 3D MFL signals with different shape defects, the characteristic values should be selected from the basic features of common defects. Therefore, in defining the characteristic

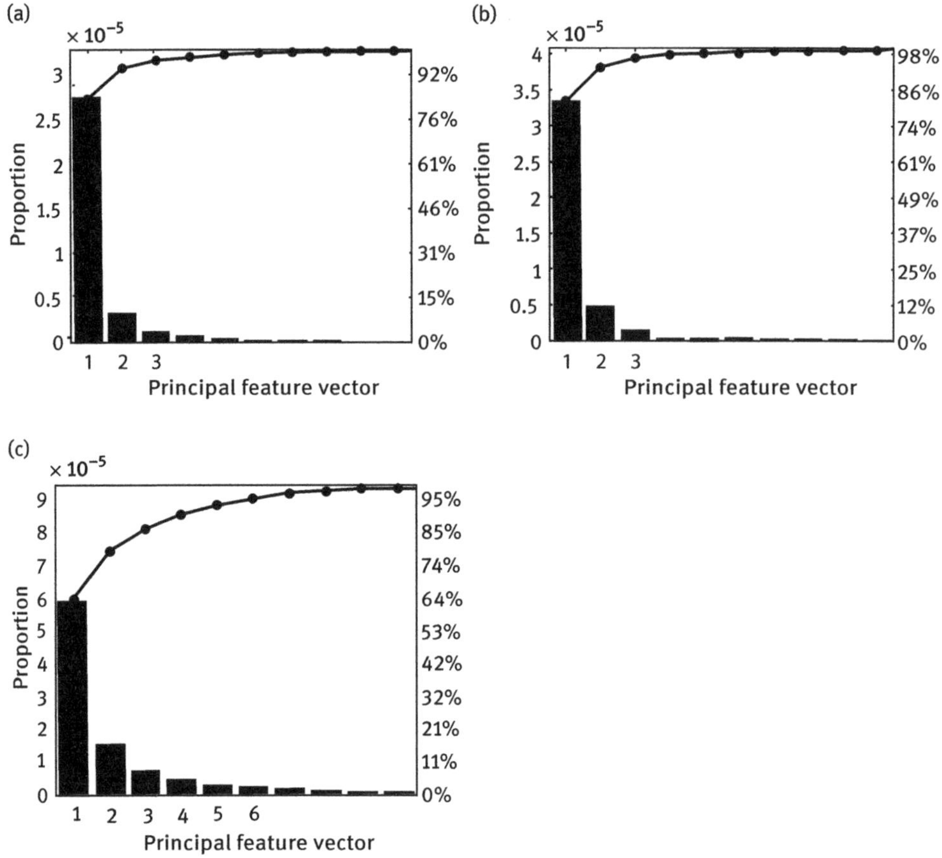

Figure 4.49: The cumulative results and the feature vector proportion of the 3D MFL signal of 28.6 mm × 7.2 mm cylindrical defect: (a) axial component; (b) radial component; and (c) circumferential component.

values, the basic characteristics reflecting the defect opening profile or depth, such as peak–valley value, the axial and circumferential spacing of peak–valley value, the "volume" and "energy" of magnetic leakage field, should be taken into account. The "volume" of the magnetic leakage field is the surface integral of the magnetic induction intensity, whose unit is $T \cdot mm^2$. The "energy" of the magnetic leakage field is the surface integral of the square of the magnetic induction intensity, whose unit is $T^2 \cdot mm^2$.

Take the 28.6 mm × 7.2 mm cambered surface defect for an example, and the characteristics of 3D MFL signal are defined. Because the axial, radial and circumferential MFL signals of the defects are different, the characteristic values of these three components are defined separately.

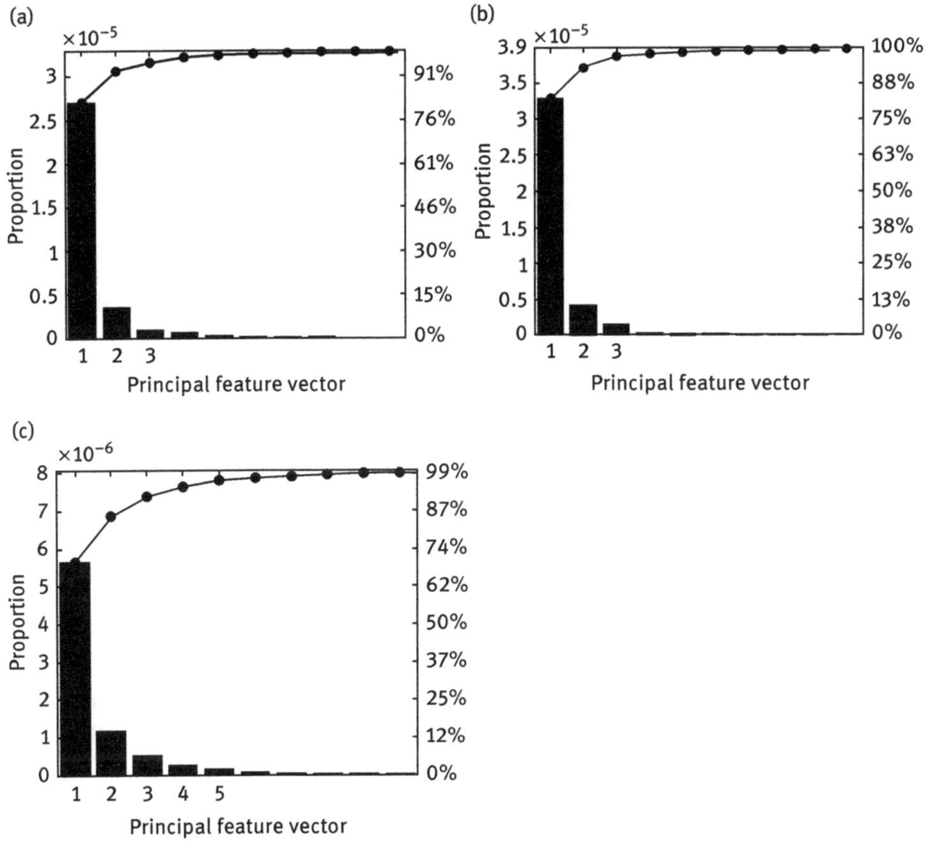

Figure 4.50: The cumulative results and the feature vector proportion of the 3D MFL signal of 28.6 mm × 14.3 mm × 7.2 mm rectangular defect: (a) axial component; (b) radial component; and (c) circumferential component.

4.3.1.1 Axial Component

Figure 4.51 shows the axial MFL signal of the cambered surface defect, which contains a peak area (including two peak points) and two valley point value. The two valley points are located at the left and right opening of the defect along the axial direction, and the peak area is located at the center of the defect opening.

First, the peak and valley points of the axial signal reflect the turning point of the bottom profile of the defect. Therefore, the peak and valley values of the axial MFL detection signal and the axial and circumferential spacing of the corresponding points are selected as the characteristic values.

Second, the amplitude of the axial component of the MFL reflects the size of the defect depth, so the volume and energy of the axial MFL detection signal is selected as the characteristic value. In addition, the axial MFL detection signal is abrupt at the

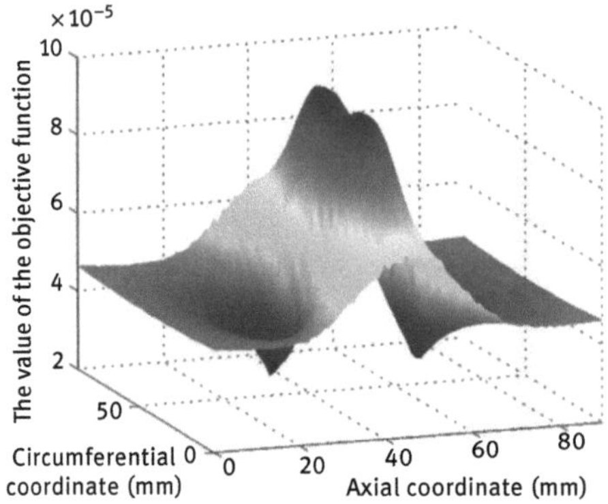

Figure 4.51: The axial MFL signal of the cambered surface defect of 28.6 mm × 7.2 mm.

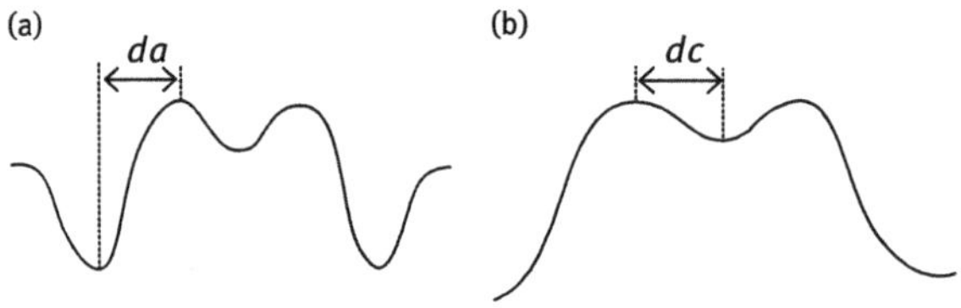

Figure 4.52: Schematic diagrams of the characteristic values of the center detection curve of the axial MFL detection signal: (a) axial center detection curve and (b) circumferential center detection curve.

boundary of the defect opening, so the amplitude of the maximum value of the signal gradient and its coordinate are selected as characteristic values.

Finally, taking into account the complex distribution of the peak area of the axial MFL detection signal, the center of the detection signal is detected along the axial direction and circumferential direction, as shown in Figure 4.52. In the center detection curve along the axial direction, take the peak–valley value and the corresponding distance d_a as the characteristic value; in the center detection curve along the circumferential direction, take the peak–valley value and the corresponding distance d_c as the characteristic value.

The characteristic values of the final defined axial MFL detection signal is provided in Table 4.8.

4.3.1.2 Radial Component

Figure 4.53 shows the radial MFL signal of the cambered surface defect that has a peak point and a valley point. The two points are, respectively, located at the left and right openings of the defects along the axial direction. So the peak and valley values, and

Table 4.8: Definition of characteristic values of the axial MFL detection signal.

Characteristic value	Unit	Definition
P_{a1}	T	Peak of the signal
P_{a2}	T	Valley of the signal
P_{a3}	mm	The axial distance between the peak and valley
P_{a4}	mm	The circumferential distance between the peak and valley
P_{a5}	T/mm	Maximum gradient
P_{a6}	mm	The axial coordinate of the gradient maximum point
P_{a7}	mm	The circumferential coordinate of the gradient maximum point
P_{a8}	$T \cdot mm^2$	Volume of magnetic leakage field
P_{a9}	$T^2 \cdot mm^2$	Energy of magnetic leakage field
P_{a10}	T	Center curve peak
P_{a11}	T	Center curve valley
P_{a12}	mm	Circumferential spacing between the valley points of the circumferential center curve
P_{a13}	T	Peak value of axial center curve
P_{a14}	T	Valley of the axial center
P_{a15}	mm	Axial distance between peak and valley of the axial center curve

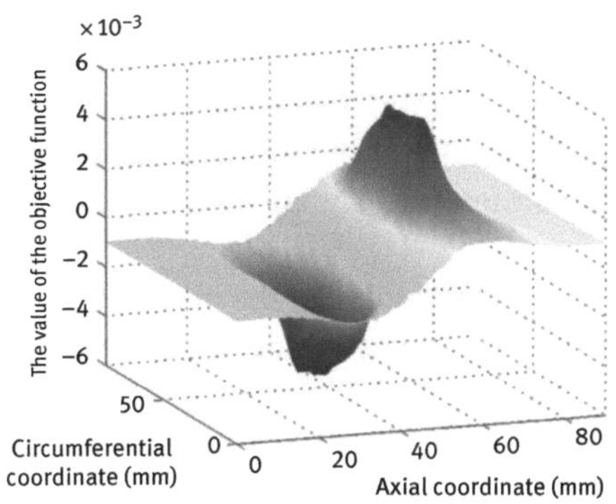

Figure 4.53: The radial MFL signal of the cambered surface defect of 28.6 mm × 7.2 mm.

the axial and circumferential spacing of the corresponding points are selected as the characteristic values. At the same time, the contour of the radial MFL signal basically reflects the opening shape of the defect. It is necessary to select the radii of the peak and the valley areas as the feature values.

In addition, the amplitude of the radial component of the MFL reflects the size of the defect depth. Therefore, the volume and energy of the radial MFL detection signal

Table 4.9: Definition of characteristic values of the radial MFL detection signal.

Characteristic value	Unit	Definition
P_{r1}	T	Peak of the signal
P_{r2}	T	Valley of the signal
P_{r3}	mm	The axial distance between the peak and valley
P_{r4}	mm	The circumferential distance between the peak and valley
P_{r5}	mm	The radius of the peak area
P_{r6}	mm	The radius of the valley area
P_{r7}	$T \cdot mm^2$	Volume of magnetic leakage field
P_{r8}	$T^2 \cdot mm^2$	Energy of magnetic leakage field

are selected as the complementary characteristic values. The final defined radial MFL detection signal characteristic values are shown in Table 4.9.

4.3.1.3 Circumferential Component

Figure 4.54 shows the circumferential MFL signal of the cambered surface defect that has two peak points and two valley points. The four points are all located at the edge of the defect opening. So the peak values, valley values and the axial and circumferential spacings of the corresponding points are selected as the characteristic values. Because the contour of the circumferential MFL signal is better consistent with the side boundary of the defect, it is necessary to select the radii of the peak area and the valley area as feature values. At the same time, the MFL detection signal is abrupt at the boundary of defect opening, so the amplitude of the gradient maximum value and its corresponding coordinate of the circumferential signal are chosen as the characteristic values.

In addition, the amplitude of the circumferential component of the magnetic leakage field reflects the defect depth, so the volume and energy of the circumferential

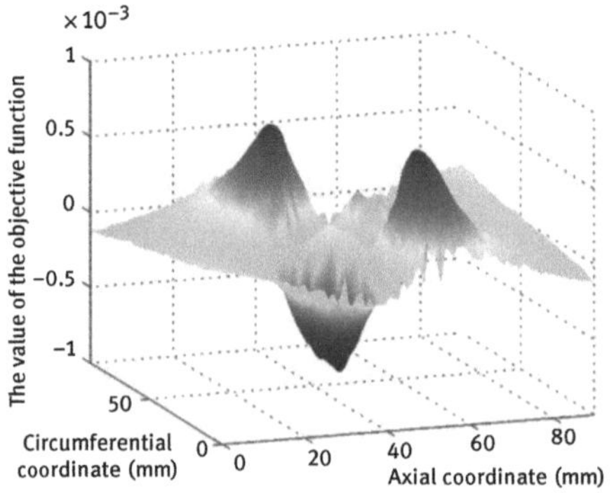

Figure 4.54: The circumferential MFL signal of the cambered surface defect of 28.6 mm × 7.2 mm.

MFL detection signal can be chosen as the complementary characteristic value. The final defined circumferential MFL detection signal characteristic values are shown in Table 4.10.

In order to reduce the data redundancy in the 3D MFL signal as far as possible, the characteristic values defined above of the 3D MFL signal were analyzed by the principal component analysis method, and the main that could effectively reflect the defect profile parameters characteristic values were extracted.

Principal component analysis is a linear data descending dimension method. Through constructing the orthogonal basis of the data space, multiple variables containing redundant data will simplify to the representation of several variables so as to reduce data redundancy and reduce the data size of the target.

The characteristic values defined above of the 3D MFL signal are 15 axial characteristic values, 8 radial characteristic values and 11 circumferential characteristic values. The basic steps of analyzing the 34 characteristic values by the principal component analysis are as follows:

1. x testing defects $D_1, D_2, \ldots, D_i, \ldots, D_{x-1}, D_x$ are used, calculate the 34 characteristic values $P_1, P_2, \ldots, P_{34}$ of 3D MFL signal corresponding to each defect. The combination of all the characteristic values is used for target matrix M:

$$M = \left\{ \begin{matrix} P_{1,1}, P_{1,2} & \ldots & P_{1,33}, P_{1,34} \\ & \ldots & \ldots \\ P_{i,1}, P_{i,2}, \ldots, & P_{i,j}, \ldots, & P_{i,33}, P_{i,34} \\ & \ldots & \ldots \\ P_{x,1}, P_{x,2} & \ldots & P_{x,33}, P_{x,34} \end{matrix} \right\} . \tag{4.23}$$

2. Calculate all feature vectors of target matrix M, calculate the component proportion of each feature vector $\mathbf{V}$, and from high to low according to the proportion, all feature vectors are sorted.

Table 4.10: Definition of characteristic values of the circumferential MFL detection signal.

Characteristic value	Unit	Definition
P_{c1}	T	Peak of the signal
P_{c2}	T	Valley of the signal
P_{c3}	mm	The axial distance between the peak and valley
P_{c4}	mm	The circumferential distance between the peak and valley
P_{c5}	mm	The radius of the peak area
P_{c6}	mm	The radius of the valley area
P_{c7}	T/mm	The maximum gradient value
P_{c8}	mm	The axial coordinate of the maximum gradient point
P_{c9}	mm	The circumferential coordinate of the maximum gradient point
P_{c10}	$T \cdot mm^2$	Volume of magnetic leakage field
P_{c11}	$T^2 \cdot mm^2$	Energy of magnetic leakage field

3. Set component proportion threshold δ for judging the main characteristic values of the 3D MFL signal. According to the above ranking result, k feature vectors $\mathbf{V}_1, \ldots, \mathbf{V}_i, \ldots, \mathbf{V}_k$ with larger component proportion are selected.ĄB
4. Select the characteristic values $P_1, \ldots, P_i, \ldots, P_k$ of the 3D MFL signal in the target matrix corresponding to the feature vectors $\mathbf{V}_1, \ldots, \mathbf{V}_i, \ldots, \mathbf{V}_k$.

Based on a set of cambered surface defects, the main characteristic values of the 3D MFL signal are extracted. The minimum value, maximum value and step size of the defect surface diameters are 14.3 mm, 28.6 mm and 1.4 mm, respectively. The minimum value, maximum value and step size of the defect depths are 1.4 mm, 7.2 mm and 1.4 mm, respectively. The total number of defects is 55.

Thirty-four main characteristic values are solved, respectively, for the 3D MFL signal of each defect, and principal component analysis is performed on the matrix composed of these characteristic values. The principal component analysis results are shown in Figure 4.55. For convenience, the analysis results are listed in the first 10 characteristic values of the largest proportion of the total composition. In order to reduce the redundant data and maintaining the highest possible accuracy, the threshold of characteristic value proportion is set as 90 %, all the 3D MFL signal can be expressed by seven main characteristic values. The selected seven main characteristic values are P_{a8}, P_{a9}, P_{a1}, P_{a2}, P_{r3}, P_{c4} and P_{a6}.

In order to further verify the above results, the main characteristic values of the 3D MFL signals are extracted based on the rectangular defects and the cylinder defects. The sizes of the two kinds of defects are shown in Table 4.11. The total number of rectangular defects is 260, and the total number of cylinders is 55.
Figures 4.56 and 4.57 are the results of the principal component analysis based on the characteristic values of the MFL signals obtained from the rectangular and cylindrical defects. For convenience, the analysis results are listed in the first 10 characteristic

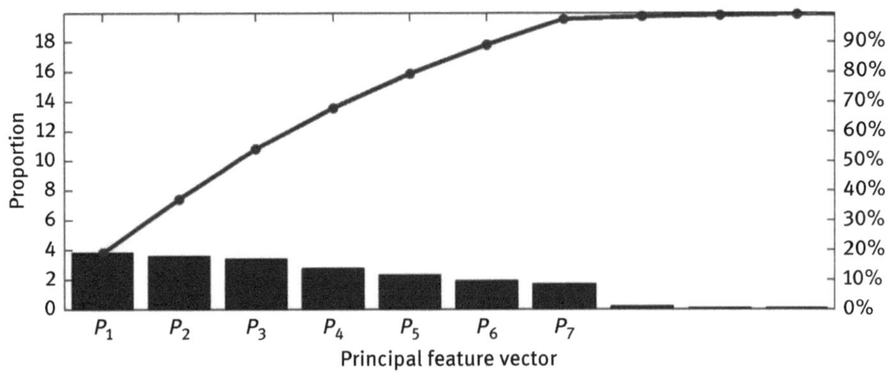

Figure 4.55: The results of principal component analysis of 3D MFL signal characteristics of cambered surface defects.

Table 4.11: Sizes of rectangular defects and cylindrical defects for main feature extraction.

Parameters	Rectangular defects			Cylinder defects	
	Length	Width	Depth	Diameter	Depth
Minimum value (mm)	14.3	7.2	1.4	14.3	1.4
Maximum value (mm)	100.1	28.6	7.2	28.6	7.2
Step size (mm)	7.2	7.2	1.4	1.4	1.4

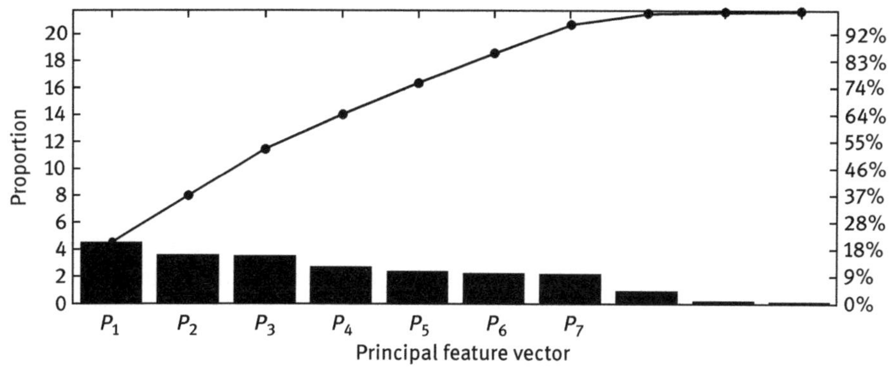

Figure 4.56: The results of principal component analysis of 3D MFL signal characteristics of rectangular defects.

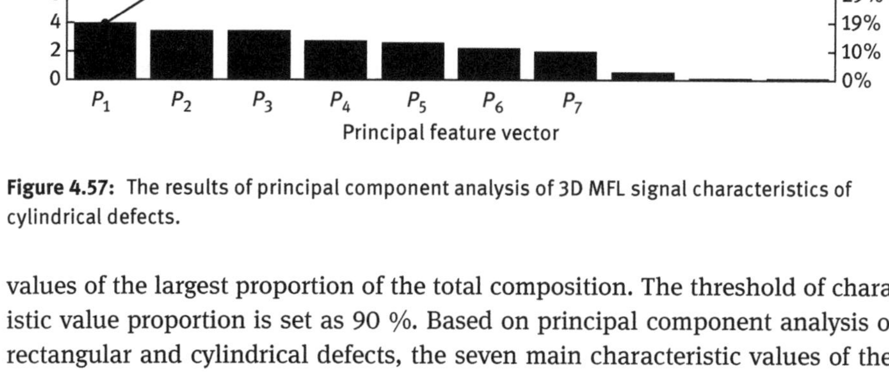

Figure 4.57: The results of principal component analysis of 3D MFL signal characteristics of cylindrical defects.

values of the largest proportion of the total composition. The threshold of characteristic value proportion is set as 90 %. Based on principal component analysis of the rectangular and cylindrical defects, the seven main characteristic values of the two defects are the same. They are P_{a8}, P_{a9}, P_{a1}, P_{a2}, P_{r3}, P_{c4} and P_{a6}, whose sequences in the two defects are different.

It can be seen from the above results, in the premise of the threshold settings of characteristic value proportion for 90 %, whether based on the cambered surface defects, rectangular defects or cylindrical defects, the main characteristic values of the defects are the seven characteristics of P_{a8}, P_{a9}, P_{a1}, P_{a2}, P_{r3}, P_{c4} and P_{a6}. Therefore, the 3D MFL signal of any defect can be effectively expressed by the seven main characteristic values.

In fact, in the inversion of the 3D profile of the defect, according to the defect open contour detection method proposed before, the open contour has been detected by the gradient information of the 3D MFL detection signal. Main characteristic value P_{a6} corresponds to the coordinate of the axial MFL signal gradient maximum point has been calculated, so we can get rid of P_{a6} in the list of the main characteristic values. Table 4.12 shows the main characteristic values of the 3D MFL signal inversion defect contour.

In the subsequent inversion process of the 3D profile of defects, the main characteristic values will replace the 3D original MFL signal as the input or output of neural network. So it can greatly reduce the number of neurons in the neural network, thereby reducing the complexity of network training and the training time.

4.3.2 Defect 3D Profile Strip Model

According to the defect opening profile detected by the previous method, the 3D contour inversion region of the defect is basically chosen. Based on this and the depth parameters, the 3D profile neural network of the defect is established.

In order to reduce the parameters of the 3D model, the method of establishing the 3D profile strip model of defect is proposed. The specific steps are as follows:

1. To determine the partition size Δw, the wall region to be solved along the circumferential direction is divided into N equal parts. Set the depths outside the

Table 4.12: Main characteristic values of three-dimensional MFL signal for defect profile inversion.

Characteristic value	Unit	Definition
P_1	$T \cdot mm^2$	Volume of axial magnetic leakage field
P_2	$T^2 \cdot mm^2$	Energy of axial magnetic leakage field
P_3	T	Peak value of axial MFL signal
P_4	T	Valley value of axial MFL signals
P_5	mm	Axial spacing between peak value and valley value of radial MFL signal
P_6	mm	Circumferential spacing between peak value and valley value of circumferential MFL signal

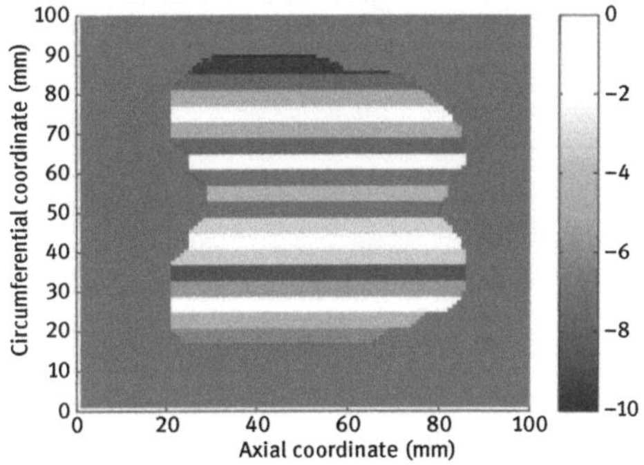

Figure 4.58: Initial partition of the defect opening profile.

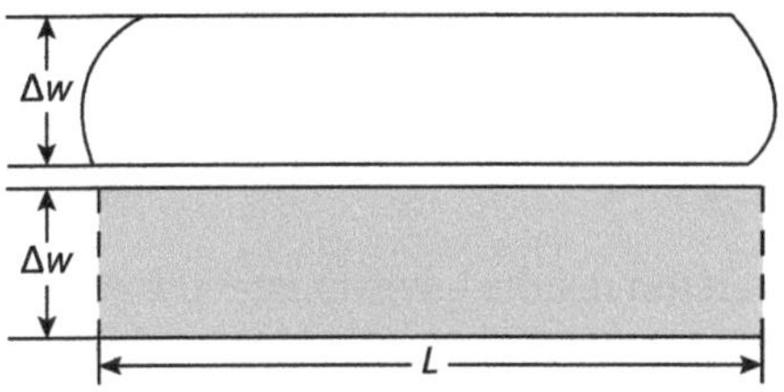

Figure 4.59: Equivalent rectangular representation of any equal part of opening defect contour.

defect opening contour as 0, and the depth of each equal parts inside the contour opening as the same depth. We can get the preliminary results as shown in Figure 4.58.

2. Calculate the equivalent length of each equal part inside the defect opening contour and represent the equal parts with equivalent rectangular, as shown in Figure 4.59. The equivalent rectangular length can be calculated as follows:

$$L = S/\Delta w, \tag{4.24}$$

where S is the area of the equal opening section and Δw is the divided size in the model along the circumferential direction.

3. Combine all the equivalent rectangles, and the opening of new contour is obtained. Determining depth d of each equal part in the radial direction, then defect 3D contour strip model M_S is obtained as shown in Figure 4.60.

In this model, the unknown parameters are the depth d for each equal part. Let L_i represent the length of ith equal part, d_i denote the depth of ith equal part, in the premise of the partition size Δw along the circumferential direction, 3D contour strip model of defects can be described as follows:

$$M_S = \left\{(L_1, d_1), \ldots, (L_i, d_i), \ldots, (L_N, d_N)\right\}. \tag{4.25}$$

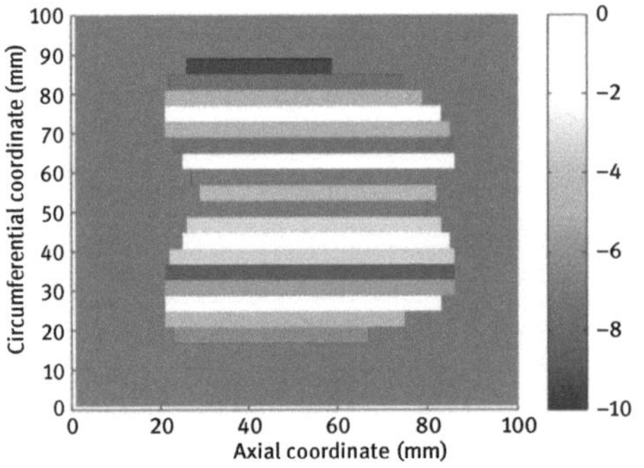

Figure 4.60: Defect 3D profile strip model.

In formula (4.25), the length and depth inside the contour of the defect are the parameters to be solved. Outside the opening contour of the defect, length and depth parameters are 0:

$$L_i = d_i = 0. \tag{4.26}$$

According to the strip model established, the inversion results have nonsmooth edges. Considering the smooth edge of actual defects, in the defect 3D profile inversion process, the inversion results based on strip model are smoothed to get smoother defect edge.

4.3.3 Forward Prediction of MFL Signal Based on RBF Neural Network

The input of the RBF neural network which is used for the forward prediction of the 3D MFL signal is based on the defect parameters of the 3D profile strip model, and the output is the main characteristic values of the extracted 3D MFL signal [110]. In the training to the forward prediction neural network, the equivalent strip models of rectangle, cambered surface and cylindrical sample defects are established. Finally, based on the training of the RBF neural network, the main characteristics of the defect MFL signal are predicted.

RBF neural network satisfies the nonlinear mapping requirements of defect 3D quantitative contour. If there are enough units in the hidden layer, a three-layer network can implement arbitrary precision approximation to any continuous function. Therefore, the RBF neural network is used to predict the 3D MFL signal.

Figure 4.61 is the basic structure of RBF neural network, including the input layer, the output layer and the hidden layer. The input vector $\mathbf{X}$ is provided with M components, and the output vector $\mathbf{Y}$ contains N components, so the complete mapping between the input and the output of the RBF neural network is provided as follows:

$$y_j = f_j(\mathbf{X}) = \sum_{i=1}^{H} W_{ij}\varphi_i(\|\mathbf{x} - \mathbf{c}_i\|), \tag{4.27}$$

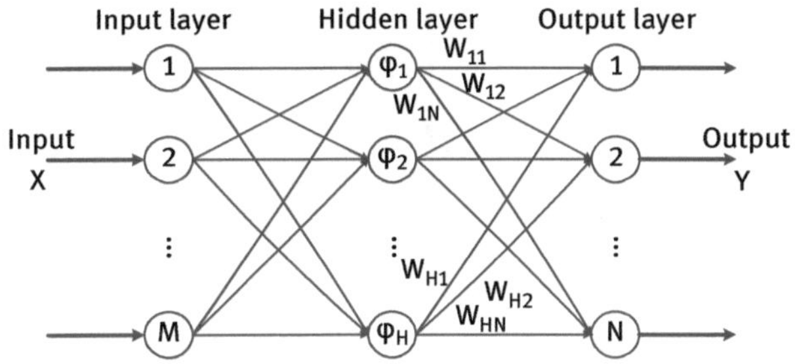

Figure 4.61: Structure of RBF neural network.

where y_j is the jth component of output vector; φ_i is the ith basis function in the hidden layer; $\mathbf{c}_i$ is the center vector of the basis function φ_i; W_{ij} is the connection weight between the ith basis function in the hidden layer and the jth component of the output vector; H is the number of neurons in the hidden layer.

In order to realize the nonlinear mapping function of network hidden layer, select the following Gauss function as the basis function of RBF neural network:

$$\phi_i(||\mathbf{x} - \mathbf{c}_i||) = \exp\left(-\frac{||\mathbf{x} - \mathbf{c}_i||^2}{2\sigma_i^2}\right), \tag{4.28}$$

where $\mathbf{c}_i$ and σ_i are the center vector and the width parameter of the Gauss basis function, respectively.

The calculation formula of the Gauss function is brought into the neural network, and the complete input–output mapping relationship of RBF neural network is obtained:

$$y_j = f_j(\mathbf{X}) = \sum_{i=1}^{H} W_{ij}\exp\left(-\frac{||\mathbf{x} - \mathbf{c}_i||^2}{2\sigma_i^2}\right), \tag{4.29}$$

where $\mathbf{c}_i$ and σ_i are the center vector and the width parameter of the Gauss basis function; W_{ij} is the connection weight between the ith basis function in the hidden layer and the jth component of the output vector; H is the number of neurons in the hidden layer.

According to the iterative training process of RBF neural network shown in Figure 4.62, it is known that the difference between the actual output and the target output can be reduced by adjusting the weights of the network connection and the basis function parameters. The mean square error between the actual output of the network and the target output is selected as the objective function in the network training process:

$$E = \frac{1}{N}\sum_{j=1}^{N}(e_j)^2 = \frac{1}{N}\sum_{j=1}^{N}(y_j - t_j)^2, \tag{4.30}$$

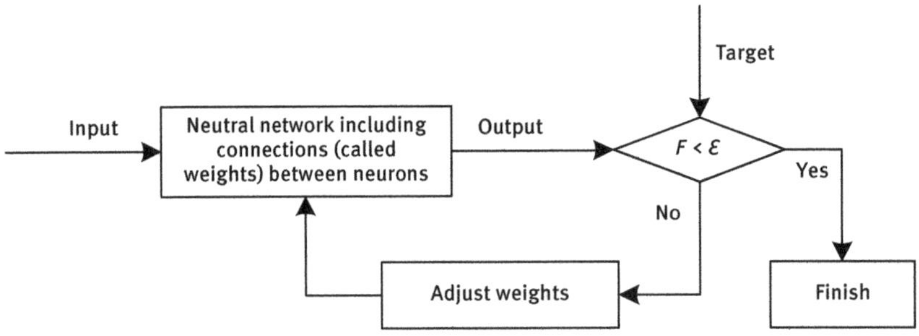

Figure 4.62: Iterative training process of RBF neural network.

where y_j and t_j are the jth components of actual output $\mathbf{Y}$ and the target output $\mathbf{T}$ of the network; N is the number of components of the output vector.

In the iterative training process of RBF neural network, the gradient descent algorithm is used to modify the connection weights between the hidden layer and the output layer of the network. It can be deduced as follows:

$$\begin{aligned} W_{ij}(k+1) &= W_{ij}(k) - \eta_W \frac{\partial E}{\partial W_{ij}} + \mu \Delta W_{ij}(k) \\ &= W_{ij}(k) - \frac{2\eta_W}{N} e_j \varphi(||x_j - c_i||) + \mu \Delta W_{ij}(k) \\ &= W_{ij}(k) - \frac{2\eta_W (y_j - t_j)}{N} \exp\left(-\frac{||\mathbf{x} - \mathbf{c}_i||^2}{2\sigma_i^2}\right) + \mu \Delta W_{ij}(k), \end{aligned} \tag{4.31}$$

where η_W is the learning rate of the algorithm, which is used to control the convergence speed of the network connection weight W_{ij} correction algorithm.
The central vector $\mathbf{c}_i$ and the width σ_i of Gauss basis function are modified by the gradient descent method, and the following iterative correction formula can be derived:

$$\mathbf{c}_i(k+1) = \mathbf{c}_i(k) - \frac{2\eta_c}{N} \sum_{j=1}^{N} \frac{W_{ij}(y_j - t_j)||\mathbf{x} - \mathbf{c}_i(k)||}{\sigma_i^2} \exp\left(-\frac{||\mathbf{x} - \mathbf{c}_i(k)||^2}{2\sigma_i^2}\right), \tag{4.32}$$

$$\sigma_i(k+1) = \sigma_i(k) - \frac{2\eta_\sigma}{N} \sum_{j=1}^{N} \frac{W_{ij}(y_j - t_j)||\mathbf{x} - \mathbf{c}_i||^2}{\sigma_i^3(k)} \exp\left(-\frac{||\mathbf{x} - \mathbf{c}_i||^2}{2\sigma_i^2(k)}\right), \tag{4.33}$$

where η_c and η_σ are, respectively, used to control the convergence speed of the center vector $\mathbf{c}_i$ and the width σ_i correction algorithm.

Based on the main characteristic values extracted from the 3D MFL signal and the established strip model, the forward prediction of MFL signals is conducted by the RBF neural network. The parameters of the 3D profile strip model of defects are the input of RBF neural network:

$$M_S = \left\{(L_1, d_1), \ldots, (L_i, d_i), \ldots, (L_N, d_N)\right\}, \tag{4.34}$$

where L_i and d_i are the length and depth of the ith equal part in the 3D contour strip model, respectively.

The output of the forward prediction RBF neural network has six main characteristic values of the 3D MFL signal, which can effectively reflect the defect profile parameters.

Figure 4.63 is an iterative training process for the forward prediction of RBF neural network. For any defect, according to 3D contour strip model, the corresponding strip model parameters M_S can be obtained. M_S is brought into the RBF neural network. A vector $\mathbf{P}_P$ consisting of six main characteristic values is obtained. Then, mean square error E between the predicted results $\mathbf{P}_P$ and vector $\mathbf{P}_R$ consisting of the main characteristic values is calculated:

$$E = \frac{1}{6}\sum_{j=1}^{6}(e_j)^2 = \frac{1}{6}\sum_{j=1}^{6}(P_{Pj} - P_{Rj})^2, \tag{4.35}$$

where P_{Pj} and P_{Rj} are the main characteristic values of the network prediction and the actual signal of jth MFL signal, respectively.

The error formula is brought into the gradient descent algorithm, and then the following iterative correction formula of the network connection weights W_{ij}, the basis function center vector $\mathbf{c}_i$ and the width σ_i are deduced:

$$W_{ij}(k+1) = W_{ij}(k) - \frac{\eta_W(P_{Pj} - P_{Rj})}{3}\exp\left(-\frac{||\mathbf{x} - \mathbf{c}_i||^2}{2\sigma_i^2}\right) + \mu\Delta W_{ij}(k), \tag{4.36}$$

$$\mathbf{c}_i(k+1) = \mathbf{c}_i(k) - \frac{\eta_c}{3}\sum_{j=1}^{6}\frac{W_{ij}(P_{Pj} - P_{Rj})||\mathbf{x} - \mathbf{c}_i(k)||}{\sigma_i^2}\exp\left(-\frac{||\mathbf{x} - \mathbf{c}_i(k)||^2}{2\sigma_i^2}\right), \tag{4.37}$$

$$\sigma_i(k+1) = \sigma_i(k) - \frac{\eta_\sigma}{3}\sum_{j=1}^{6}\frac{W_{ij}(P_{Pj} - P_{Rj})||\mathbf{x} - \mathbf{c}_i||^2}{\sigma_i^3(k)}\exp\left(-\frac{||\mathbf{x} - \mathbf{c}_i||^2}{2\sigma_i^2(k)}\right), \tag{4.38}$$

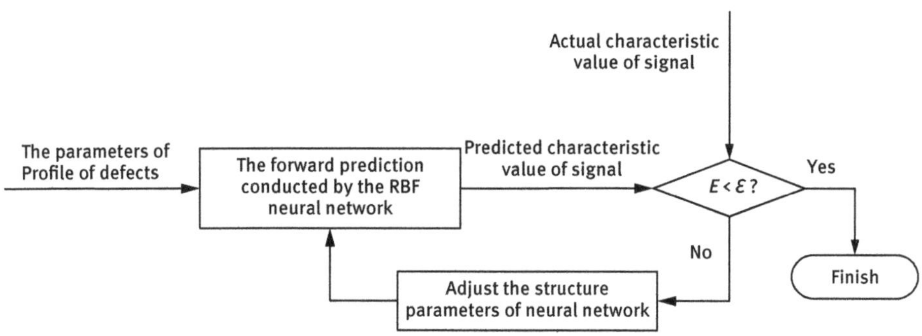

Figure 4.63: The iterative training process of forward prediction RBF neural network.

where η_W, η_c and η_σ are used to control the convergence speed of the modified algorithm of the network connection weight W_{ij}, basis function center vector $\mathbf{c}_i$ and width σ_i.

In order to train the RBF neural network, a 3D MFL detection signal database based on sample defect is established. The sample defects include rectangular defects, cambered surface defects and cylindrical defects. However, the actual contour of the three kinds of samples is not completely consistent with the 3D profile model. Therefore, the equivalent strip model is established to correspond to the actual contour.

4.3.3.1 Equivalent Model of Rectangular Defect

In the case of determining the size of the partition Δw, it is necessary to divide the rectangular defect along the circumferential direction of the pipeline, and then the equivalent defect 3D profile strip model can be established quickly.

If the defect is divided into N equal parts according to the size of division Δw, the bottom of each part is in an equal high plane, and the length and width of the opening rectangular are L and Δw, respectively, so there is no need for additional equivalent processing and computing. Equivalent 3D profile strip model M_{SR} of rectangular defects can be obtained by the direct combination of various parts. Based on the equivalent strip model, the $L \times W \times H$ rectangular defect in the case of partition size Δw can be expressed as

$$M_{SR} = \left\{(L_1, H_1), \ldots, (L_i, H_i), \ldots, (L_k, H_k)\right\}, \tag{4.39}$$

where the length of each equal part is $L_i = L$, depth is $H_i = H$.

If the defect cannot be divided into N equal parts according to the size of division Δw, the parts inside the defect do not need additional processing, but the parts at the edge of the defect need equivalent treatment. At the edge of the defect, the length and depth of the rectangular block are still L and H. Set the width as W_X. The length and the width of the rectangular block are L and Δw, respectively, and its equivalent depth is

$$H_X = HW_X/\Delta w. \tag{4.40}$$

The final equivalent 3D profile strip model M_{SR} of rectangular defects is shown in Figure 4.64.

4.3.3.2 Equivalent Model of Cylindrical Defects

In the case of determining the size of the partition Δw, the cylindrical defect is divided along the circumferential direction of the pipe, and then the equivalent defect 3D profile strip model is established according to the following steps.

No matter whether the defect is just divided into N equal parts according to the partition Δw, all equal parts need equivalent treatment. The cylindrical defect depth

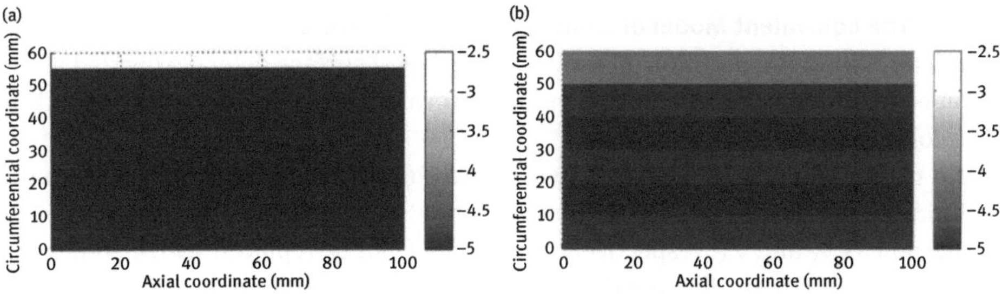

Figure 4.64: Sketch map of equivalent 3D contour strip model of rectangular defects: (a) real defect and (b) equivalent model.

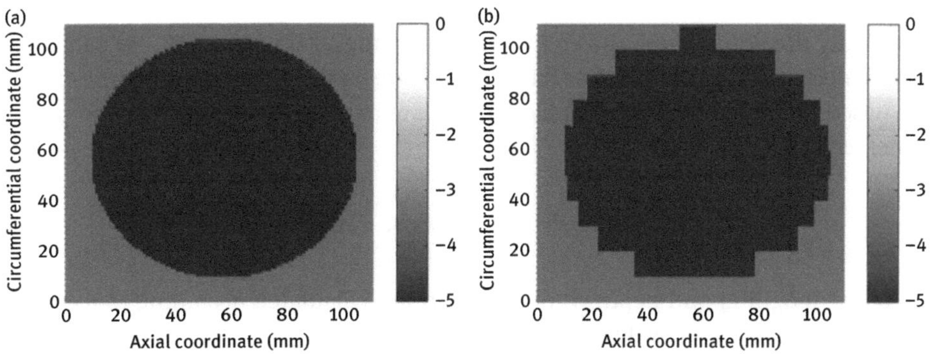

Figure 4.65: Sketch map of equivalent 3D contour strip model of cylindrical defects: (a) real defect and (b) equivalent model.

is H, the arbitrary division area is S_X, the division is replaced by an equivalent cuboid whose width and depth are Δw and H, respectively. So the length of the cuboid can be obtained:

$$L_X = S_X/\Delta w. \tag{4.41}$$

Combining the equal parts, the equivalent 3D contour strip model M_{SC} of cylindrical defect can be obtained as shown in Figure 4.65.

In the case of partition size Δw, based on the equivalent strip model, the cylindrical defect k can be expressed as

$$M_{SC} = \left\{(L_1, H_1), \ldots, (L_i, H_i), \ldots, (L_k, H_k)\right\}, \tag{4.42}$$

where the depth of each part is $H_i = H$, the length L_i is calculated according to formula (4.41).

4.3.3.3 The Equivalent Model of Cambered Surface Defects

After determining the partition size Δw, the cambered surface defect is divided along the pipe circumferential direction, then the equivalent strip model is built according to the following steps.

No matter whether the defect is just divided into N equal parts according to the partition Δw, all equal parts need equivalent treatment. Set the area and volume of the division as S_X and V_X, respectively, the division can be replaced with a equivalent cuboid whose width Δw, the length L_X and depth H_X of cuboid can be obtained:

$$L_X = S_X/\Delta w, \tag{4.43}$$

$$H_X = V_X/S_X. \tag{4.44}$$

Combining the equal parts, the equivalent 3D contour strip model M_{SA} of cambered surface defect can be obtained as shown in Figure 4.66. In the case of partition size Δw, based on the equivalent strip model, the cambered surface defect $D \times H$ can be expressed as

$$M_{SA} = \left\{(L_1, H_1), \ldots, (L_i, H_i), \ldots, (L_k, H_k)\right\}, \tag{4.45}$$

where the length L_i and depth H_i of each part are calculated according to formulas (4.43) and (4.44), respectively.

In order to train the forward prediction RBF neural network, the sample defect database on the pipe of 457 mm diameter and 14.3 mm wall thickness is established. Table 4.13 lists the size parameters of the defect samples. The defect types in the sample database include rectangular defects, cambered surface defects and cylindrical defects, their numbers are 260, 55 and 55, respectively.

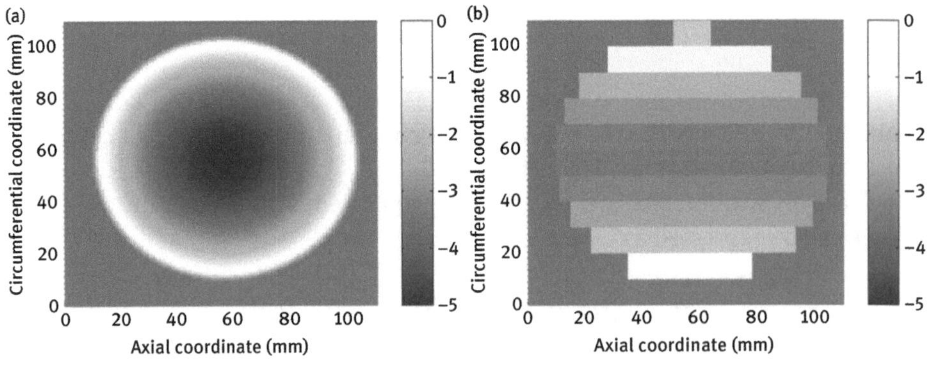

Figure 4.66: Sketch map of equivalent 3D contour strip model of cambered surface defects: (a) real defect and (b) equivalent model.

Table 4.13: Size parameters of training sample defect of neural network.

Parameters	Rectangular defects			Cambered surface defects		Cylindrical defects	
	Length	Width	Depth	Diameter	Depth	Diameter	Depth
Minimum value (mm)	14.3	7.2	1.4	14.3	1.4	14.3	1.4
Maximum value (mm)	100.1	28.6	7.2	28.6	7.2	28.6	7.2
Step size (mm)	7.2	7.2	1.4	1.4	1.4	1.4	1.4

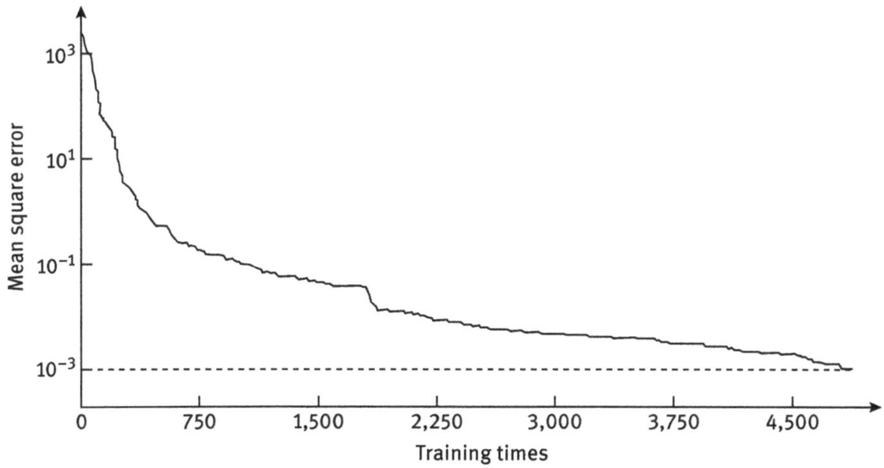

Figure 4.67: Error convergence curve of forward prediction RBF neural network training process.

About 70 % and 15 % defect samples are randomly selected for neural network training and verification, respectively. The remaining 15 % defect samples are used for testing and analysis of the trained RBF neural network.

The gradient descent algorithm is used to train the forward prediction RBF neural network, and get the error convergence curve of the training process as shown in Figure 4.67. From this curve, we can know that in the training process of the forward prediction RBF neural network, the iterative correction method based on the gradient descent algorithm makes the forecast output of the network to approach the target output at a faster speed. After 5,000 times of training, the network converges to the desired error value, which can achieve the accuracy requirements of the main characteristic value of the MFL signal.

If the trained RBF neural network still cannot meet the accuracy requirements, through changing the number of neurons and the algorithm for network training, increasing the number of testing data to adjust the neural network, the desired prediction accuracy can be reached. In the RBF neural network that can meet the requirements, based on the parameters of the 3D profile strip model, the main characteristic value of the 3D MFL signal is predicted.

Based on the trained forward prediction RBF neural network, defects MFL signal forward prediction is tested using defects not belong to the sample library to validate the proposed RBF neural network prediction precision, sensitivity and applicability to practical irregular defects.

Using rectangular defects, cambered surface defects and cylindrical defects not in the sample database, forward predictions are tested to verify the accuracy of the forward prediction RBF neural network. Based on the equivalent strip model of defects, the main characteristic values of the 3D MFL signal are predicted by the forward prediction RBF neural network. The error between the predicted result and the main characteristic values of the actual MFL signal is calculated.

In order to verify the prediction accuracy of RBF neural network to the main characteristic values of MFL signal, the prediction error of the main characteristic value is defined as follows:

$$E = \sqrt{\sum_{j=1}^{6}(P_{Pj} - P_{Rj})^2 / \sum_{j=1}^{6}(P_{Rj})^2}, \tag{4.46}$$

where P_{Pi} and P_{Ri} are the network prediction and the main characteristic value of the actual MFL signal, respectively.

The prediction error of the main characteristic value of the 3D MFL signal is shown in Table 4.14. According to the statistical results, we can know that the prediction error of the main characteristic value of the 3D MFL signal is not more than 10 % based on the forward prediction RBF neural network. This shows that the established RBF neural network has high prediction accuracy.

In order to verify the sensitivity to the defect profile parameters, 1 %, 2 % and 5 % random noise are added in the parameters of defect 3D contour strip model, and the main characteristic values of defect MFL signal are predicted. The prediction error of the main characteristic value of the 3D MFL signal with different noise level is shown in Table 4.15.

Table 4.14: Prediction error of main characteristic value of 3D MFL signal.

Defect shape	Defect sizes (mm)	Prediction error (%)
Rectangle	30 × 15 × 5	8.7
Rectangle	40 × 20 × 10	8.4
Cambered surface	25 × 5	9.6
Cambered surface	30 × 10	9.1
Cylinder	25 × 5	9.2
Cylinder	30 × 10	8.9

Table 4.15: Prediction error of main characteristic value of three-dimensional MFL signal under different noise levels.

Defect shape	Defect sizes (mm)	1 % noise (%)	2 % noise (%)	5 % noise (%)
Rectangle	30 × 15 × 5	10.4	16.7	23.5
Rectangle	40 × 20 × 10	9.8	15.4	21.9
Cambered surface	25 × 5	11.7	17.3	26.2
Cambered A surface	30 × 10	11.1	16.5	22.7
Cylinder	25 × 5	11.3	16.9	23.4
Cylinder	30 × 10	10.6	17.1	24.3

From the above error statistics, prediction results are compared with the case of no noise, and the prediction error of the main characteristic value of 3D MFL signal significantly increases with the increase of noise level. This shows that the established forward-predictive RBF neural network has high sensitivity to the parameters of the defect strip model. When the parameters of the defect model change, the main characteristic value of the MFL signal can be modified quickly.

In order to further verify the applicability of irregular defects in actual pipeline, the main characteristic value of the irregular defect MFL signal in Figure 4.68 is predicted using the established forward RBF neural network, then the prediction error between the result and the actual MFL signal main characteristic value is calculated. Based on calculation, the error between the predicted result of RBF neural network and the true value of MFL signal is 14.1 %. Although the error is larger than the prediction error of the rule defect, it is still in the acceptable range. It is proved that the established RBF neural network is a forward model to predict the irregular defects in the actual pipeline. Therefore, the established forward prediction RBF neural network can be used to predict the main characteristic values of MFL signal defects with arbitrary shapes.

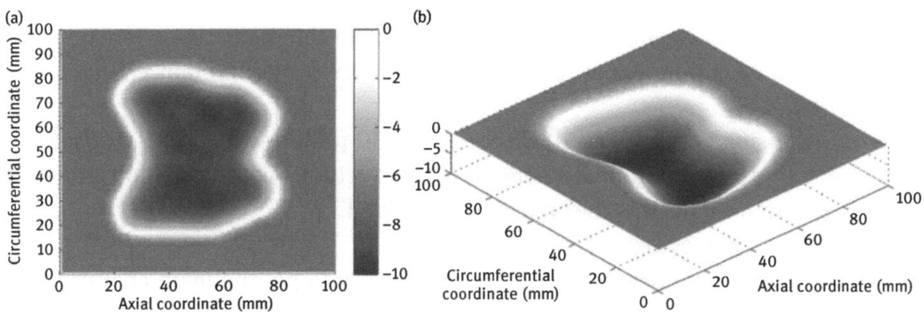

Figure 4.68: An irregular defect for the forward-predictive test of the main characteristic value of MFL signal: (a) flat display and (b) 3D display.

4.3.4 Iterative Inversion of RBF Neural Network for 3D Profile of Defects

Because the inversion of 3D profile of defects is an ill-conditioned inverse problem of MFL, and the types of training defect samples only include rectangular defects, cambered surface defects and cylindrical defects, in neural network direct inversion methods the existing problem of low inversion precision and generalization ability is insufficient, which cannot be directly applied to the inversion of defect 3D contour [111].

Taking into account that the iterative method has high inversion precision and strong generalization ability, the neural network iterative inversion method is proposed. The basic idea of the proposed inversion method is, first, training MFL forward-predictive neural network, and then embedding it into the iterative loop as the forward model, inversing defect 3D profile by iterative correction method.

So the neural network can be used to solve the problem of the MFL forward problem and not the ill-conditioned MFL inverse problem, so it can get a high precision of prediction. In the aspect of inversion speed, the neural network that can predict MFL quickly replaces the original finite element calculation in the iterative method. The speed of the MFL forward problem can be improved greatly. In addition, the mapping relationship between the input and output data of neural network is clear, so differential calculation can be easily carried out. Gradient descent method can be applied to the iterative correction of the defect profile parameters, so a faster defect 3D profile inversion speed can be obtained.

Figure 4.69 shows the basic flow of the iterative inversion method embedding the forward prediction RBF neural network for defect 3D profile. Among them, the forward prediction RBF neural network is used for the forward prediction from defect profile to MFL signal, the input data is the defect parameters based on 3D contour strip model and the output data is the main characteristic value of the 3D MFL signal that can effectively reflect defect contour parameters.

The basic steps of the iterative inversion of 3D profile neural network from the 3D MFL detection signal can be obtained from Figure 4.69.

1. Detect the defect opening profiles, and extracting the main characteristic value of 3D MFL detection signal, comprising the corresponding main characteristic vector **PR**.
2. Setting division width W, establish the defect 3D profile strip model M_S based on the detecting open contour S, where the length vector **L** of each partition is known, and the depth vector **H** is the amount to be solved.
3. Set the initial defect depth vector **H**.
4. Forward predicting based on RBF neural network to get the main characteristic value vector **PP**.
5. Calculate the error E between **PP** and **PR**. If E is less than the threshold δ, the iterative inversion is ended, and the final contour of the defect is obtained after smoothing; otherwise, the defect depth vector **H** is modified and jump to step 4.

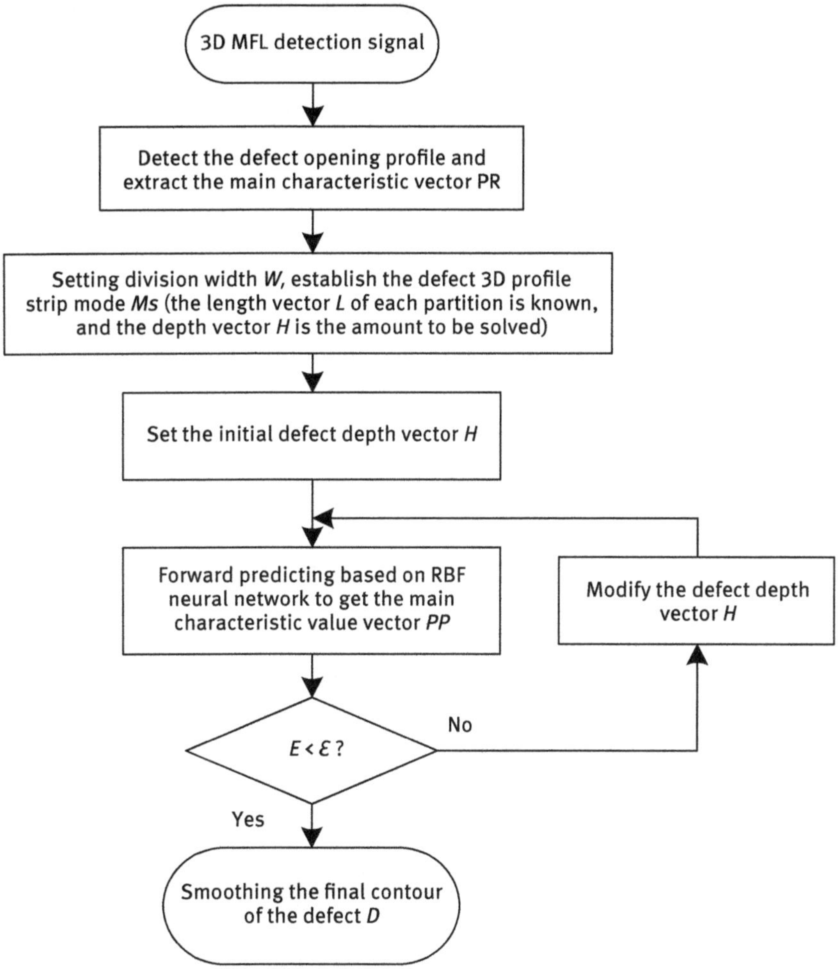

Figure 4.69: The basic flow of the iterative inversion method of RBF neural network with defect 3D profile.

After training based on the sample defect, the connection weights W_{ij}, the center vector $\mathbf{c}_i$ of Gauss basis function and the width σ_i of the forward prediction RBF neural network are known. Therefore, based on the mapping relationship between the input and output data of the forward prediction RBF neural network, the relationship expressions are determined as follows:

$$y_j = f_j(\mathbf{X}) = \sum_{i=1}^{H} W_{ij} \exp\left(-\frac{\|\mathbf{x} - \mathbf{c}_i\|^2}{2\sigma_i^2}\right), \tag{4.47}$$

where **X** is the defect parameter vector for the input of the 3D profile strip model; y_j is the jth component of the output vector **Y**, which represents the jth characteristic value of the MFL signal.

In order to correct the defect parameter vector, the error E between the predicted results **PP** and the actual value **PR** of the main characteristic vector is defined as follows:

$$E = \frac{1}{6}\sum_{j=1}^{6}(e_j)^2 = \frac{1}{6}\sum_{j=1}^{6}(P_{Pj} - P_{Rj})^2, \tag{4.48}$$

where P_{Pj} and P_{Rj} are the jth main characteristic values of the predicted MFL signal and the actual MFL signal, respectively.

By formulas (4.47) and (4.48), the first-order derivation of the error E to any parameter x_i of the defect is obtained:

$$\frac{\partial E}{\partial x_i} = \frac{1}{3}\sum_{j=1}^{6}\left[(P_{Pj} - P_{Rj})\sum_{i=1}^{H}\frac{W_{ij}(c_i - x_i)}{\sigma_i^2}\exp\left(-\frac{\|\mathbf{x} - \mathbf{c}_i\|^2}{2\sigma_i^2}\right)\right]. \tag{4.49}$$

In the process of iterative inversion of defect 3D profile, based on the gravity descent algorithm, the defect parameters **X** are iteratively modified by the first-order derivation, and the following correction formula can be deduced:

$$\begin{cases} x_i(k+1) = x_i(k) + \eta\left(-\dfrac{\partial E}{\partial x_i(k)}\right) \\ \qquad = x_i(k) - \dfrac{\eta}{3}\sum\limits_{j=1}^{6}\left\{(P_{Pj} - P_{Rj})\sum\limits_{i=1}^{H}\dfrac{W_{ij}[c_i - x_i(k)]}{\sigma_i^2}\exp\left(-\dfrac{\|\mathbf{x}(k) - \mathbf{c}_i\|^2}{2\sigma_i^2}\right)\right\}. \end{cases} \tag{4.50}$$

If only using the gravity descent algorithm, iterative inversion process may fall into the local optima, and thus cannot converge to the global optimal defect parameter vector. Therefore, the virtual temperature variable T is defined, and the simulated annealing conditions are introduced to enlarge the search space of the algorithm. In each iteration, when the defect parameter vector is updated from $\mathbf{X}^k$ to $\mathbf{X}^{k+1}$, the new defect parameter $\mathbf{X}^{k+1}$ is accepted only when the following formula is true:

$$\exp\left(\frac{E^k - E^{k+1}}{T}\right) \geq \text{rand}(0, 1), \tag{4.51}$$

where E^k and E^{k+1} are the prediction error of the defect parameters $\mathbf{X}^k$ and $\mathbf{X}^{k+1}$, respectively.

From the above formula, when $E^k > E^{k+1}$, the new defect profile parameters are better than the existing parameters, and $\mathbf{X}^k$ must be updated to $\mathbf{X}^{k+1}$; when $E^k < E^{k+1}$, the new defect profile parameters are not improved compared to the existing parameters, and $\mathbf{X}^k$ is updated to $\mathbf{X}^{k+1}$ with a certain probability. In this case, the iterative inversion process can get rid of the local optimal solution so as to continue to search the global optimum.

In order to speed up the global convergence of the iterative inversion process, when the iteration process is carried out a certain number of times, it can be regarded as the equilibrium state in the current temperature T. The temperature parameter T is modified by the following formula:

$$T_{k+1} = \alpha \cdot T_k, \quad 0 < \alpha < 1, \tag{4.51}$$

where the value α is usually between 0.8 and 0.95.

The termination criteria of the iterative inversion method of neural network are the following three: The error E is less than a certain threshold δ_E; temperature T is less than a certain threshold δ_T; and the corresponding error of the defect profile is not changed in the continuous k searches. When any of the above criterion is established, the iterative inversion process is terminated. Therefore, the defect profile

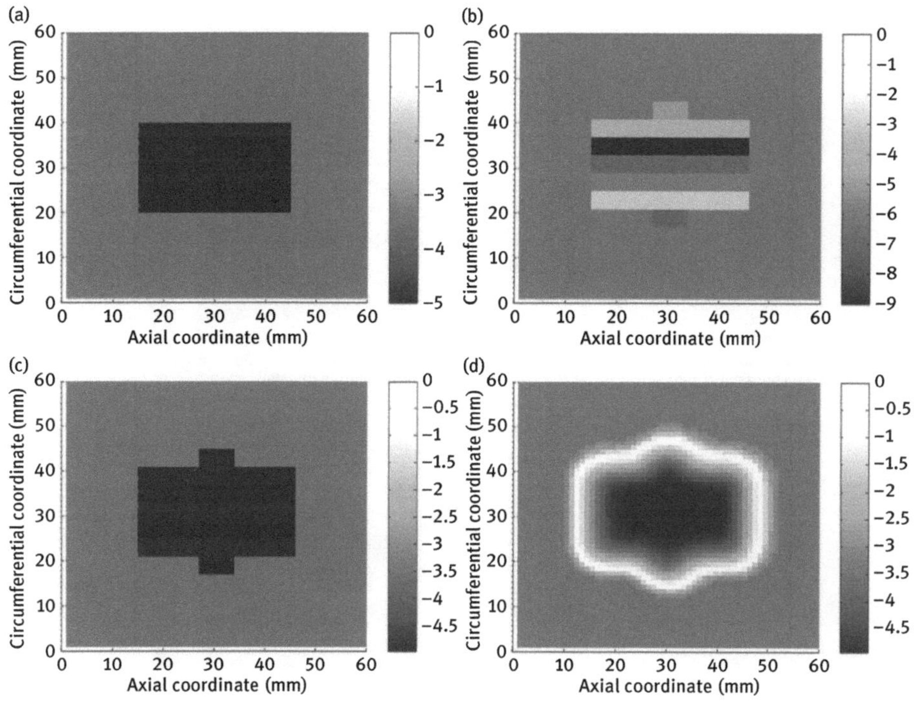

Figure 4.70: Iterative inversion of RBF neural network for rectangular defect profile of 30 mm × 20 mm × 5 mm: (a) real contour; (b) strip model based on the recognition results of open profile; (c) inversion result of strip model; and (d) final defect profile.

is obtained based on the defect parameters strip model, and then the appropriate smoothing process is carried out in order to get a smooth edge.

The precision, antidisturbance ability and the generalization ability of the RBF neural network iterative inversion method were studied by using the 3D profile inversion test of the defect.

Based on RBF neural network iterative inversion method, the 3D profiles of 30 mm × 20 mm × 5 mm rectangular defect, 30 mm × 5 mm cambered surface defect and 30 mm × 5 mm cylindrical defect are inversed, respectively. The results are shown in Figures 4.70–4.72. It can be known that there is a slight error in the opening shape of the inversion results of the test defects due to the equivalent treatment of the outermost edges of the strip model. However, in terms of the position and depth of the maximum point, the inversed contour and the real contour are still in good agreement.

Table 4.16 shows the statistical results of the error of 3D profile inversion. From the table, it can be seen that the RMS errors of 3D inversions of the rectangular, cambered surface and cylindrical defects are 15.7 %, 12.4 % and 13.2 %, respectively. Compared to that of the stochastic search iterative inversion method, the error of the neural network iterative inversion method is slightly larger, but the inversion time almost can be neglected.

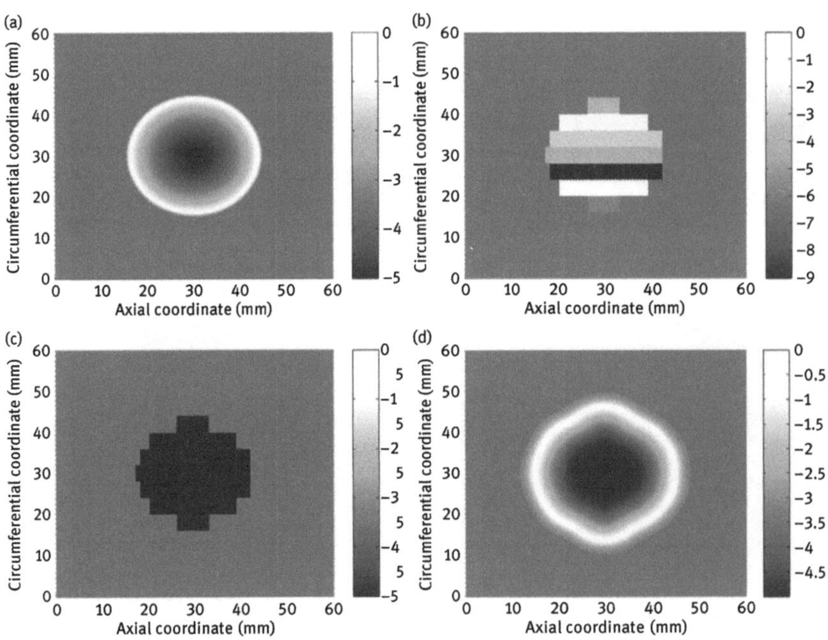

Figure 4.71: Iterative inversion of RBF neural network for cambered surface defect profile of 30 mm × 5 mm: (a) real contour; (b) strip model based on the recognition results of open profile; (c) inversion result of strip model; and (d) final defect profile.

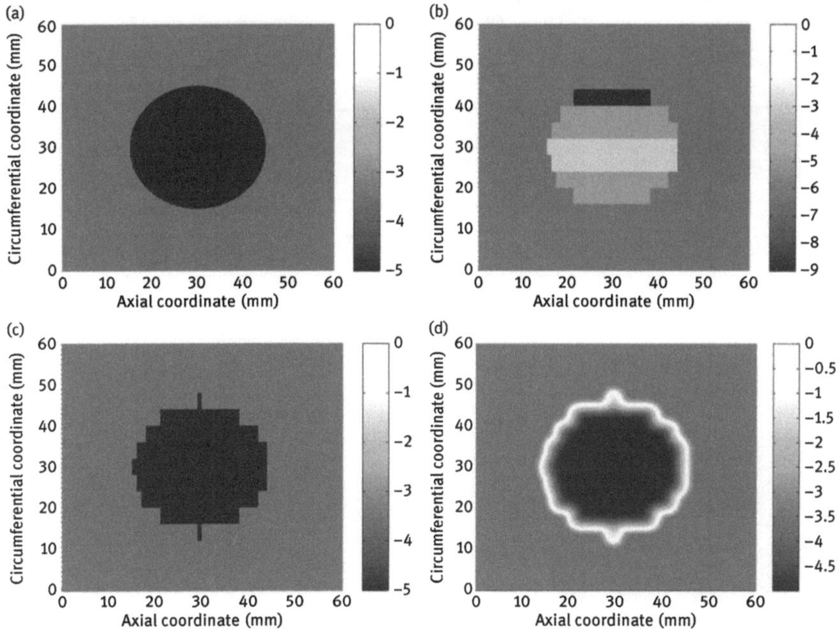

Figure 4.72: Iterative inversion of RBF neural network for cylindrical defect profile of 30 mm × 5 mm: (a) real contour; (b) strip model based on the recognition results of open profile; (c) inversion result of strip model; and (d) final defect profile.

Table 4.16: Inversion error of defect 3D profile.

Defect shape	Dimension (mm)	E_m (%)	E_p (%)	E_n (%)
Rectangle	30 × 20 × 5	15.7	6.4	6.2
Arc	30 × 5	12.4	5.5	5.6
Cylinder	30 × 5	13.2	6.1	6.5

About 1 %, 2 % and 5 % random noises were added in the 3D MFL signals, then 3D defect profile inversion test was preceded using RBF neural network iterative inversion method. The inversion results are shown in Figures 4.73–4.75. It can be seen that with the increase of the noise level, the difference between the inversion contour and the real contour is gradually increasing.

Table 4.17 shows the inversion error of the various test defects in different noise levels. Compared with Table 4.16, we can know that the error of 3D contour inversion based on RBF neural network iterative inversion method is significantly increased when the noise disturbance is added to the 3D MFL signal. In the case of 30 mm × 20 mm × 5 mm rectangular defect, the RMS error of the inversion was 16.4 %

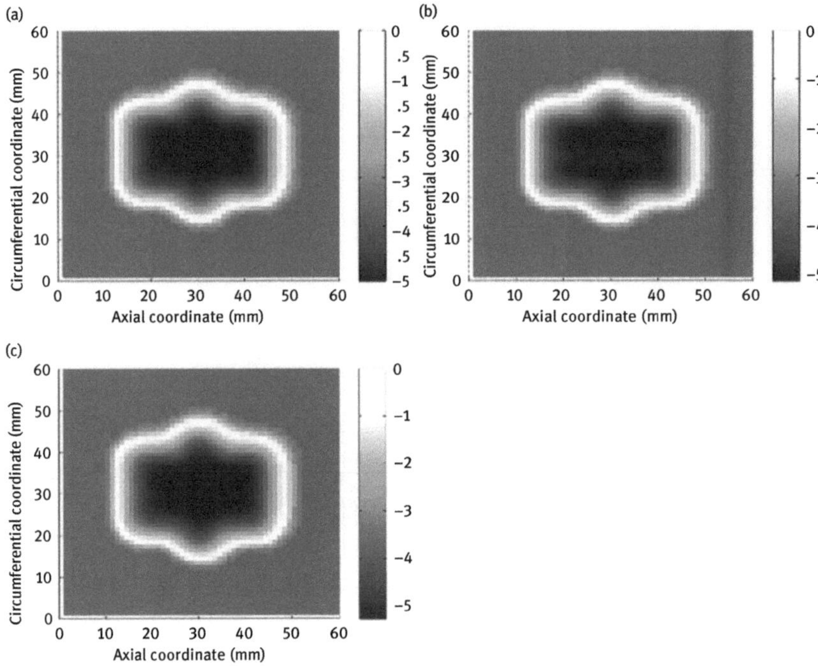

Figure 4.73: RBF neural network iterative inversion results of 30 mm × 20 mm × 5 mm rectangular defect profile under different noises: (a) 1 % noise; (b) 2 % noise; and (c) 5 % noise.

at the 1 % noise level; however, the RMS error of the inversion was 28.5 % at the 5 % noise level.

Compared with the inversion results of the random search iterative inversion method, the inversion error of the neural network iterative inversion method is larger, and the inversion error increases with the increase of the noise level. This shows that the neural network inversion iteration results are affected by noise largely; the anti-interference ability of the method is relatively weak.

The shapes of the defects in the actual oil and gas pipelines vary greatly and are very different from those of rectangular, cambered surface and cylindrical defects samples used for neural network training. Therefore, based on the irregular defect shown in Figure 4.76, the generalization ability of the neural network iterative inversion method for the unknown shape defects is studied.

From the inversion results shown in Figure 4.76, the opening identification result of the experimental defect is basically consistent with the true opening profile, and the final inversion contour is similar to the real defect profile. The RMS error of the 3D profile inversion results is 21.4 %. Taking into account the rapidity of the iterative inversion of neural networks, the error is still within an acceptable range. However, compared with the stochastic search iterative inversion method, in the same recognized opening conditions, the inversion error of neural network iterative inversion method is relatively increased by 72.6 %. This shows that the neural network iterative

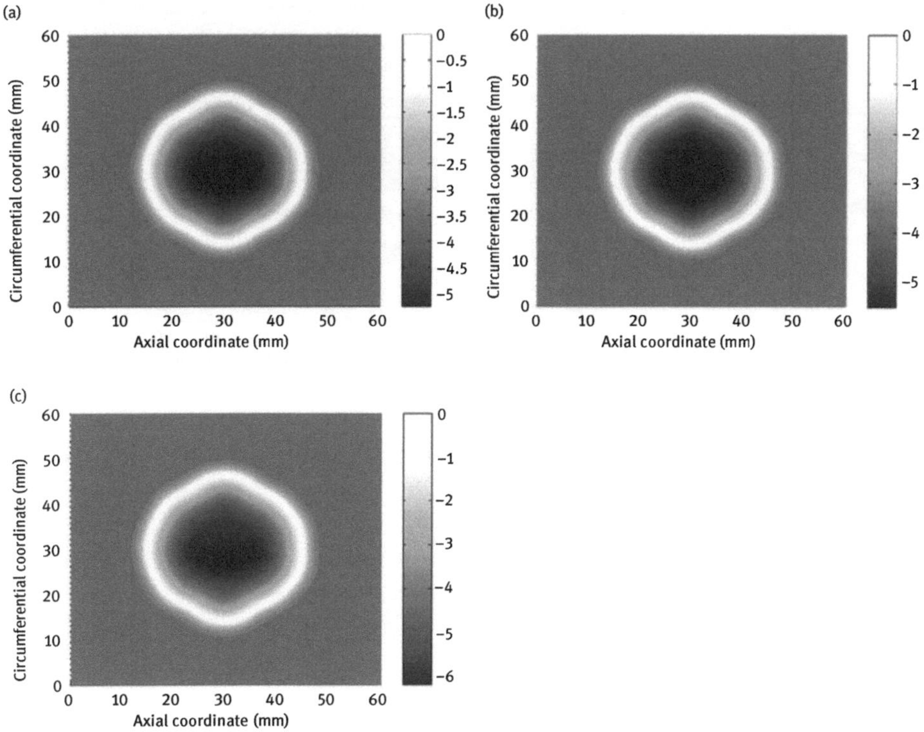

Figure 4.74: RBF neural network iterative inversion results of 30 mm × 5 mm cambered surface defect profile under different noises: (a) 1 % noise; (b) 2 % noise; and (c) 5 % noise.

inversion method is relatively weak to the generalization ability of the irregular defects in the actual pipeline.

4.4 Multistage Successive Approximation Inversion Method for Defect 3D Profile

The stochastic search iterative inversion method and the neural network iterative inversion method for the 3D profile of defects have their own advantages and disadvantages [3]. The stochastic search iterative inversion method has strong anti-interference ability and applicability of actual defects, but it needs to consume a lot of time to obtain a high inversion precision, and the inversion results will also be affected significantly by the initial defect outline. The neural network iterative inversion method can inverse defect fastly, but its anti-interference ability and generalization ability of the unknown shape defect is relatively weak.

In order to compensate the advantages and disadvantages of these two methods, this section explores on the combination of the two methods and puts forward the

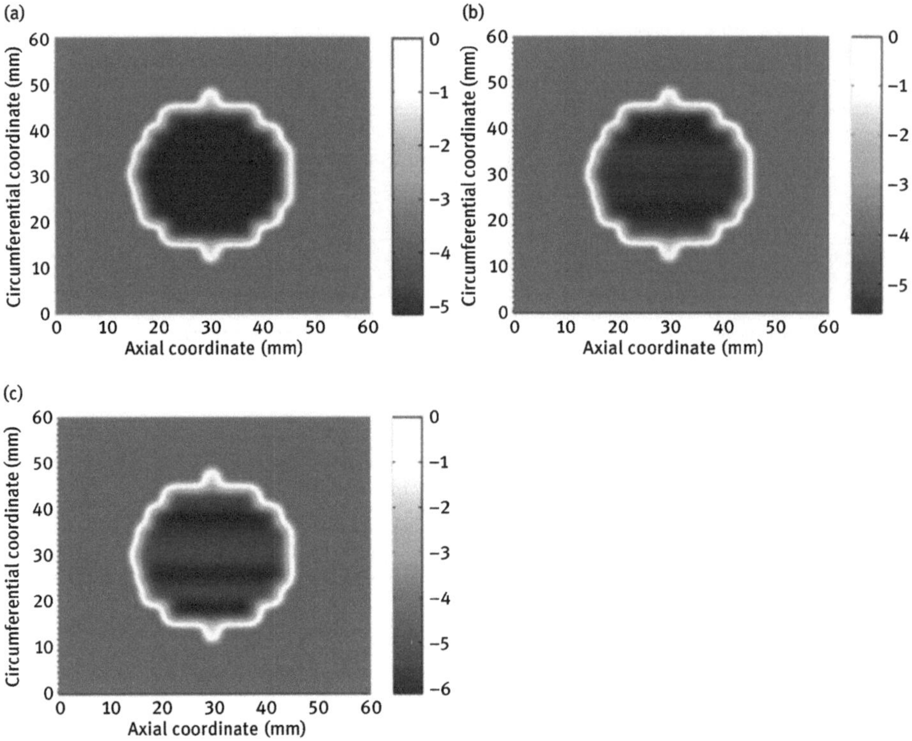

Figure 4.75: RBF neural network iterative inversion results of 30 mm × 5 mm cylindrical defect profile under different noises: (a) 1 % noise; (b) 2 % noise; and (c) 5 % noise.

Table 4.17: Inversion error of 3D profile of defects at different noise levels.

Defect (mm)	1 % noise (%)			2 % noise (%)			5 % noise (%)		
	E_m	E_p	E_n	E_m	E_p	E_n	E_m	E_p	E_n
Rectangular defect 30 × 20 × 5	16.4	8.3	7.8	19.1	14.4	14.1	28.5	29.5	26.3
Cambered surface defect 30 × 5	13.6	7.5	6.9	16.4	13.2	11.9	24.6	26.4	27.3
Cylindrical defect 30 × 5	14.8	7.2	8.1	18.0	12.5	13.8	26.1	27.2	26.4

stepwise refinement strategy of 3D mesh model of defects and the stepwise refinement strategy of the finite element mesh size, which forms a 3D contour multistage successive approximation inversion method with better comprehensive performances. Finally, through 3D profile inversion test of the actual oil and gas pipeline defects, the effect of the inversion method is verified and compared with the random search iterative inversion method and neural networks iterative inversion method.

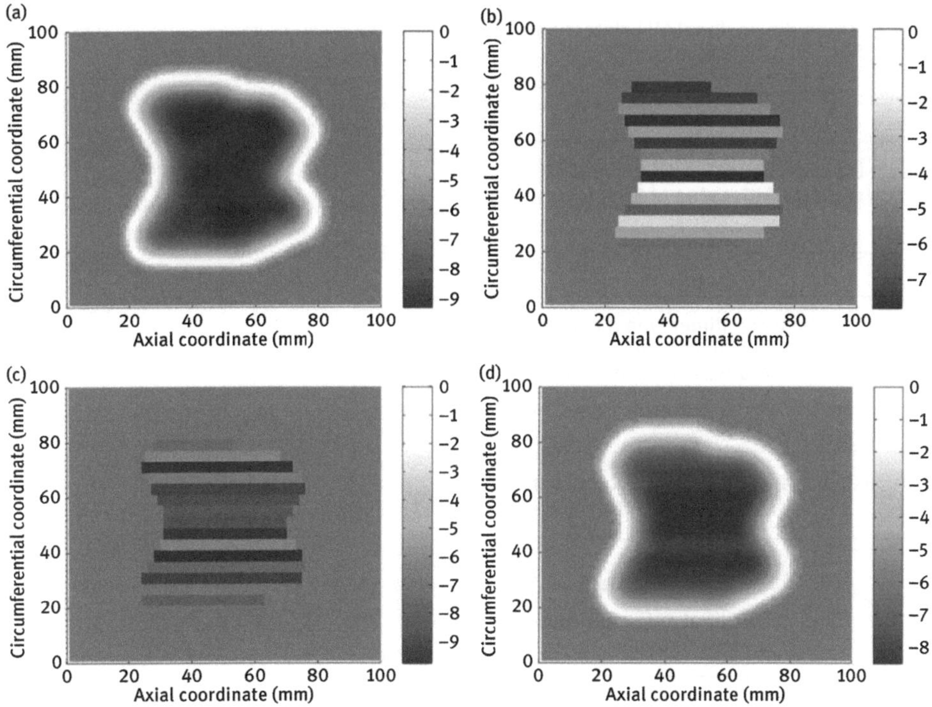

Figure 4.76: Iterative inversion of RBF neural network for irregular defects: (a) real contour; (b) strip model based on the recognition results of open profile; (c) inversion result of strip model; and (d) final defect profile.

4.4.1 Multistage Inversion of Defect Profiles

In order to obtain satisfactory inversion results, an appropriate initial defect profile is needed in the stochastic search iterative inversion method. The neural network iterative inversion method is used to obtain the defect profile, which can be used as the initial defect profile of the stochastic search iterative inversion method. Thus, based on the initial defect opening contour recognition results predicted by the neural networks iterative inversion method, 3D contour of defects is searched in global range with high precision using stochastic search iterative inversion method.

Based on the above ideas, a multistage hybrid inversion method of defect 3D contour is proposed. The basic process is shown in Figure 4.77.

According to the process of the proposed composite inversion method, after obtaining the 3D MFL signal of any defect, the following steps are processed:

1. According to the 3D MFL detection signal, identify the defect opening profile S.
2. Based on the opening contour S, a 3D contour strip model M_S of the defect is established.

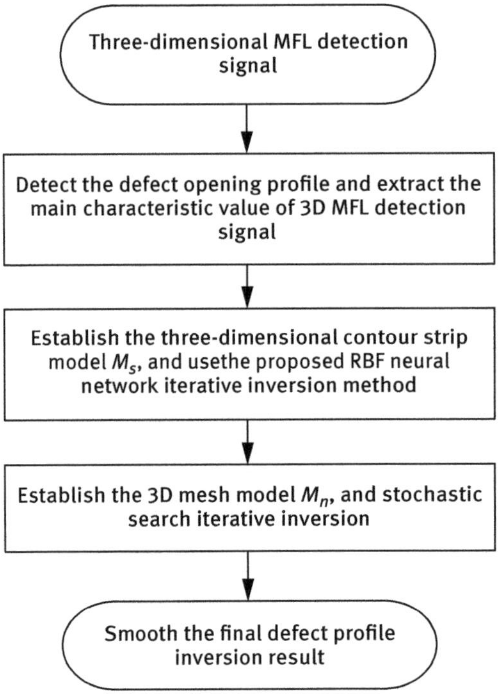

Figure 4.77: Flowchart of multistage composite inversion method for defect 3D profile.

3. By using the proposed RBF neural network iterative inversion method, the inversion result D_s of defect profile is obtained.
4. On the basis of D_s, a 3D mesh model of the defect is established, and the initial defect profile D_0 of the next step is obtained.
5. Taking the D_0 as the initial defect profile, a more accurate inversion result D_n of the defect profile is obtained by the stochastic search iterative inversion method.
6. The final defect profile inversion result D is obtained by appropriate smoothing of D_n.

A key of the composite inversion method is to establish a mesh model M_n from the inversion result D_s of the strip model and obtain the initial defect profile D_0.

Take the schematic diagram of Figure 4.78 as a reference, after getting the inversion result of the defect contour strip model D_s, the defect 3D mesh model M_n will be obtained by the following conversion steps, and the initial defect profile D_0 is calculated:

1. The 3D contour mesh model of the defect is determined by dividing the pipe along the axial and the circumferential directions. The D_s plane is divided into $N_1 \times N_2$ lattice.
2. To select all the overlapping grids with D_s, the area of defect 3D profile inversion to be solved is obtained.

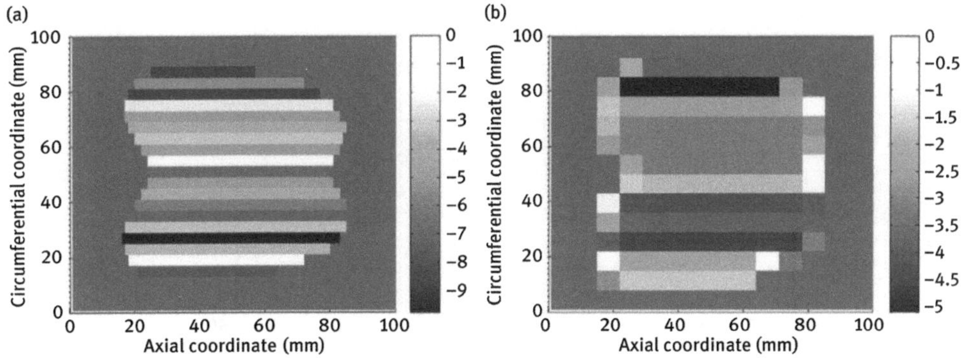

Figure 4.78: The transformation from the strip model to the mesh model: (a) inversion results of defect 3D profile strip model and (b) the mesh model of the defect 3D profile.

3. The pipe wall is divided into N_3 layer along the radial direction and adds depth parameter d to all the meshes to be solved. The 3D profile mesh model is established.
4. According to the depth value of D_S, the equivalent depths of all grids in M_n are established, and the initial defect profile D_0 of the stochastic search iterative inversion is obtained.

In the mesh model M_n of the defect 3D profile, the unknown parameters are the depth d of each grid. Because the pipe wall is divided into N_3 layers, the feasible region of the depth value of any grid is

$$D = \left\{d|0 \le d_{i,j} \le N_3\,(i = 1, 2, \ldots, N_1; j = 1, 2, \ldots, N_2)\right\}. \tag{4.52}$$

The expression of any defect in the mesh model can be obtained by making $d_{i,j}$ as the mesh depth of the ith row and the row jth column in the 3D profile mesh model:

$$M_n = \begin{Bmatrix} d_{1,1}, & d_{1,2}, \ldots, d_{1,j}, \ldots, d_{1,N_2-1}, & d_{1,N_2} \\ & \cdots & \\ d_{i,1}, & d_{i,2}, \ldots, d_{i,j}, \ldots, d_{i,N_2-1}, & d_{i,N_2} \\ & \cdots & \\ d_{N_1,1}, & d_{N_1,2}, \ldots, d_{N_1,j}, \ldots, d_{N_1,N_2-1}, & d_{N_1,N_2} \end{Bmatrix}. \tag{4.53}$$

According to the depth of D_S, the equivalent depths of all grids in the initial defect profile D_0 are determined. For any grid X within the D_0, if it is overlapping with n regions of the D_S, the calculation formula of the equivalent depth D_X of the grid X can be obtained:

$$D_X = \sum_{i=1}^{n} S_i d_i / S_X, \tag{4.54}$$

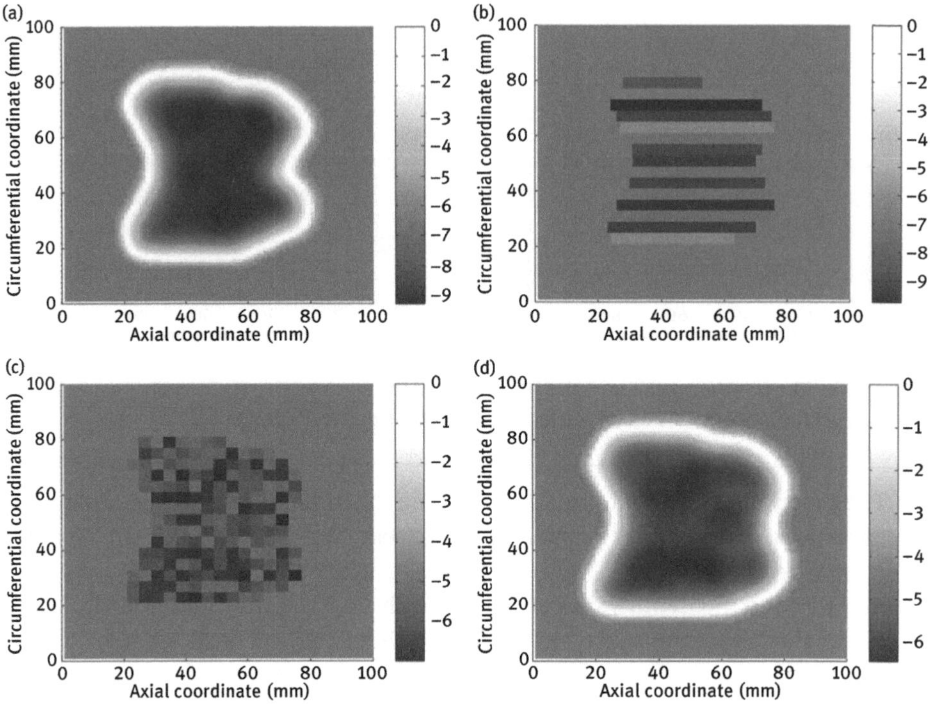

Figure 4.79: Multistage inversion of irregular defect 3D profile: (a) real defect profile; (b) the inversion result obtained by the iterative inversion method of RBF neural network based on the 3D contour strip model; (c) the inversion result obtained by the stochastic search iterative inversion method based on the 3D mesh model; and (d) the final inversion profile.

where $S_1, S_2, \ldots, S_n$ are the overlap areas between the grid X and n overlapping regions in region D_s, $d_1, d_2, \ldots, d_n$ are the corresponding depths, S_X is the area of grid X.

Based on the irregular defect shown in Figure 4.79, the 3D profile inversion test is carried out to verify the performance of the multistage hybrid inversion method. The diameter and wall thickness of the test pipe are 457 mm and 14.3 mm. The partition sizes in the strip model along the circumferential and radial directions of the pipe are 2 and 1.4 mm, respectively. The partition sizes for the mesh model along the pipe axial, circumferential and radial directions are 2, 2 and 1.4 mm, respectively. The iteration number of the stochastic search iterative inversion algorithm is set up to 4,000.

In the defect tests of the multistage hybrid inversion process, the first stage is based on strip model and RBF neural network iterative inversion method, and the second stage is based on mesh model and stochastic search iterative inversion method. From Figure 4.79, the inversion result in the first stage based on strip model is closer to the real defect outline. Therefore, it can avoid the influence of the initial defect profile, which is not suitable for the second stage.

The RMS error of the inversion defect profile in the first stage is 21.4 %, and the error of the final inversion defect profile is 10.3 %. If the iterative number is fixed at

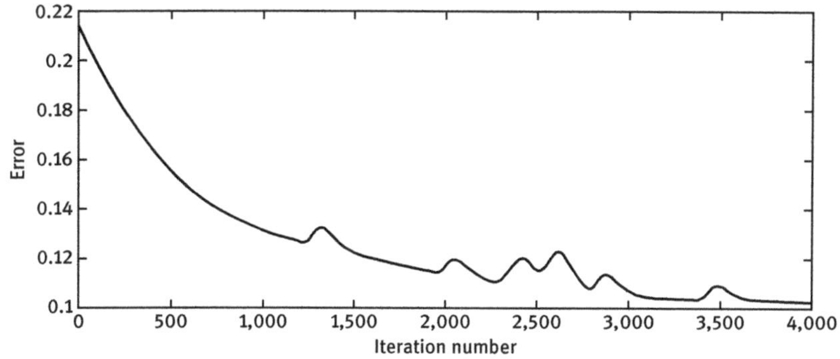

Figure 4.80: The error convergence curve of multistage hybrid inversion in the second stage.

4,000, compared to that of the stochastic search iterative inversion method, the defect profile inversion error of the multistage hybrid inversion method is reduced obviously. It is also shown that the applicability of irregular defects in actual pipeline is stronger.

Figure 4.80 is the error convergence curve in the second stage of the multistage inversion. Compared with the error convergence curve of stochastic search iterative inversion method, which is shown in Figure 4.46, the initial defect contour obtained by the iterative inversion of neural network, whose error is 21.4 %, is significantly better than the initial defect profile of random selection. After the adoption of multistage hybrid inversion method, under the same number of iterations, the inversion error of defect 3D profile is significantly reduced. It confirms that the multistage inversion method can improve the inversion results of the defect profile.

4.4.2 Progressive Refinement of the Defect Mesh Model

In the 3D profile of defect multistage hybrid inversion method, the second stage of the stochastic search iterative inversion guarantees the high-precision defect profile inversion results. The inversion results are directly influenced by the dimension of the mesh model of the defect 3D profile. If the mesh size is too large, the iterative inversion can quickly complete the entire space search, but the error between the inversion contour and the real defect profile is larger. On the contrary, if the partition size is too small, it can get the inversion results closer to the real defect, but because of the larger solution space, the time of stochastic search iterative inversion has increased dramatically.

In order to strike a balance between the inversion error and the time consumption, a dynamic partitioning strategy is proposed, which is shown in Figure 4.81. So in the initial stage, the mesh model is divided into large size, and the time consumed by the iterative inversion is short. Based on the inversion results, the partition size is reduced to refine the division of the 3D mesh model, and the iterative inversion is proceeded

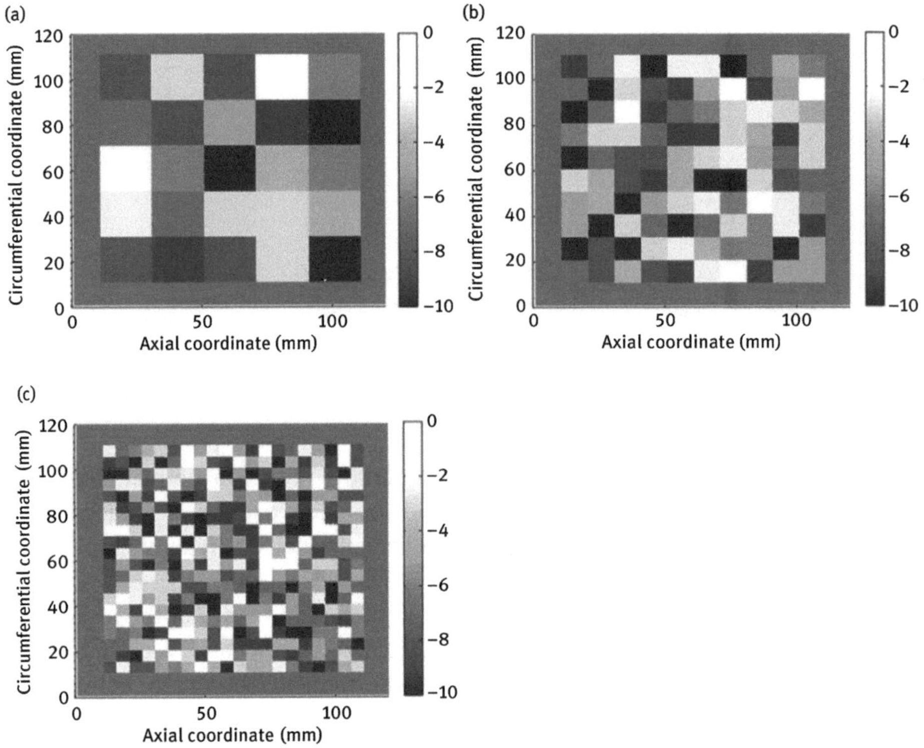

Figure 4.81: Sketch map of progressive refinement of 3D contour mesh model: (a) basic partition size; (b) partition size reduced to 1/2; and (c) partition size reduced to 1/4.

again, which can further improve the 3D contour inversion precision of defects. On the whole, the method can obtain the accurate defect profile inversion results as the dividing dimension decreases.

After the progressive refinement strategy of the mesh model of the defect 3D mesh model is introduced, the process of the multistage hybrid inversion method of the defect is revised, as shown in Figure 4.82. According to the revised process, after obtaining the inversion result D_S of defect profile based on strip model, the following steps are processed:

1. Determine the mesh sizes along the circumferential, axial and radial directions.
2. On the basis of D_S, a defect 3D mesh model M_n is established, and the initial defect profile D_0 is obtained.
3. The inversion result D_n of defect profile is obtained by stochastic searching.
4. If the time is finished or the mesh model is small enough, the D_n is smoothed, and the final inversion result D is obtained.
5. Otherwise, reduce the mesh size of the mesh model, jump to step 3 and continue to perform.

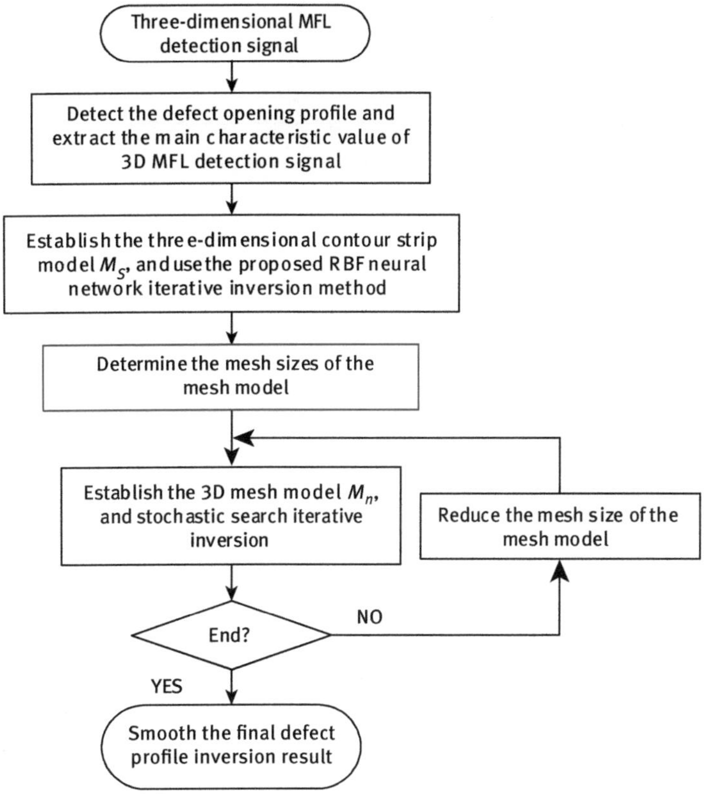

Figure 4.82: Flowchart of the multistage hybrid inversion method of the defect 3D profile after the progressive refinement of the mesh model.

In order to verify the effect of progressive refinement of the mesh model, the 3D profile inversion test is carried out based on an irregular defect shown in Figure 4.83. The diameter and wall thickness of the test pipe are 457 and 14.3 mm, respectively.

In the first stage of the hybrid inversion process, the strip model is divided into 2 and 1.4 mm in the circumferential direction and the radial direction; in the second stage, the mesh model is divided 3 times. In order to compare the inversion results without the progressive refinement strategy in the mesh model, the numbers of iterations in the three partition sizes are set to 500, 1,000 and 2,500, respectively, and the total number of iterations is still 4,000 (Table 4.18).

Figure 4.83 shows the multistage hybrid inversion results of an irregular defect. It can be known that there is still a large error between the inversion result and the real defect profile after the first division of the mesh model. However, with the gradual refinement of the mesh model, the defect profile obtained by the inversion method is gradually approaching the true defect profile. After smoothing the inversion results

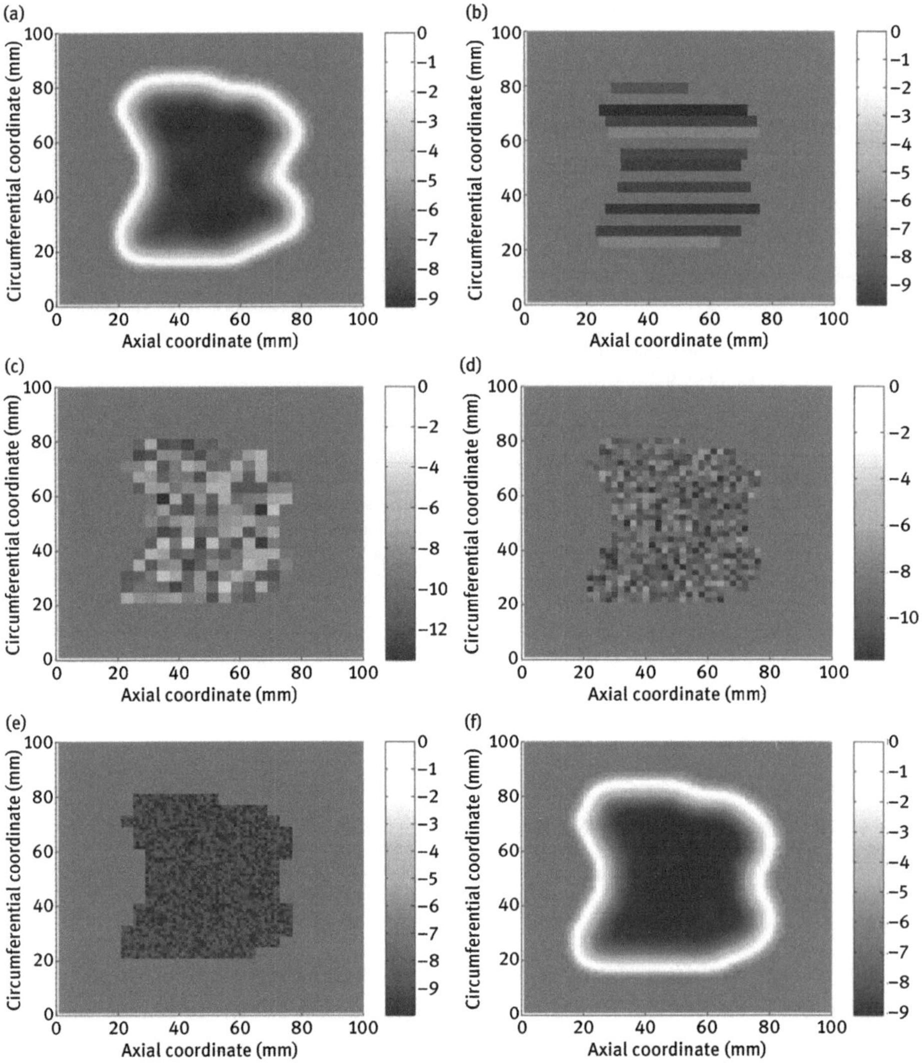

Figure 4.83: Three-dimensional profile multistage inversion of irregular defects by progressive refinement of mesh model: (a) real defect profile; (b) inversion results based on strip model; (c) the inversion results after the first division of the mesh model; (d) the inversion results after the second division of the mesh model; (e) the inversion results after the third division of the mesh model; and (f) the final inversion results.

Table 4.18: Three mesh sizes and their corresponding iteration times.

Parameter	First division	Second division	Third division
Division size in the mesh model (mm)	8 × 8 × 5.6	4 × 4 × 2.8	2 × 2 × 1.4
Iteration number	500	1,000	2,500

of the third division, the inversion result of the defect profile is very close to the real defect profile.

After calculation, the RMS error of the final 3D profile is 8.6 %. Compared with the inversion without introducing mesh model refinement strategy, the time of 4,000 iterations did not change. However, after the gradual refinement of the mesh model, the final inversion error of the test defect is reduced from 10.3 % to 8.6 %.

After the gradual refinement of the mesh model, the error convergence curve of the second stage of the experimental defects in the multistage hybrid inversion is shown in Figure 4.84. The error convergence rates in the first and the second divisions of the mesh model are obviously faster than the error convergence rate of the third division. Compared to the error convergence of the mesh model without refinement, the error convergence rate is faster and the final inversion error is smaller. Therefore, the mesh model can effectively speed up the error convergence rate of the iterative inversion of the defect profile.

4.4.3 Progressive Refinement of Division Size of Finite Element Model

In the multistage inversion process of the defect 3D profile, the precision and speed of the stochastic search iterative inversion in the second stage are not only affected by the mesh size of the 3D profile of the defect but also by the division size of finite element forward calculation model. When the finite element mesh size is large, the finite element calculation time is shortened, the defect contour in the iterative inversion can also be obtained quickly, but the MFL signal calculation error is larger, which can cause larger inversion error, and even lead to nonconvergence. Instead, when the finite element mesh size is too small, although accurate MFL signal calculation results can be obtained, the time of one finite element calculation is too long; thus, the total time consumed by the stochastic search iterative inversion has increased dramatically.

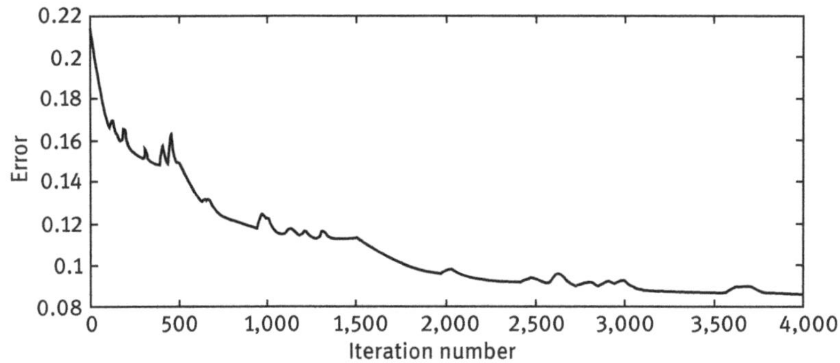

Figure 4.84: The error convergence curve in the second stage of the multistage hybrid inversion after the gradual refinement of the mesh model.

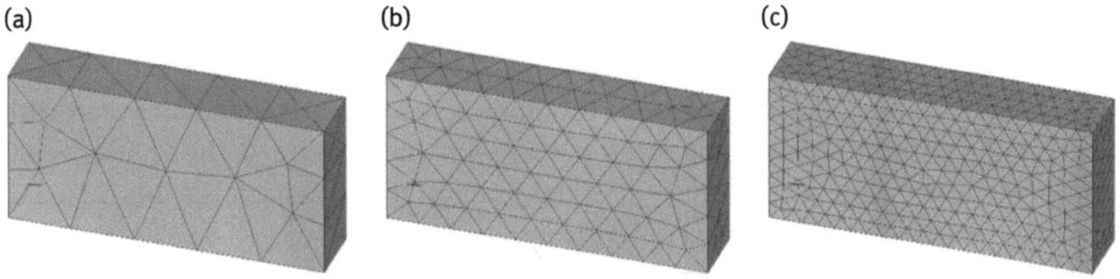

Figure 4.85: The sketch map of mesh progressive refinement of finite element calculation: (a) basic partition size; (b) partition size reduced to ½; and (c) partition size reduced to 1/4.

Similar to the progressive refinement strategy of the defect mesh model, in order to achieve the balance between the inversion error and time consumption, a progressive refinement method for the finite element calculation is presented, and the finite element forward calculation model is refined by gradually decreasing the size of the mesh, as shown in Figure 4.85.

After using the method, in the initial stage of the iterative inversion of random search, the finite element mesh is large, and the time consumed by the iteration inversion is short. On the basis of the inversion results, by reducing the mesh of the finite element calculation, the finite element forward calculation model is refined, which can obtain more accurate 3D profile inversion results. In a certain period of time, with the mesh reduction of the finite element calculation, more accurate 3D contour inversion results can be obtained.

The progressive refinement method of the finite element mesh is adopted to the multistage hybrid inversion method of the defect 3D profile. Finally, the final multistage successive approximation inversion method of the defect 3D contour is formed as shown in Figure 4.86.

According to the above process, in the iterative inversion process of the defect 3D profile, the mesh model and the finite element calculation model are refined simultaneously. After obtaining the inversion result D_s of the defect profile based on strip model, the following steps are carried out:

1. Determine the mesh size of the forward finite element calculation, as well as the partition sizes of the mesh model along the circumferential, axial and radial directions.
2. On the basis of D_s, the defect 3D contour mesh model M_n is established, and the corresponding initial defect profile D_0 is obtained.
3. The inversion results of defect profile D_n are obtained by random searching.
4. If the time is over, the finite element mesh is small enough, or the size of the mesh model is small enough, D_n is smoothed to get the final defect profile D.
5. Otherwise, decrease the finite element mesh size and the partition size of the defect mesh model, and skip to step 3 to continue.

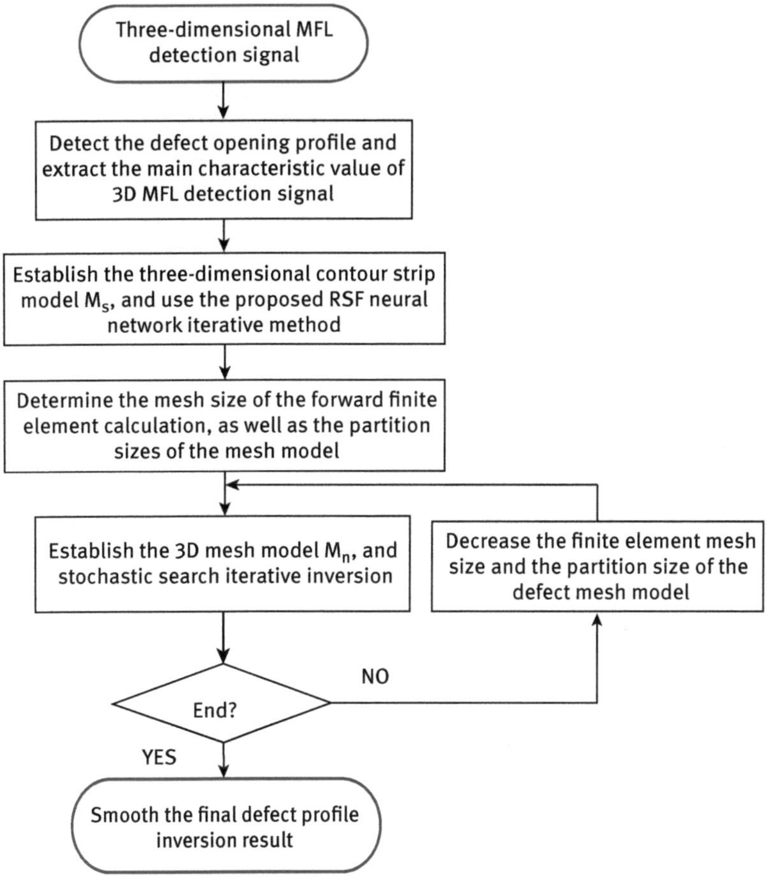

Figure 4.86: The flowchart of the multistage successive approximation inversion method for the defect 3D contour.

Based on the multistage successive approximation inversion method, the 3D profile of an irregular defect in Figure 4.87 is retrieved. The outer diameter and wall thickness of the pipeline are 457 and 14.3 mm, respectively. The pipe circumferential and radial partition dimensions of the strip model used in the first stage of neural network iterative inversion are 2 and 1.4 mm, respectively.

In order to verify the effectiveness of the stepwise refinement strategy of the finite element method, the mesh model and the finite element calculation model are divided 3 times in the second stage of the stochastic search iterative inversion process. Table 4.19 lists the three partition sizes and the corresponding numbers to terminate the iterations. The three iteration numbers are set to 500, 1,000 and 2,500, respectively. The total number of iterations is still 4,000.

Figure 4.87 shows the result of multistage successive approximation inversion for irregular defects. The result shows that the error between the inversion result and the

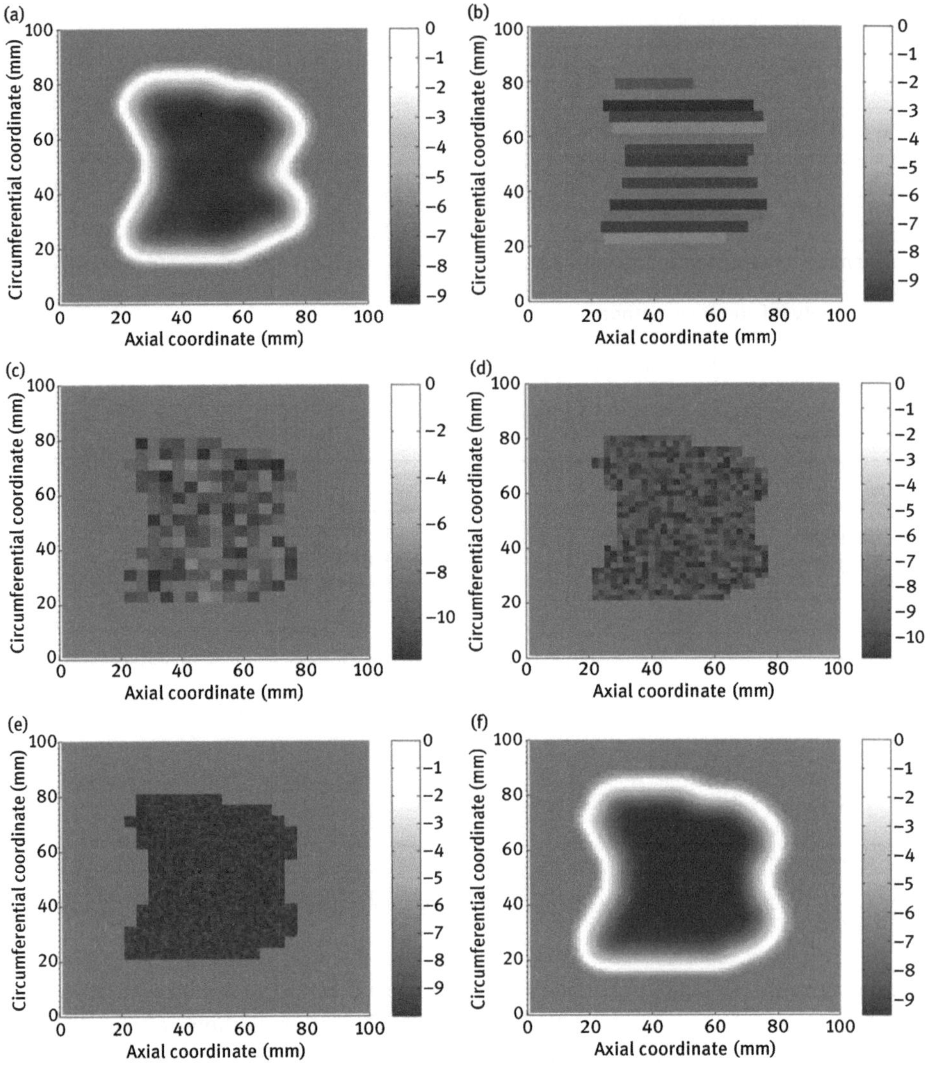

Figure 4.87: Multistage successive approximation inversion for 3D profile of irregular defects: (a) real defect profile; (b) inversion results based on strip model; (c) the inversion results after the first division of the mesh model; (d) the inversion results after the second division of the mesh model; (e) the inversion results after the third division of the mesh model; and (f) the final inversion results.

Table 4.19: Three kinds of partition sizes and corresponding iteration times.

Parameter	First division	Second division	Third division
Defect mesh model (mm)	8 × 8 × 5.6	4 × 4 × 2.8	2 × 2 × 1.4
Finite element calculation model (mm)	4	2	1
Iteration number	500	1,000	2,500

real defect profile obtained from the first division is large. However, after the gradual refinement strategy, with the progressive refinement of the mesh model and finite element mesh size, the defect profile obtained by the inversion method is fast approaching to the true defect profile. After smoothing treatment, the final inversion result of the defect 3D profile has been obtained, which is very close to the true defect profile. At the same time, although the total number of iterations is 4,000, there are 500 times finite element calculations in the mesh sizes enlarged 4 times to the original sizes and another 1,000 times finite element calculations in the mesh sizes enlarged 2 times to the original sizes; hence, the total iterative inversion time will be greatly reduced.

The final inversion error of the irregular defect test is 9.1 %, and the consumption time is about 170 h. Compared with the inversion results by the refinement of defect mesh model and without any finite element model refinement, the total time consumption of defect profile inversion is reduced from 200 to 170 h, and the corresponding inversion error is increased from 8.6 % to 9.1 %. Compared with the inversion results without introducing any refinement strategy, the total time of the multistage successive approximation inversion method is reduced from 200 to 170 h, and the corresponding inversion error is also reduced from 10.3 % to 9.1 %.

Figure 4.88 shows the error convergence curve of the multistage successive approximation in the second stage. Compared with the error convergence process without introducing finite element model refinement strategy, the convergence speed of the multistage inversion method is faster, and the final inversion error is smaller. In the fixed number of iterations, the progressive refinement strategy for the mesh of finite element calculation can reduce the inversion error of the defect 3D profile; in a certain inversion error, the strategy can effectively reduce the time of the iterative inversion process.

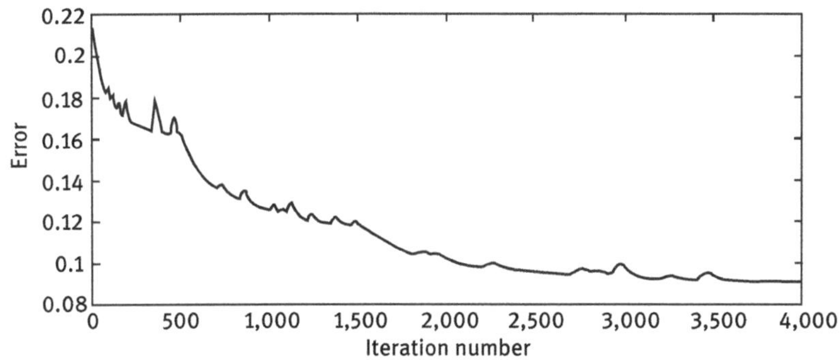

Figure 4.88: The error convergence curve of the second stage of the multistage successive approximation inversion.

4.4.4 Three-Dimensional Profile Inversion Test of Actual Defect

An actual oil and gas pipeline is detected, and then defect 3D profile inversions are tested based on the tested data. The performance of the multistage successive approximation inversion method will further validate through the actual defects testing, and it is compared with the stochastic search iterative inversion method and the neural network iterative inversion method.

The 3D MFL detector used for oil and gas pipeline is shown in Figure 4.89. The detector along the circumferential direction is evenly distributed with 160 magnetic probes. The sampling intervals along the axial direction and the circumferential direction of the pipe are 3.3 and 8.4 mm. The diameter of the pipeline is 457 mm. The depths of the pipeline are 14.3, 9.5 and 7.1 mm, respectively. There are 100 standard artificial rectangular, cambered surface and cylindrical defects on the pipe wall.
As an example, Figure 4.90 shows the axial MFL detection signal thumbnail, in which the axial and circumferential detection points are 8,000 and 160, respectively, corresponding to the sampling interval is 3.3 and 8.4 mm. It can be known from the detection signal, in the stainless steel pipe and the connection flange between the various pipes, the MFL detection signal is in the saturation state. In the other parts, the artificial standard defects have a significant MFL detection signal.

From the pipe with wall thickness of 14.3 mm, 28.6 mm × 14.3 mm × 7.2 mm rectangular defect, 28.6 mm × 7.2 mm cambered surface defect and 28.6 mm × 7.2 mm cylindrical defect are selected as examples. And then, based on the multistage

Figure 4.89: The three-dimensional MFL inner detector for oil and gas pipeline and the test pipe.

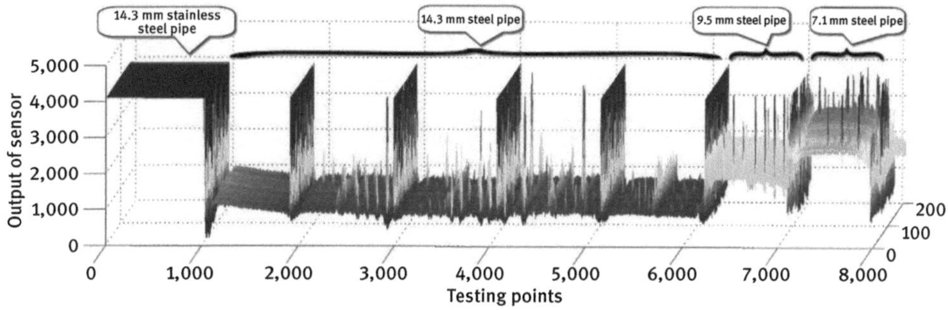

Figure 4.90: The axial MFL detection signal thumbnail of the testing pipe.

Table 4.20: Three partition sizes and the corresponding iteration times of the inversion process in the second stage.

Parameter	First division	Second division	Third division
Defect mesh model (mm)	8 × 8 × 5.6	4 × 4 × 2.8	2 × 2 × 1.4
Finite element calculation model (mm)	4	2	1
Iteration number	300	500	1,200

successive approximation inversion method, the inversion experiment of the defect 3D profile is carried out.

In the application of multistage successive approximation inversion method, the division dimensions along the circumferential and radial directions in the first stage of the strip model are 2 and 1.4 mm. In the second stage of stochastic search iterative inversion, the mesh model and finite element model are divided 3 times. The division sizes and corresponding iteration numbers are shown in Table 4.20. The numbers of iterations in the three partition sizes are set to 300, 500 and 1,200, respectively, and the total number of iterations is 2,000.

The multistage successive approximation inversion method is applied to inverse the 3D profile of the three sample defects. Figures 4.91–4.93 give the processes and the final results of the multistage successive approximation inversion of the three sample defects, respectively. It can be seen that the inversion results of the neural network iterative inversion in the first stage, although there is a large error to the real defect outlines, basically reflect the characteristics of the overall defect contours; thus, they prepared good initial defect profiles for the inversion of the second stage. In the second stage of the stochastic search iterative inversion process, there is still a large error between the results obtained after the first division and the real defect profile. However, with the gradual refinement of the mesh model and finite element model of the defect, the defect profile of the inversion is gradually approaching to the real defect

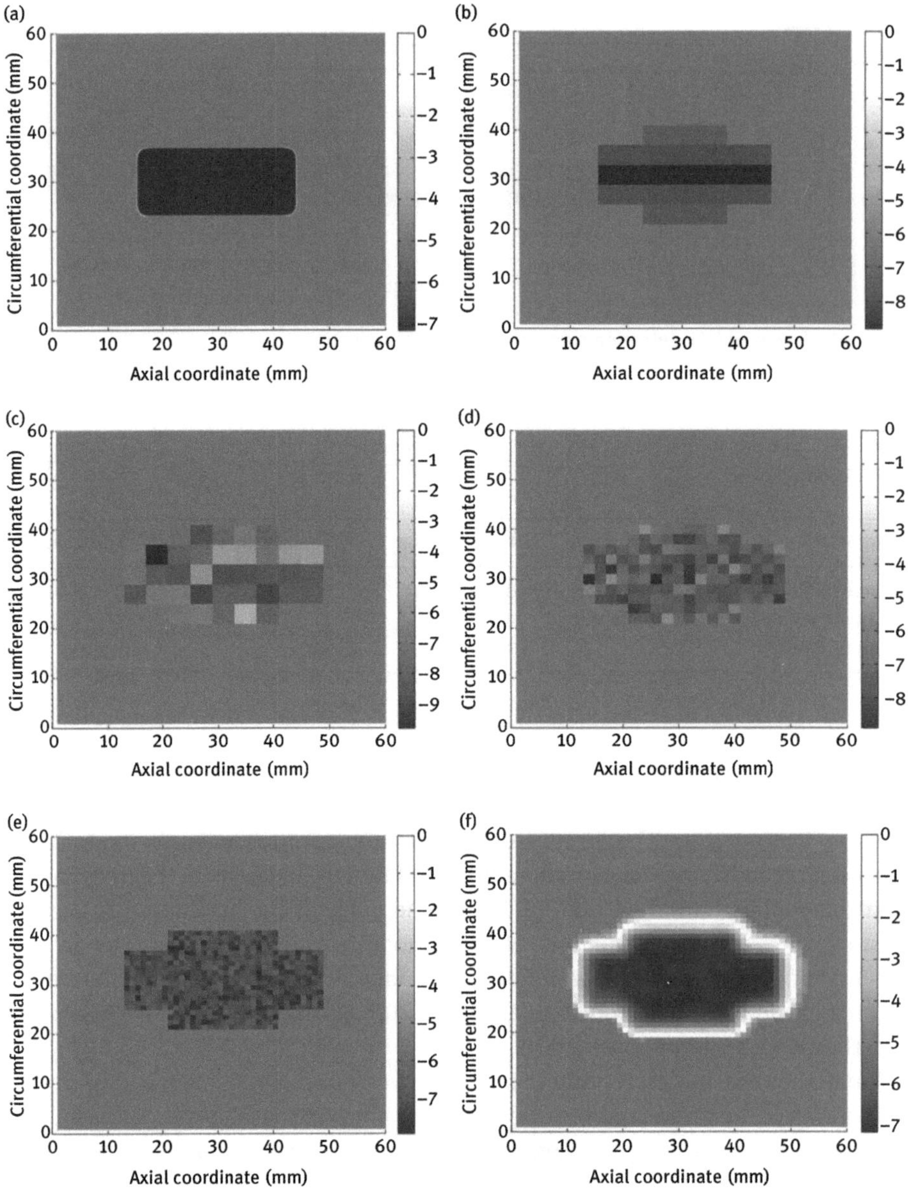

Figure 4.91: Multistage successive approximation inversion for 3D profile of actual 28.6 mm × 14.3 mm × 7.2 mm rectangular defect: (a) real defect profile; (b) inversion results based on strip model; (c) the inversion results after the first division of the mesh model; (d) the inversion results after the second division of the mesh model; (e) the inversion results after the third division of the mesh model; and (f) the final inversion results.

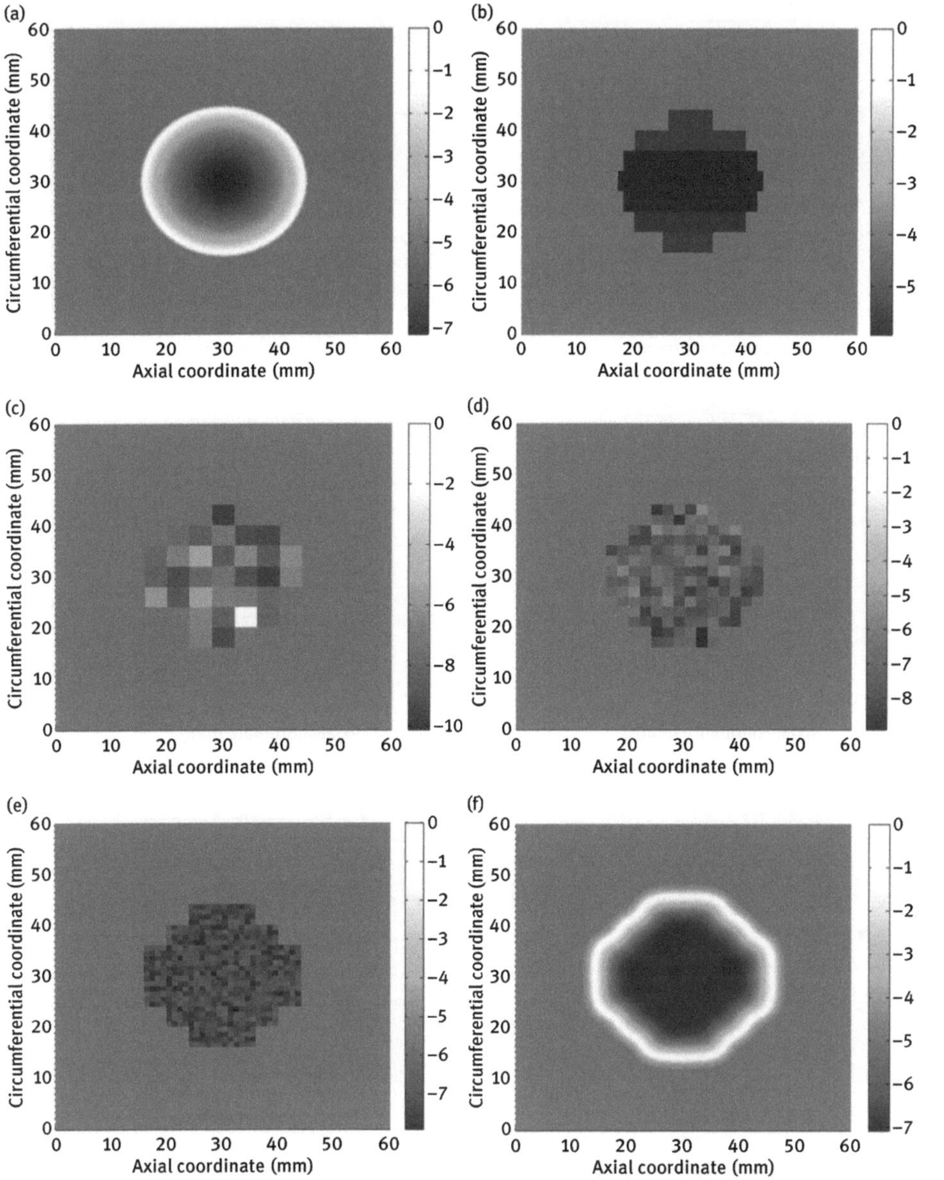

Figure 4.92: Multistage successive approximation inversion for 3D profile of actual 28.6 mm × 7.2 mm cambered surface defect: (a) real defect profile; (b) inversion results based on strip model; (c) the inversion results after the first division of the mesh model; (d) the inversion results after the second division of the mesh model; (e) the inversion results after the third division of the mesh model; and (f) the final inversion results.

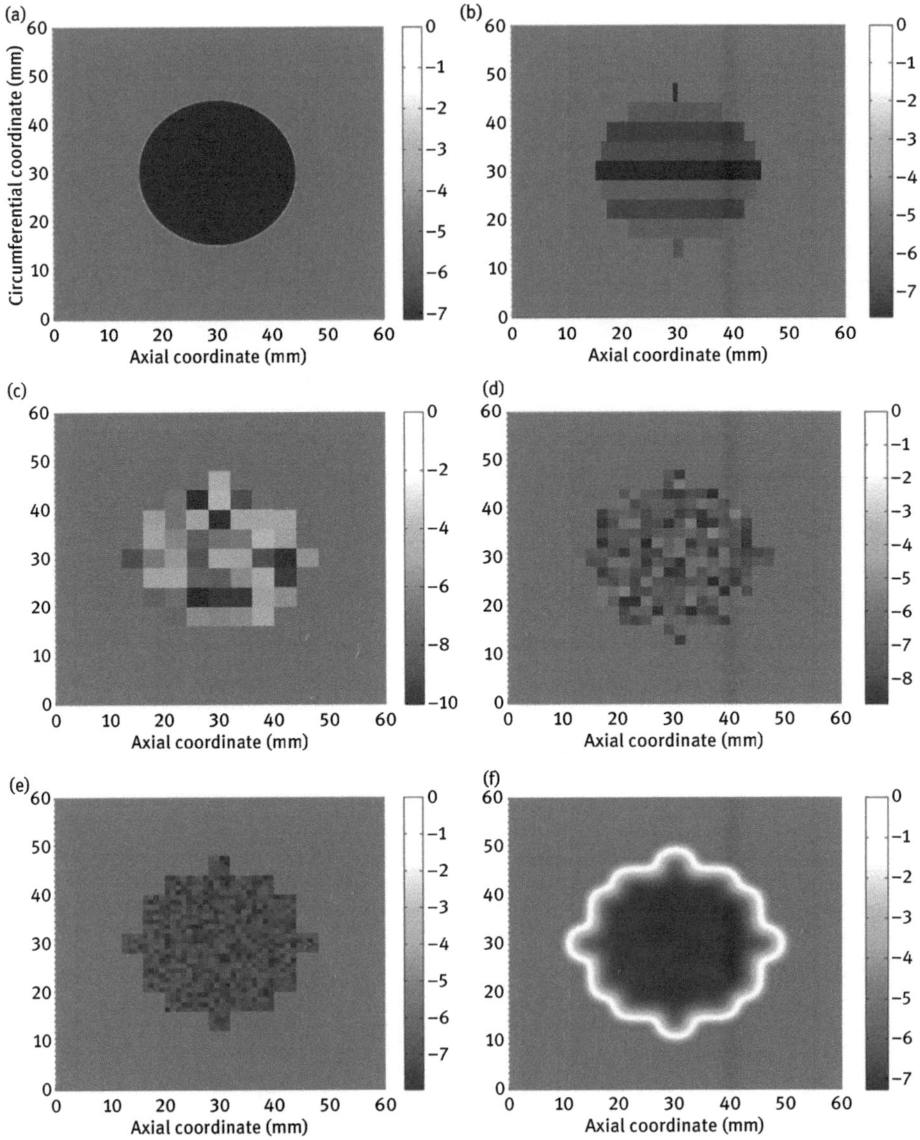

Figure 4.93: Multistage successive approximation inversion for 3D profile of actual 28.6 mm × 7.2 mm cylindrical defect: (a) real defect profile; (b) inversion results based on strip model; (c) the inversion results after the first division of the mesh model; (d) the inversion results after the second division of the mesh model; (e) the inversion results after the third division of the mesh model; and (f) the final inversion results.

Table 4.21: The inversion errors and the time of 3D contour retrieval based on different inversion methods.

Defect (mm)	Stochastic search		Neural network	Multistage successive approximation				
	E (%)	*T* (h)	*E* (%)	E_1 (%)	E_{21} (%)	E_{22} (%)	E_{23} (%)	*T* (h)
Rectangular defect 28.6 × 14.3 × 7.2	13.1	200	18.2	18.2	13.3	10.1	8.7	170
Cambered surface defect 28.6 × 7.2	11.4	200	14.7	14.7	12.6	9.5	8.2	170
Cylindrical defect 28.6 × 7.2	12.2	200	15.5	15.5	12.9	9.8	8.4	170

profile. After smoothing the results of the third division, the inversion results are very close to the true defect profiles.

Besides this, the stochastic search iterative inversion method and the neural network iterative inversion method are also used in the inversion experiments. The inversion errors and the time of the three kinds of methods are shown in Table 4.21. The time of the neural network iterative inversion method can be neglected, and the errors of the multistage successive approximation inversion method include E_1, E_{21}, E_{22} and E_{23}, which indicate the inversion errors of the first stage and the inversion errors of the three division sizes in the second stage, respectively.

The different stages of the multistage successive approximation inversion process are analyzed. The errors of the first phase are 15.5 %, 14.7 % and 18.2 %, respectively; in the process of iteration in the second stage, with the gradual refinement of defect mesh model and finite element model, the error between the inversed profile and the real profile decreases gradually. The final inversion errors after the third division have been reduced to 8.7 %, 8.2 % and 8.4 %, which shows that the inversion results at this time are very close to the real defect outlines.

In the multistage successive approximation inversion method experiments, the ultimate inversion errors of rectangular defect, cambered surface defect and cylindrical defect are 8.7 %, 8.2 % and 8.4 %, respectively. Their corresponding inversion time is 170 h. Compared with the results of the neural network iterative inversion method, the inversion error of the multistage successive approximation inversion method is reduced by 52.2 %, 44.2 % and 45.8 %, respectively. Compared with the results of the stochastic search iterative inversion method, the total time of the multistage successive approximation inversion method is reduced by 15 %, and the inversion error is reduced by 33.6 %, 28.1 % and 31.1 %, respectively. Thus, compared with the iterative inversion of neural networks method and stochastic search iterative inversion method, defect 3D contour of multistage successive approximation inversion method in the aspects of retrieval speed and precision obtained obvious improvement.

4.5 Influence and Correction of Sampling Precision on the 3D Profile Inversion of Defects

Compared with the 3D MFL theoretical signal, the sampling precision of the oil and gas pipeline 3D MFL data along the axial and circumferential directions of the pipe is low [112]. When the MFL testing is performed on the actual oil and gas pipelines, it is needed to study the influence of the low-signal sampling precision on the inversion results of the 3D profile of the defects.

4.5.1 The Influence of Sampling Precision

Taking the oil and gas pipeline MFL detector-typed TH-PipeScan as an example, the sampling intervals along the axial direction and circumferential direction of the pipeline are 3.3 mm and 8.4 mm. For comparison, 28.6 mm × 14.3 mm × 7.2 mm rectangular defect, 28.6 mm × 7.2 mm cambered surface defect and 28.6 mm × 7.2 mm cylindrical defect are still selected as examples. The diameter of the pipe is 457 mm and the depth of the pipe is 14.3 mm.

By modifying the sampling interval of the axial and circumferential directions of the pipe in the finite element simulation model, the simulation results under different sampling intervals were obtained, and then the defect profile inversion test was carried out. The sampling intervals of the 3D MFL simulation signals along the axial direction and circumferential direction of the pipeline are of two kinds: the first sampling intervals are 3.3 and 8.4 mm, and the second sampling intervals are 1.1 and 2.1 mm. After getting the MFL simulation signals of the two sampling intervals, 3D contour defect inversions of the multistage successive approximation inversion method are tested.

As an example, Figures 4.94 and 4.95 show the 3D profile inversion results of 28.6 mm × 14.3 mm × 7.2 mm rectangular defect under two different sampling intervals. Comparison of Figures 4.94 and 4.91 shows that under the same sampling interval, based on the simulated MFL signal and the measured MFL signal, the difference between the 3D profile inversion results is not very large. A comparison between Figures 4.94 and 4.95 shows that when the sampling interval decreases along the axial direction and circumferential direction of the pipeline, the inversion results in each stage of 3D defect contour will be significantly improved.

The errors of 3D defect contours in multistage successive approximation inversion results are shown in Table 4.22. When the sampling intervals along the axial direction and the circumferential direction of the pipe are 3.3 and 8.4 mm, the error of the 3D profile between the simulated MFL signal and the measured MFL signal is not very different. When the sampling interval along the axial direction and circumferential direction of the pipe was reduced to 1.1 and 2.1 mm, the defect 3D inversion error decreased obviously. The statistical results show that after reducing the sampling

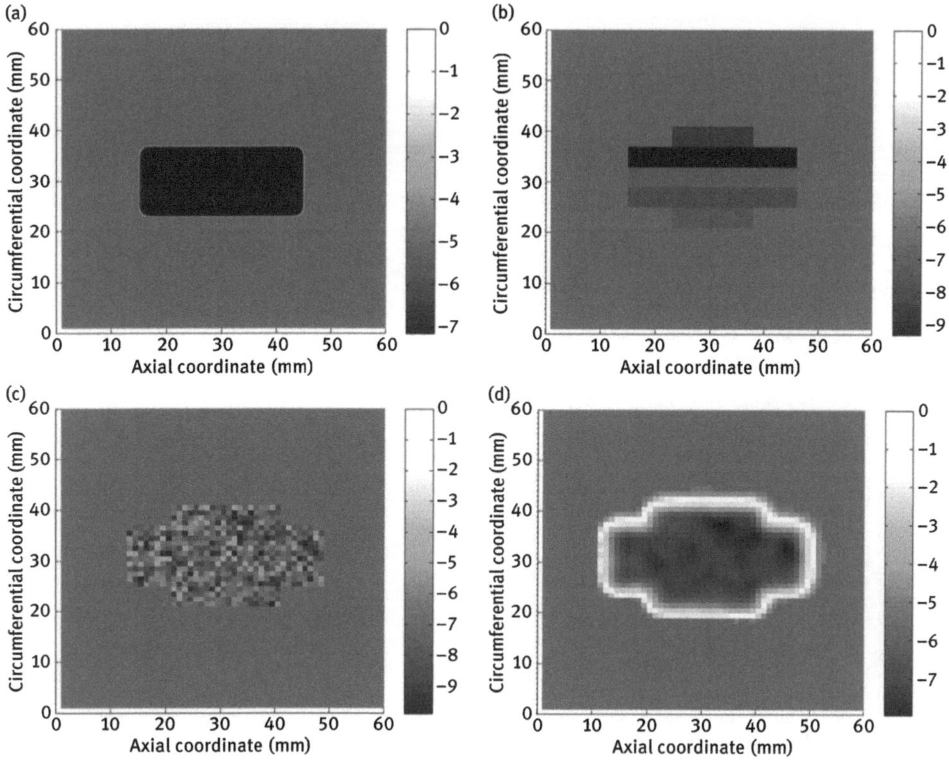

Figure 4.94: Three-dimensional profile inversion results of 28.6 mm × 14.3 mm × 7.2 mm rectangular defect (the axial and circumferential sampling intervals are 3.3 and 8.4 mm): (a) real defect profile; (b) inversion results based on strip model; (c) the inversion results after the third division of the mesh model; and (d) the final inversion results.

interval, the inversion results can be improved significantly by using the multistage successive approximation inversion method.

The defect profile inversion results above showed that the sampling precision will affect the result of defect 3D contour inversion. When the sampling intervals along the axial direction and circumferential direction of the pipeline increase, the amount of 3D MFL information reduces and the inversion error of defect 3D contour will increase. Therefore, in the actual oil and gas pipeline leakage detection, because the actual detector sampling precision is low, 3D MFL detection signal should be corrected, which will improve the accuracy of defect profile inversion.

4.5.2 Interpolation Correction Method for Three-Dimensional MFL Signal

The sampling accuracy along the pipe axial and circumferential directions in the MFL detector is low. Therefore, the information should be added simultaneously to the 3D

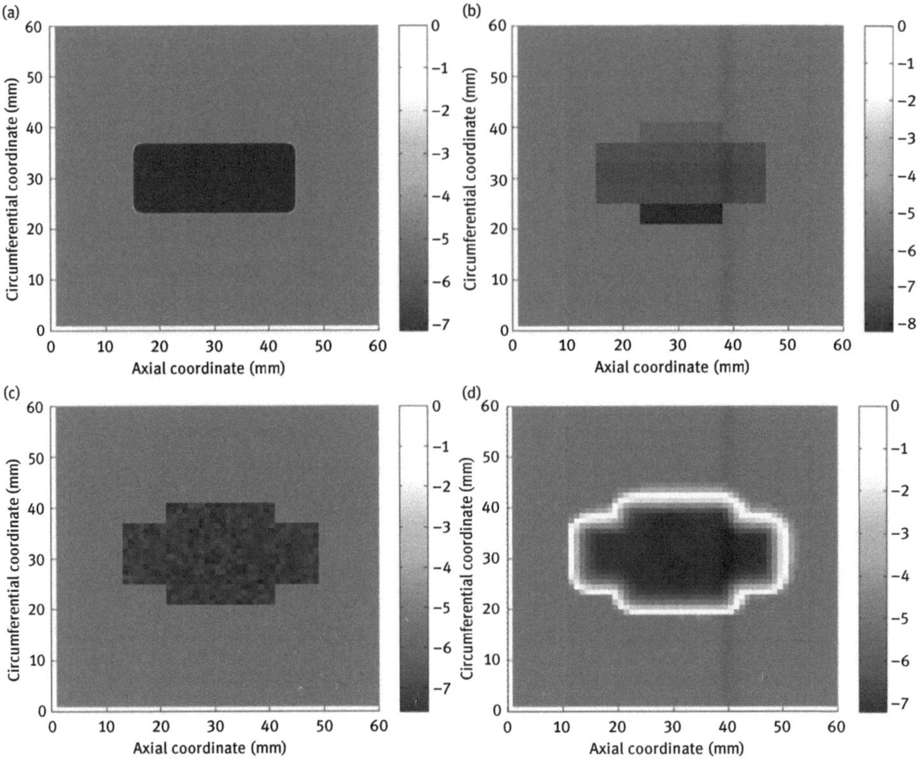

Figure 4.95: Three-dimensional profile inversion results of 28.6 mm × 14.3 mm × 7.2 mm rectangular defect (the axial and circumferential sampling intervals are 1.1 and 2.1 mm): (a) real defect profile; (b) inversion results based on strip model; (c) the inversion results after the third division of the mesh model; and (d) the final inversion results.

Table 4.22: Errors of multistage successive approximation inversions of defects in different cases.

Defect (mm)	Error based on the measured signal (the first sampling interval) (%)	Error based on simulation signal (the first sampling interval) (%)	Error based on simulation signal (the second sampling interval) (%)
Rectangular defect 28.6 × 14.3 × 7.2	8.7	8.6	7.7
Cambered surface defect 28.6 × 7.2	8.2	8.1	7.3
Cylindrical defect 28.6 × 7.2	8.4	8.2	7.4

MFL detecting signals along the axial direction and circumferential directions. At the same time, the defect 3D profile inversion method needs to detect the gradient information of the 3D MFL to find the defect profile. Therefore, the interpolation correction method should not destroy the differential value of each point in the original MFL detection signal.

The interpolation curve obtained by the three-spline interpolation method has the first-order and the second-order differential continuity, which can satisfy the above requirements. Therefore, based on the three-spline interpolation method, the 3D MFL detection signal is interpolated in the circumferential adjacent sensors and between the adjacent acquisition points, respectively. By constructing virtual sensor and virtual acquisition point, the sampling precision of the actual MFL detector is improved.

For a single detection curve in the 3D MFL detection signal of any defect, set its sampling interval as δ, the coordinate sequence of the original detection point as $x_0, x_1, \ldots, x_n$, and the corresponding detection signal sequence as $y_0, y_1, \ldots, y_n$. Within the interval between two adjacent detection points x_i and x_{i+1}, the three polynomial curve can be expressed as follows:

$$S_i(x) = a_i + b_i(x - x_i) + c_i(x - x_i)^2 + d_i(x - x_i)^3. \tag{4.55}$$

By the continuity of the interpolating function curve, we can know that

$$S_i(x_i) = y_i, \tag{4.56}$$
$$S_i(x_{i+1}) = y_{i+1}. \tag{4.57}$$

From the first- and second-order differential continuity of the interpolation function curve, we can know that

$$S_i'(x_{i+1}) = S_{i+1}'(x_{i+1}), \tag{4.58}$$
$$S_i''(x_{i+1}) = S_{i+1}''(x_{i+1}), \tag{4.59}$$

The first- and second-order differential of the three-order polynomial curves are obtained:

$$S_i'(x) = b_i + 2c_i(x - x_i) + 3d_i(x - x_i)^2, \tag{4.60}$$
$$S_i''(x) = 2c_i + 6d_i(x - x_i). \tag{4.61}$$

The sampling interval δ is brought into the above formulas, it follows that

$$a_i = y_i, \tag{4.62}$$
$$a_i + \delta b_i + \delta^2 c_i + \delta^3 d_i = y_{i+1}. \tag{4.63}$$

By formula (4.58), the following can be introduced:

$$S_i'(x_{i+1}) = b_i + 2c_i(x_{i+1} - x_i) + 3d_i(x_{i+1} - x_i)^2 = b_i + 2c_i\delta + 3d_i\delta^2, \tag{4.64}$$

$$S_{i+1}'(x_{i+1}) = b_{i+1} + 2c_i(x_{i+1} - x_{i+1}) + 3d_i(x_{i+1} - x_{i+1})^2 = b_{i+1}. \tag{4.65}$$

Then

$$b_i + 2c_i\delta + 3d_i\delta^2 - b_{i+1} = 0. \tag{4.66}$$

By formula (4.59), the following can be introduced:

$$2c_i + 6d_i\delta - 2c_{i+1} = 0. \tag{4.67}$$

Setting $m_i = S_i''(x_i) = 2c_i$, it can be introduced by the above formula:

$$d_i = \frac{m_{i+1} - m_i}{6\delta}. \tag{4.68}$$

When c_i and d_i are brought into formula (4.67), we can get

$$b_i = \frac{y_{i+1} - y_i}{\delta} - \frac{\delta m_i}{2} - \frac{\delta}{6}(m_{i+1} - m_i). \tag{4.69}$$

When b_i, c_i and d_i are brought into formula (4.66), we can get

$$\delta m_i + 4\delta m_{i+1} + \delta m_{i+2} = \frac{6}{\delta}\left(y_{i+2} - 2y_{i+1} + y_i\right). \tag{4.70}$$

Because the defect signal curve is horizontal near the two ends and its first derivative is zero, the two ends of the signal can be regarded as a fixed boundary. Set $S_0'(x_0) = S_{n-1}'(x_n) = 0$, then

$$2\delta m_0 + \delta m_1 = \frac{6\left(y_1 - y_0\right)}{\delta}, \tag{4.71}$$

$$\delta m_{n-1} + 2\delta m_n = \frac{6\left(y_{n-1} - y_n\right)}{\delta}. \tag{4.72}$$

This result is brought into the above equations, and the final solution is obtained:

$$\begin{bmatrix} 2\delta & \delta & 0 & \cdots & \cdots & 0 \\ \delta & 2\delta & \delta & 0 & \cdots & \cdots \\ 0 & \delta & 2\delta & \delta & 0 & \cdots \\ \cdots & 0 & \cdots & \cdots & \cdots & 0 \\ 0 & \cdots & 0 & \delta & 2\delta & \delta \\ 0 & \cdots & \cdots & 0 & \delta & 2\delta \end{bmatrix} \begin{bmatrix} m_1 \\ m_2 \\ m_3 \\ m_4 \\ \cdots \\ m_n \end{bmatrix} = 6 \begin{bmatrix} 0 \\ \frac{y_2 - 2y_1 + y_0}{\delta} \\ \frac{y_3 - 2y_2 + y_1}{\delta} \\ \frac{y_4 - 2y_3 + y_2}{\delta} \\ \cdots \\ 0 \end{bmatrix}. \tag{4.73}$$

In summary, for any MFL detecting signal, the curve interpolation can be carried out along the pipeline first, and then interpolation along the circumferential direction of the pipe. For any signal curve, in order to carry out the three-spline interpolation, the sampling interval δ is brought into the matrix equations, and the second-order differential of the interpolation curve is obtained: $m_i = S_i''(x_i) = 2c_i$. The unknown coefficients of the three-spline curves can be calculated by the following formula:

$$a_i = y_i, \tag{4.74}$$

$$b_i = \frac{y_{i+1} - y_i}{\delta} - \frac{\delta m_i}{2} - \frac{\delta}{6}(m_{i+1} - m_i), \tag{4.75}$$

$$c_i = \frac{m_i}{2}, \tag{4.76}$$

$$d_i = \frac{m_{i+1} - m_i}{6\delta}. \tag{4.77}$$

Finally, in any two adjacent detection points between x_i and x_{i+1} of the original MFL detection signal, the three-polynomial curve is the following:

$$S_i(x) = a_i + b_i(x - x_i) + c_i(x - x_i)^2 + d_i(x - x_i)^3. \tag{4.78}$$

4.5.3 Test Verification

The test signal was based on the axial MFL detection of 28.6 mm × 7.2 mm × 14.3 mm rectangular defect. The interpolation corrections of the axial center curve and the circumferential center curve were carried out on the basis of the three-spline interpolation algorithm, respectively. The sampling intervals of the axial and circumferential original signals are 3.3 and 8.4 mm, respectively. Along the pipe axial direction, insert two virtual sampling points between adjacent sampling points, the equivalent of the sampling interval is reduced to 1.1 mm; along the circumferential direction of the pipe, three virtual sampling points are inserted between the adjacent sampling points, which is equivalent to reduce the sampling interval to 2.1 mm.

The interpolation correction results are shown in Figure 4.96. Regardless of the axial or circumferential detection curve, after the correction of the original detection signal based on the three-spline interpolation algorithm, new values are added between the adjacent signal points and the modified detection curve becomes smooth. It makes up the deficiency of the axial and circumferential sampling precision of the detector.

The axial MFL detection signal of the rectangular defect is interpolated along the pipe axial direction and circumferential direction, respectively. The correction results are shown in Figure 4.97. Among them, two virtual sampling points are inserted between the adjacent axial sampling points, which is equivalent to reducing the sampling interval to 1.1 mm. Three virtual sampling points are inserted between

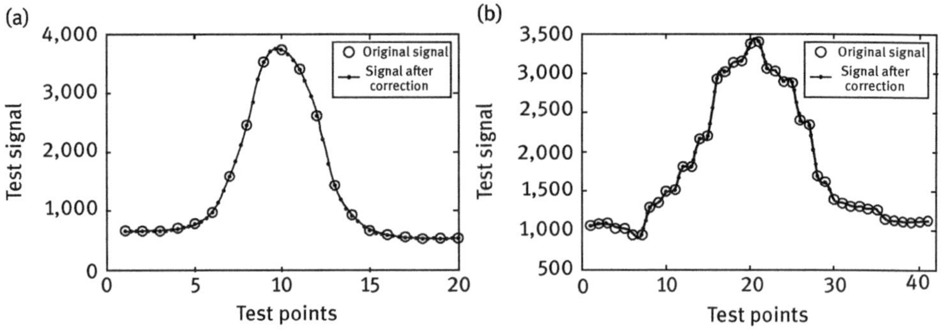

Figure 4.96: Interpolation correction of the detection curve in the axial MFL signal of 28.6 mm × 14.3 mm × 7.2 mm rectangular defect: (a) axial center curve and (b) circumferential center curve.

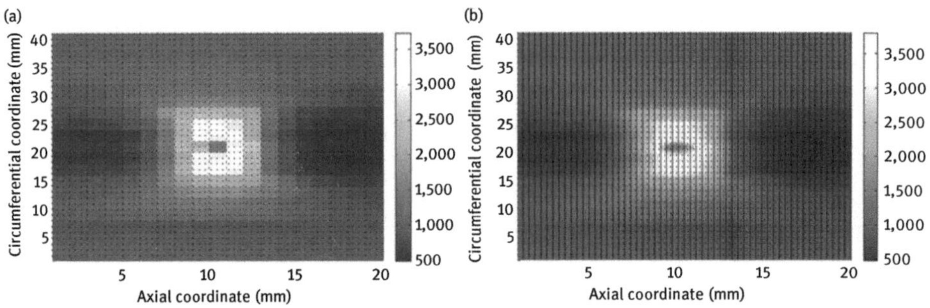

Figure 4.97: Interpolation correction of the axial MFL detection signal in 28.6 mm × 14.3 mm × 7.2 mm rectangular defect: (a) original detection signal and (b) corrected signal.

the adjacent circumferential sampling points, which is equivalent to reducing the sampling interval to 2.1 mm. In contrast, we can know that the missing data can be added between the adjacent detection points of the original MFL detection signal through the interpolation correction, so the detection signal loss caused by the low sampling precision of the detector can be reduced.

In order to verify the validity of the interpolation method, based on the original MFL detection signal and the corrected signal, the defect 3D profile inversion tests are carried out. The example defects are the 28.6 mm × 14.3 mm × 7.2 mm rectangular defect, the 28.6 mm × 7.2 mm cambered surface defect and the 28.6 mm × 7.2 mm cylindrical defect in a pipe with diameter 457 mm and wall thickness 14.3 mm.

The sampling intervals along the axial and circumferential directions are 3.3 and 8.4 mm, respectively. During the test, the original MFL signal is interpolated by 2 times. In the first amendment, two virtual axial sampling and three virtual circumferential sampling points are inserted between the adjacent points, which are equivalent to the axial and circumferential sampling intervals reducing from 1.1 and

2.1 mm. In the second amendment, five virtual axial sampling and seven virtual circumferential sampling points are inserted between the adjacent points, which are equivalent to the axial and circumferential sampling intervals reducing from 0.55 and 1.05 mm.

As an example, Figure 4.98 shows the interpolation correction results of the axial MFL detection signal of 28.6 mm × 7.2 mm cambered surface defect. From the figure, we can see that the corrected MFL signal is more natural than the original MFL signal. Compared with the first amendment, the equivalent sampling interval of the second correction is smaller, and the transition of the MFL signal image is more smooth.

Based on the original MFL detection signal, the first correction signal and the second correction signal, respectively, the 3D profiles of the defect are inversed, and the inversion errors are shown in Table 4.23. According to the results, the inversion error of the 3D profile of the defect is obviously reduced when the original MFL detection signal is interpolated along the axial direction and the circumferential direction of the pipe, and the results of the second correction of the signal is better than the first one. With 28.6 mm × 14.3 mm × 7.2 mm rectangular defect as an example, the defect

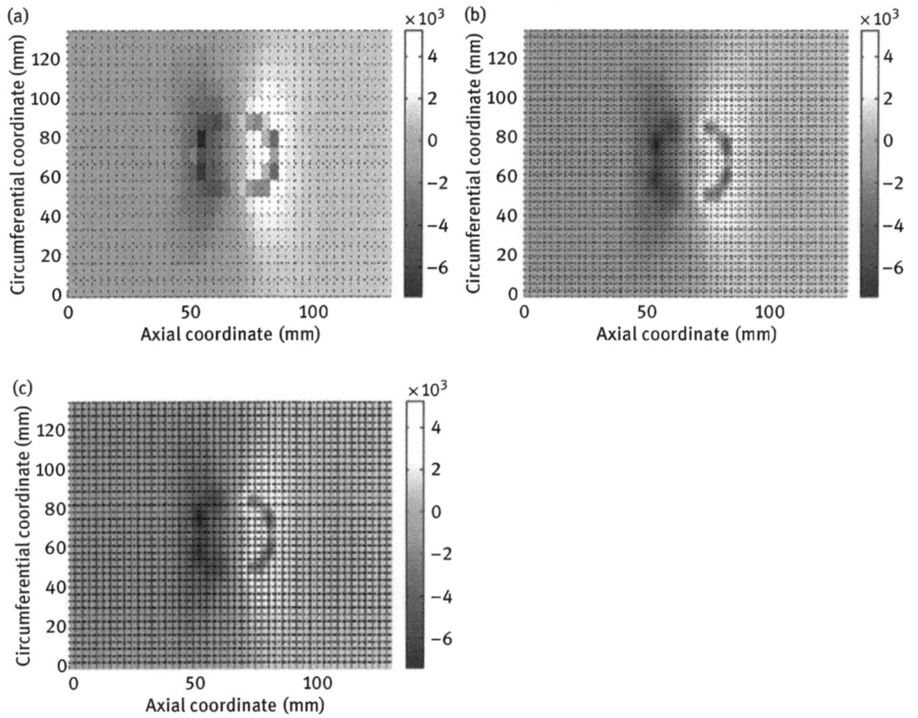

Figure 4.98: The interpolation correction result of the axial MFL test signal of the 28.6 mm × 7.2 mm cambered surface defect: (a) original signal; (b) the first corrected signal; and (c) the second corrected signal.

Table 4.23: The inversion errors of 3D profile based on the original MFL signal and the corrected signal.

Defect (mm)	Original signal	The first corrected signal	The second corrected signal
Rectangular defect 28.6 × 14.3 × 7.2	8.7	7.9	7.6
Cambered surface defect 28.6 × 7.2	8.2	7.5	7.3
Cylindrical defect 28.6 × 7.2	8.4	7.6	7.5

3D contour inversion error, based on the original MFL detection signal, is 8.7 %; the inversion error is reduced to 7.9 % after the first correction; the inversion error of the second correction is further reduced to 7.6 %.

MFL signal interpolation correction and 3D profile inversion test results show that the correction method based on cubic spline interpolation algorithm can add information to the 3D MFL detection signal, and thus reduce the defect profile inversion error caused by the low sampling accuracy of the MFL detector. Therefore, in MFL detection, due to the low sampling accuracy, the interpolation correction of 3D MFL detection signal can be carried out along the axial direction and circumferential direction to improve the accuracy of contour defect 3D inversion.

5 Three-Dimensional MFL Imaging Detection

5.1 Characteristics of Three-Dimensional MFL Signal

With reference to the classification criteria given in the national standard GB/T 27699–2011, according to the length and width of the defect, the metal loss defects are divided into six kinds: the pinhole, the horizontal slot, the horizontal groove, the tangential slot, the tangential groove and the pit. The direction of scanning (at the same time as the magnetization direction) of the magnetic flux leakage (MFL) detector is used as the direction of the defect length. The direction perpendicular to the scanning direction is used as the direction of the defect width and t is the thickness of the material to be tested. The classification criteria for defects are shown in Figure 5.1.

L indicates the defect length, W indicates defect width, so pinholes are defined as $W < t, L < t$; pits are defined as $t < L < 3t$, $t < W < 3t$; horizontal slots are defined as $W < t, L > t$; horizontal grooves are defined as $t < W < 3t$, $L > 3t$; tangential slots are defined as $W > t, L < t$; tangential grooves are defined as $t < L < 3t$, $W > 3t$.

The wides and lengths of pinholes and pits are similar or equal, the peak area and valley region of the leakage magnetic signal show similar shapes, and so the two types of defects are considered as pits in this book. The characteristics of the three-dimensional (3D) MFL signal of horizontal slot is similar to that of the horizontal groove; therefore, these two types of defects are classified as a kind of horizontal groove defect. The characteristics of the 3D MFL signal of tangential slot are similar to that of the tangential groove defect; therefore, the two kinds of defects are classified as a kind of tangential groove defect.

In order to obtain more information about the defect size and improve the quantitative accuracy of defects, the 3D leakage magnetic signals of the three types of defects were analyzed, respectively, and the signal parameters that can reflect the characteristic of 3D MFL signal are defined.

5.1.1 Parameter Definition of Pits

Figure 5.2 shows the 3D maps of the magnetic leakage signals for 24 mm × 24 mm × 2.4 mm pit. Based on different characteristics of the horizontal component, tangential component and normal component of the 3D MFL signal, the parameters of the 3D MFL signal of the pit are defined. The parameters of the three components of the signal are represented as P_h, P_t and P_v, respectively.

As shown in Figure 5.2(a), the horizontal component of the MFL signal includes a peak area and two valley values. The peak area, whose peak position can reflect the extension of the tangential direction of the pit to a certain extent, has two maxima and a minimum. Each valley region contains a minimum value. Because the magnetization is along the horizontal direction, the horizontal MFL component will change

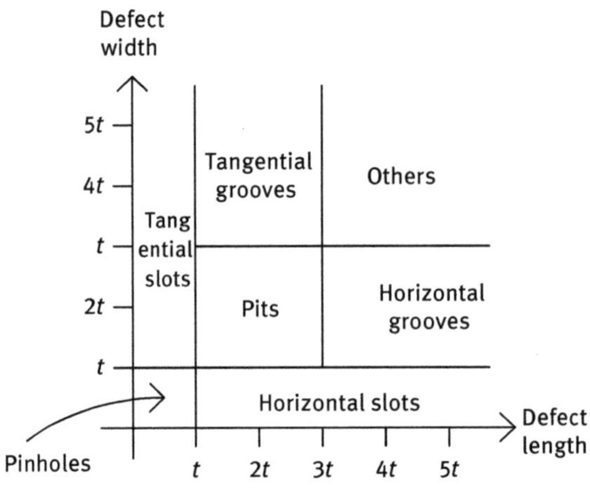

Figure 5.1: Standard for classification of metal loss defects.

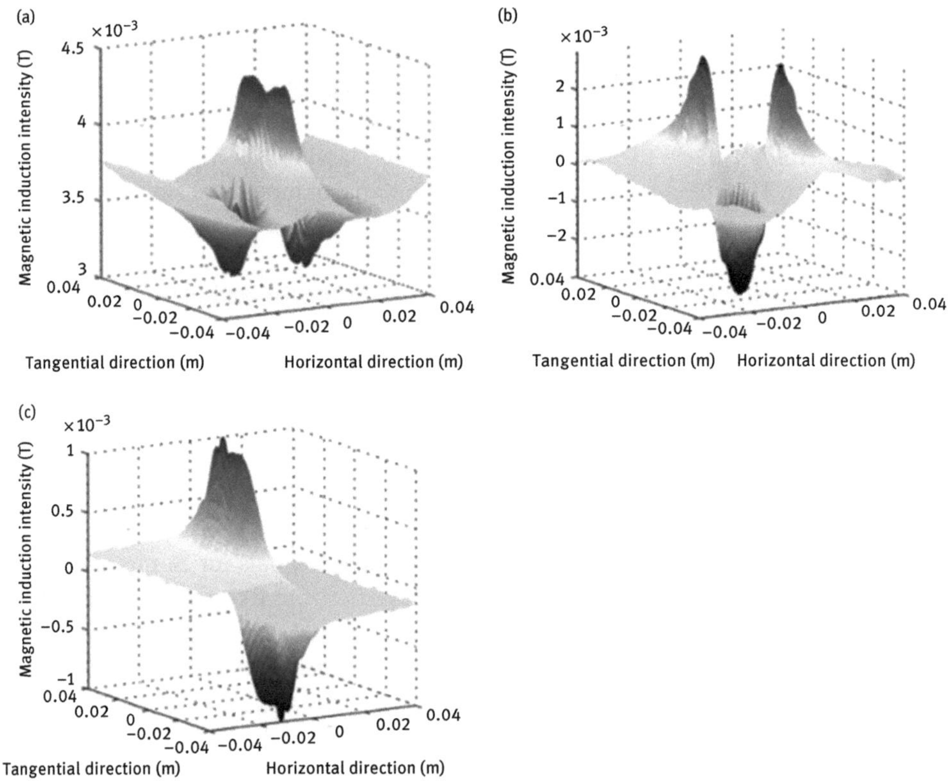

Figure 5.2: Three-dimensional magnetic flux leakage signal of 24 mm × 24 mm × 2.4 mm pit: (a) the horizontal component; (b) the tangential component; and (c) the normal component.

suddenly at the edge of the defect, which reflects the defect horizontal extension; therefore, the horizontal coordinate of the peak value of the first-order differential signal is extracted.

According to the above analysis of the horizontal component of the MFL signal, a total of 15 signal parameters are defined, as shown in Table 5.1.

From Figure 5.2(b), the tangential component of the pit MFL signal includes two peak regions and two valley regions. Each peak area contains a maximum value, and each valley area contains a minimal value. They all reflect the defect depth information. The position of the peak can reflect the position of the defect in a certain degree. Beyond that, the MFL signal will change suddenly at the edge of the defect, so the location of the mutation point can be used to characterize the diameter of the defect profile. In view of the MFL tangential component, a total of nine signal parameters are defined, as shown in Table 5.2.

Table 5.1: Parameter definition of the MFL horizontal component of pits.

Parameter	Definition	Unit	Parameter	Definition	Unit
P_{h1}	Peak value	T	P_{h9}	Gradient peak horizontal coordinate	mm
P_{h2}	Peak tangential coordinate	mm	P_{h10}	Gradient peak tangential coordinate	mm
P_{h3}	Valley value	T	P_{h11}	Intensity integral	$T \times mm^2$
P_{h4}	Valley horizontal coordinate	mm	P_{h12}	Intensity-squared integral	$T^2 \times mm^2$
P_{h5}	Peak–valley value	T	P_{h13}	Horizontal peak	T
P_{h6}	Number of points in the peak area		P_{h14}	Horizontal differential peak value	T/m
P_{h7}	Number of points in the valley area		P_{h15}	Horizontal coordinate of horizontal first-order differential peak	mm
P_{h8}	Gradient peak	T/mm			

Table 5.2: Parameter definition of the MFL tangential component of pits.

Parameter	Definition	Unit
P_{t1}	Peak value	T
P_{t2}	Peak horizontal coordinate	mm
P_{t3}	Peak tangential coordinate	mm
P_{t4}	Peak radius	mm
P_{t5}	Gradient peak	T/mm
P_{t6}	Gradient peak horizontal coordinate	mm
P_{t7}	Gradient peak tangential coordinate	mm
P_{t8}	Intensity integral	$T \times mm^2$
P_{t9}	Intensity-squared integral	$T^2 \times mm^2$

Table 5.3: Parameter definition of the MFL normal component of pits.

Parameter	Definition	Unit
S	Peak value	T
P_{v2}	Peak horizontal coordinate	mm
P_{v3}	Number of points in the peak area	
P_{v4}	Number of points in the valley area	
P_{v5}	Intensity integral	$\mathrm{T} \times \mathrm{mm}^2$
P_{v6}	Intensity squared integral	$\mathrm{T}^2 \times \mathrm{mm}^2$
P_{v7}	Horizontal differential peak value	T/m
P_{v8}	Gradient peak horizontal coordinate	mm

As shown in Figure 5.2(c), the normal component of the pit MFL signal includes a peak region and a valley region, which reflects the depth information of the pit. The position of the peak and valley of the signal reflects the position of the pit to a certain extent. In addition, the position of the signal mutation point is characterized by the pit diameter, so the peak value of the first-order differential signal in the horizontal axis is extracted. According to the normal component of the MFL signal, a total of eight signal parameters are defined, as shown in Table 5.3.

5.1.2 Parameter Definition of Horizontal Grooves

Figure 5.3 shows the 3D maps of the magnetic leakage signals for 48 mm × 24 mm × 1.2 mm horizontal groove. In view of the different characteristics of the horizontal component, tangential component and normal component, the parameters of the 3D MFL signal of the horizontal groove are defined. The parameters of the three components of the signal are represented as H_{h}, H_{t} and H_{v}, respectively.

From Figure 5.3(a), we can know that the MFL horizontal component of the horizontal groove defect has two peaks, and the span of the two peaks can be characterized by the extension of the defect in the horizontal direction. When the feature value is extracted, the signal parameters such as the span of the signal peak area and the horizontal coordinate of the peak value of the first-order differential tangential signal are added. In view of the horizontal component of the horizontal groove MFL signal, the 16 signal parameters are defined in Table 5.4.

From Figure 5.3(b), it is known that the 3D MFL tangential component of the horizontal groove is different from that of the pit, which includes not only two peaks but also two maxima. The horizontal and tangential coordinates of the maximum reflect the length and width of the defect, accordingly, when the signal parameters of the 3D MFL signal tangential component of the horizontal groove are selected, the parameters that are related to the maximum are added. In addition, the difference between the length and width of the horizontal groove is large, which results in the situation that the peak radius of the 3D MFL tangential component no longer reflects

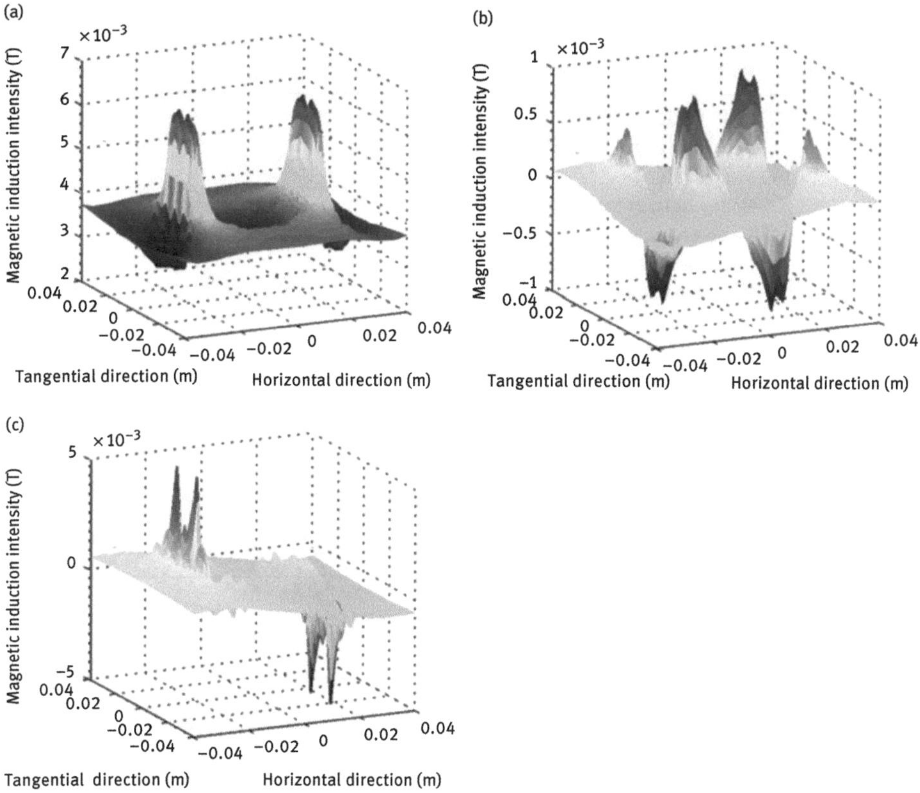

Figure 5.3: Three-dimensional magnetic flux leakage signal of 48 mm × 24 mm × 1.2 mm horizontal groove: (a) the horizontal component; (b) the tangential component; and (c) the normal component.

defect geometric parameters. So when the signal parameters are defined, the peak radius is not used. In view of the tangential component of the MFL signal, the 12 signal parameters are defined in Table 5.5.

Figure 5.3(c) shows that compared to the pit, the normal component of groove MFL signal no longer contains peak area, but has two obvious peak points. In the definition of signal parameters, the number of points in the peak area is no longer chosen; however, according to the signal characteristics, the horizontal and tangential coordinates of the signal peak are extracted as the signal parameters. In view of the normal component of the horizontal groove MFL signal, seven signal parameters are defined in Table 5.6.

5.1.3 Parameter Definition of Tangential Grooves

Figure 5.4 shows the 3D maps of the magnetic leakage signals for 24 mm × 48 mm × 1.2 mm tangential groove. According to the different characteristics of the

Table 5.4: Parameter definition of the MFL horizontal component of horizontal grooves.

Parameter	Definition	Unit
H_{h1}	Peak value	T
H_{h2}	Peak horizontal coordinate	mm
H_{h3}	Valley value	T
H_{h4}	Valley horizontal coordinate	mm
H_{h5}	Peak–valley value	T
H_{h6}	Intermediate minimum	T
H_{h7}	Peak area span	mm
H_{h8}	Valley area span	mm
H_{h9}	Intensity integral	$T \times mm^2$
H_{h10}	Intensity squared integral	$T^2 \times mm^2$
H_{h11}	Horizontal peak	T
H_{h12}	Horizontal coordinate of horizontal peak	mm
H_{h13}	Horizontal differential peak value	T/m
H_{h14}	Horizontal coordinate of horizontal first-order differential peak	mm
H_{h15}	Tangential differential peak value	T/m
H_{h16}	Horizontal coordinate of tangential first-order differential peak	mm

Table 5.5: Parameter definition of the MFL tangential component of horizontal grooves.

Parameter	Definition	Unit
H_{t1}	Peak value	T
H_{t2}	Peak horizontal coordinate	mm
H_{t3}	Peak tangential coordinate	mm
H_{t4}	Signal maximum	T
H_{t5}	Maximum horizontal coordinate	mm
H_{t6}	Maximum tangential coordinate	mm
H_{t7}	Peak radius	mm
H_{t8}	Gradient peak	T/mm
H_{t9}	Horizontal coordinate of gradient peak	mm
H_{t10}	Tangential coordinate of gradient peak	mm
H_{t11}	Intensity integral	$T \times mm^2$
H_{t12}	Intensity-squared integral	$T^2 \times mm^2$

Table 5.6: Parameter definition of the MFL normal component of horizontal grooves.

Parameter	Definition	Unit
H_{v1}	Peak value	T
H_{v2}	Peak horizontal coordinate	mm
H_{v3}	Peak tangential coordinate	mm
H_{v4}	Intensity integral	$T \times mm^2$
H_{v5}	Intensity-squared integral	$T^2 \times mm^2$
H_{v6}	Horizontal differential peak value	T/m
H_{v7}	Horizontal coordinate of horizontal first-order differential peak	mm

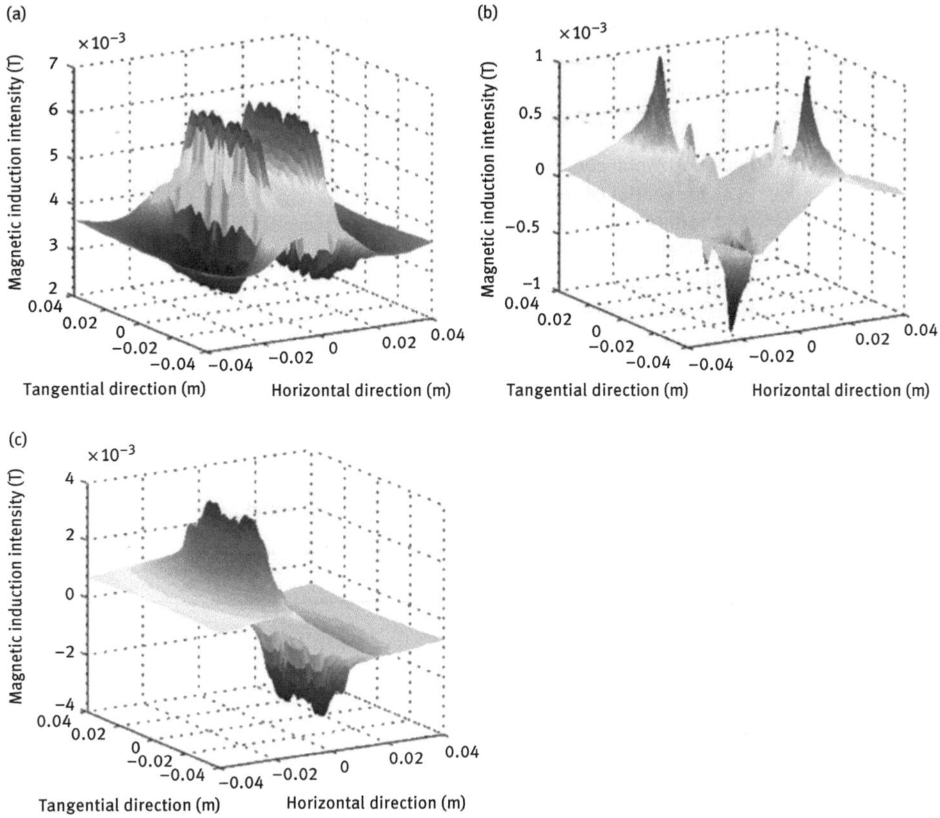

Figure 5.4: Three-dimensional magnetic flux leakage signal of 24 mm × 48 mm × 1.2 mm tangential groove: (a) the horizontal component; (b) the tangential component; and (c) the normal component.

Table 5.7: Parameter definition of the MFL horizontal component of tangential grooves.

Parameter	**Definition**	**Unit**
T_{h1}	Peak value	T
T_{h2}	Peak horizontal coordinate	mm
T_{h3}	Valley value	T
T_{h4}	Valley horizontal coordinate	mm
T_{h5}	Peak–valley value	T
T_{h6}	Intermediate minimum	T
T_{h7}	Peak area span	mm
T_{h8}	Peak–valley span	mm
T_{h9}	Peak area spacing	mm
T_{h10}	Intensity integral	$T \times mm^2$
T_{h11}	Intensity-squared integral	$T^2 \times mm^2$
T_{h12}	Horizontal differential peak value	T/m
T_{h13}	Horizontal coordinate of horizontal first-order differential peak	mm

Table 5.8: Parameter definition of the MFL tangential component of tangential grooves.

Parameter	Definition	Unit
T_{t1}	Peak value	T
T_{t2}	Peak horizontal coordinate	mm
T_{t3}	Peak tangential coordinate	mm
T_{t4}	Peak radius	mm
T_{t5}	Maximum	T
T_{t6}	Gradient peak	T/mm
T_{t7}	Horizontal coordinate of gradient peak	mm
T_{t8}	Tangential coordinate of gradient peak	mm
T_{t9}	Intensity integral	$T \times mm^2$
T_{t10}	Intensity-squared integral	$T^2 \times mm^2$

Table 5.9: Parameter definition of the MFL normal component of tangential grooves.

Parameter	Definition	Unit
T_{v1}	Peak value	T
T_{v2}	Peak horizontal coordinate	mm
T_{v3}	Peak area span	mm
T_{v4}	Span between peak area and valley area	mm
T_{v5}	Intensity integral	$T \times mm^2$
T_{v6}	Intensity-squared integral	$T^2 \times mm^2$
T_{v7}	Horizontal peak	T
T_{v8}	Horizontal coordinate of horizontal peak	mm
T_{v9}	Horizontal differential peak value	T/m
T_{v10}	Horizontal coordinate of horizontal first-order differential peak	mm

horizontal, tangential and normal components of the MFL signal, the parameters of 3D MFL signal of the tangential groove defects are defined. The parameters of the three components of the signal are represented as T_h, T_t and T_v, respectively. The results are given in Tables 5.7, 5.8 and 5.9.

5.2 Defect Classification Quantitative Method under Complete MFL Signal

Neural network is usually used to quantify the defect under a unified network, without taking into account the MFL signal differences induced by the different shapes of defects, which will lead to a large error in the quantization. In order to improve the quantitative accuracy, first, the defects are classified based on the 3D MFL signal, then for different types of defects, the corresponding neural network is trained.

First, according to the level of pits, grooves and defects of the three types of grooves, the radial basis function (RBF) neural network is established for pits, horizontal defects and tangential defects classification. Due to the difference of 3D MFL signals of different kinds of defects, the method by extracting the characteristic values of the input signal, it is difficult to get the appropriate input signal for the defect classification neural network. So the grid average processing method of the horizontal component of the MFL signal is proposed, which is used to obtain the input signal of the RBF classification network. And the classification accuracy of the established RBF classification network is verified by the test of the defect classification.

Second, according to the level of pits, grooves and three defects, the backpropagation (BP) neural networks for defect geometry quantization are constructed for pits, horizontal grooves and tangential grooves, respectively. In order to control the complexity of the constructed BP neural network, and to save the training time of the BP neural network, in the process of network training, the Bayesian algorithm is introduced to correct the inter-layer connection weights and the super-parameters of the network.

Finally, through the test on the performance of defect quantification, the proposed neural network training time, defect quantization accuracy and anti-interference ability are verified.

5.2.1 Defect Classification Method Based on RBF Neural Network

In the neural network model, a single neuron can only realize the linear transformation from the input signal to the output signal. In order to realize the nonlinear mapping between the input data and the output data, the RBF neural network is used to classify the defects. The RBF is used as the transfer function of the RBF neural network, the mapping of the input data from the nonlinear space to the linear space is completed and the final output data is obtained by the linear transformation of the weight. Under the premise of sufficient number of hidden layer neurons, the transformation of the data from nonlinear space to linear space can realize the defects classification of MFL testing.

In the structure of RBF neural network, there is no weight connection between the input and the hidden layers; only the RBF is used to transmit the information; and the information transfer between the hidden layer and the output layer depends on the weight. Let $\mathbf{I} = (I_1, \ldots, I_m)$ be the input vector of the neural network, O the output signal, $\varphi = (\varphi_1, \ldots, \varphi_m)$ the RBF between the hidden layer and the input layer, $\mathbf{c} = (\mathbf{c}_1, \ldots, \mathbf{c}_m)$ the RBF center, $w = (w_1, \ldots, w_m)$ the weights of the hidden layer and the output layer of the neural network, n the number of neurons in the hidden layer, θ a constant, the output function of the RBF neural network can be calculated:

$$\hat{O} = \sum_{i=1}^{n} w_i \varphi_i(\|\mathbf{I} - \mathbf{c}_i\|) + \theta. \tag{5.1}$$

The Gauss function of formula (5.2) is chosen as the RBF:

$$\varphi\left(\|\mathbf{I}-\mathbf{c}\|\right)=\exp\left(-\frac{\|\mathbf{I}-\mathbf{c}\|^2}{2\sigma^2}\right), \tag{5.2}$$

where σ is the standard deviation of the Gauss function. The selected Gauss function has the following characteristics: the function value at the node center is very large, and the function value decreases rapidly from the center.

In the training of RBF neural network, the difference between the actual output value and the target output value is reduced by adjusting the weight w, the center $\mathbf{c}$ of the RBF and the standard deviation σ.

The mean square error is used as the objective function in neural network training:

$$E=\sum_{j=1}^{N}\frac{1}{N}\left(\hat{O}_j-O_j\right)^2=\sum_{j=1}^{N}\frac{1}{N}\left(\sum_{i=1}^{n}w_{i,j}\varphi_i(\|\mathbf{I}-\mathbf{c}_i\|)+\theta-O_j\right)^2. \tag{5.3}$$

In the training process of RBF neural network, the weight, the center of RBF and the standard deviation are corrected based on the steepest descent algorithm. Thus, the change of network parameters corresponding to the Lth output signal in the neural network training process is obtained:

$$\Delta w_{i,L}=-\eta_w\frac{\partial E}{\partial w_{i,L}}=-\frac{2\eta_w}{N}\left(\sum_{i=1}^{n}w_{i,L}\varphi_i\left(\|I-c_i\|\right)+\theta-O_L\right)\varphi_i\left(\|I-c_i\|\right), \tag{5.4}$$

$$\begin{aligned}\Delta c_i&=-\eta_c\frac{\partial E}{\partial c_i}\\&=-\frac{2\eta_c}{N\sigma^2}\sum_{j=1}^{N}\left[w_{i,j}\left(\sum_{i=1}^{n}w_{i,j}\varphi_i\left(\|\mathbf{I}-\mathbf{c}_i\|\right)+\theta-O_L\right)\varphi_i\left(\|\mathbf{I}-\mathbf{c}_i\|\right)\|\mathbf{I}-\mathbf{c}_i\|\right],\end{aligned} \tag{5.5}$$

$$\begin{aligned}\Delta\sigma_i&=-\eta_\sigma\frac{\partial E}{\partial\sigma_i}\\&=-\frac{2\eta_\sigma}{N\sigma_i^3}\sum_{j=1}^{N}\left[w_{i,j}\left(\sum_{i=1}^{n}w_{i,j}\varphi_i\left(\|\mathbf{I}-\mathbf{c}_i\|\right)+\theta-O_L\right)\varphi_i\left(\|\mathbf{I}-\mathbf{c}_i\|\right)\|\mathbf{I}-\mathbf{c}_i\|^2\right],\end{aligned} \tag{5.6}$$

where η_w, η_c and η_σ are learning rate for each parameter of the neural network, which are used to adjust the network parameters and the step size of the correction algorithm adaptively. If the learning rate is too small, it will lead to slow convergence of the network, and the network training time is too long. If the learning rate is too large, the network training process may be a shock, which results in poor network stability.

According to the classification criteria of the defects mentioned above, the classified objects of the trained RBF neural network include pits, horizontal grooves and

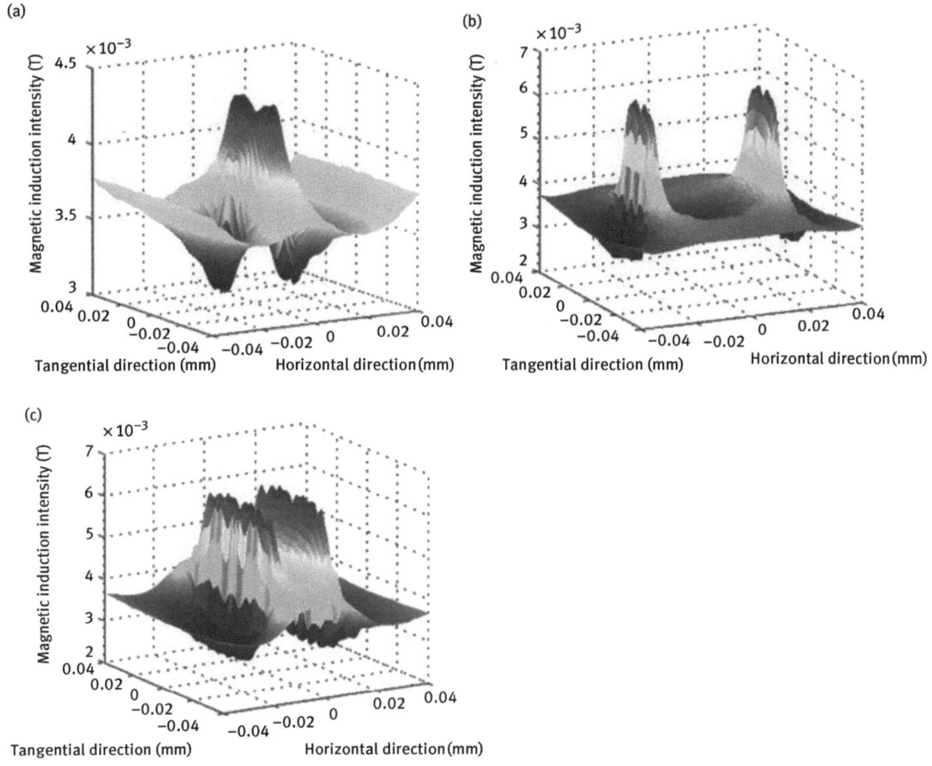

Figure 5.5: The horizontal component of defect MFL signal: (a) the pit; (b) the horizontal groove; and (c) the tangential groove.

tangential grooves. Figure 5.5 shows an example of the horizontal component of the 3D MFL signal for these three types of defects. Section 5.1 has pointed out that the horizontal component characteristics of MFL signals for the three kinds of defects have obvious differences. Therefore, in the classification of the three types of defects, only the characteristic of the horizontal component of MFL signal is used as the input signal of the neural network.

As shown in Figure 5.5, there is a large difference between the horizontal components of the pit, the horizontal groove and the tangential groove. When selecting the input signals of RBF neural network, it is difficult to obtain uniform and accurate reflection of the characteristic value of the defect species by the method of extracting the characteristic value. Therefore, a grid average processing method is proposed to extract the input signals of the RBF classification network.

According to the proposed grid average processing method, the signal is divided into grids, and the average value of the signal in each grid is calculated as the value of the grid. Figure 5.6 shows the grid average processing results of the horizontal components of the pit, the horizontal groove and the tangential groove MFL signals.

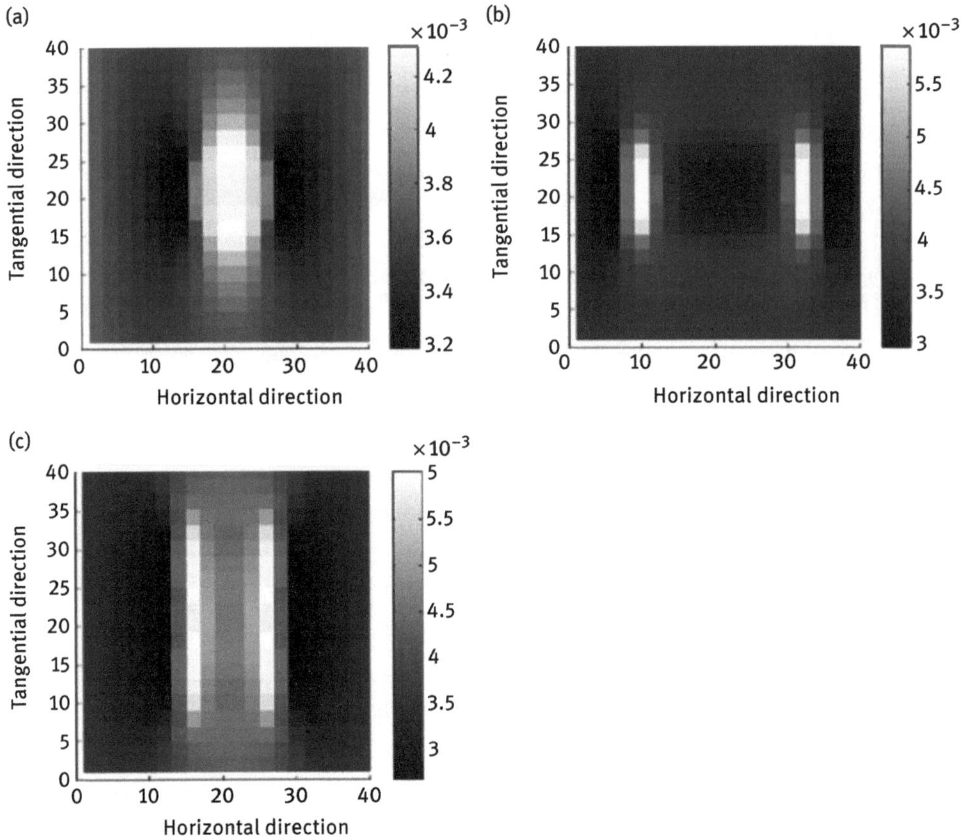

Figure 5.6: The grid average processing results of the horizontal components of defect MFL signals: (a) the pit; (b) the horizontal groove; and (c) the tangential groove.

The proposed method is used to process the MFL signal in the horizontal direction, which can reduce the amount of data in the signal under the precondition of retaining the main features of the signal. As shown in Figure 5.6, the difference between the three types of grid peak concentrated areas is larger. Therefore, in accordance with the grid size, 20 grid locations with largest grid values are extracted as the input signal of RBF neural network.

A sample library containing 90 pits, 150 horizontal grooves and 150 tangential grooves is established. The training samples of the neural network selected were 70 pits, 70 horizontal grooves and 70 tangential grooves, respectively. In addition, 20 pits, 20 horizontal grooves and 20 tangential grooves were selected as the test samples, which were used to test the classification accuracy of the classification neural network.

The extracted grid location is used as the input signal of the RBF neural network. The [0, 0, 1] [0, 0, 1] and [0, 0, 1] are used to represent the defect of the pit,

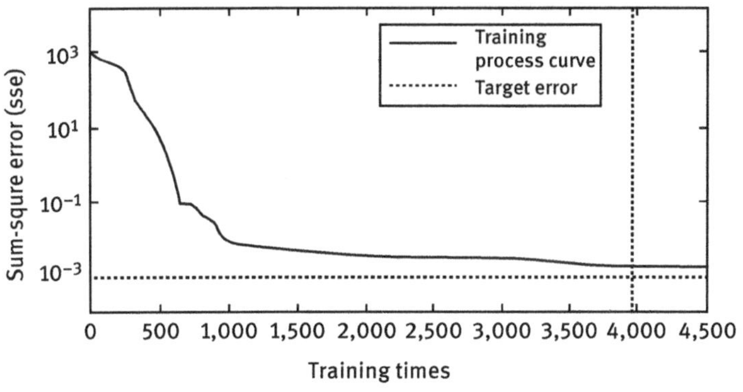

Figure 5.7: RBF neural network training process curve.

the horizontal groove and the tangential groove, and as the output signal of the RBF neural network. The target error of RBF neural network training is set to 0.001. When the training results meet the preset target error, the training of RBF neural network will be complete.

The training process curve of RBF neural network is given in Figure 5.7. We can see that, after 4,556 times of training, the constructed RBF neural network achieves the preset accuracy requirements.

The trained classification neural network is verified with 60 test samples. Table 5.10 presents the defect classification results of six test samples.

Taking the defect classification result of test sample no. 1 for example, the actual output of RBF neural network is [0.8267, 0.1280, 0.0454]. The approximate result of defect classification is [0, 0, 1]. Accordingly, the defect is classified as a pit. The same classification of test sample no. 28 was classified. It is a horizontal groove, however, the classification result is a pit, which means there is an error in the network classification.

Figure 5.8 shows the results of defect classification for 60 test samples. From the graph, we can see that in the 60 test samples, the classification result of five samples is wrong, so the correct rate of classification is 91.7 %. This shows that the trained RBF neural network can realize the classification of the three defect types.

Table 5.10: The defect classification results of six test samples.

Defect no.	Actual types	Ideal output parameter	Actual output parameter	Estimated types
1	Pit	[0, 0, 1]	[0.8267, 0.1280, 0.0454]	Pit
3	Pit	[0, 0, 1]	[−0.1329, 1.0390, 0.0938]	Horizontal groove
21	Horizontal groove	[0, 0, 1]	[−0.0771, 0.9951, 0.0821]	Horizontal groove
28	Horizontal groove	[0, 0, 1]	[0.8012, 0.0526, 0.1462]	Pit
41	Tangential groove	[0, 0, 1]	[0.1385, −0.0332, 0.8947]	Tangential groove
52	Tangential groove	[0, 0, 1]	[0.8650, 0.0276, 0.1074]	Pit

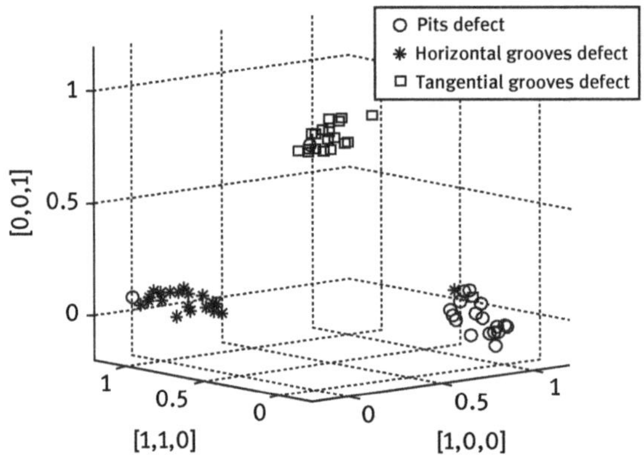

Figure 5.8: The results of defect classification for 60 test samples.

5.2.2 Defect Quantization Method Based on BP Neural Network

BP neural network has strong nonlinear mapping ability. Under the premise of sufficient number of neurons, the BP neural network can be infinitely close to any nonlinear function. Therefore, it can be used to establish the nonlinear mapping relationship between the characteristic value of the 3D MFL signal and the geometric size of the defect.

In BP neural network, set $I = (I_1, \ldots, I_m)$ as input signal, $O = (O_1, \ldots, O_n)$ as output signal, $D = (I, O)$ as training sample, w_1 as the connection weight between the input layer and the hidden layer, w_2 as the connection weight between the hidden layer and the output layer, $W = (w_1, w_2)$ as the overall weight, and the number of neurons in the hidden layer is represented by p, then the output function of the BP neural network can be calculated:

$$\widehat{O_n} = \sum_{i=1}^{p} w_2(i) \tanh\left(\sum_{j=1}^{m} w_1(j) I_j + \theta_1(j)\right) + \theta_2. \tag{5.7}$$

The BP neural network is trained based on the principle of least error, that is, by reducing the error between the actual output and the target output value, the weights are adjusted. Therefore, the objective function of the BP neural network in the training process can be obtained:

$$\begin{aligned} E_D &= \sum_{n=1}^{N} \frac{1}{N} \left(\widehat{O_n} - O_n\right)^2 \\ &= \sum_{n=1}^{N} \frac{1}{N} \left\{ \left[\sum_{i=1}^{p} w_{2,n}(i) \tanh\left(\sum_{j=1}^{m} w_{1,n}(j) I_j + \theta_1(j)\right) + \theta_2 \right] - O_n \right\}^2. \end{aligned} \tag{5.8}$$

In the training process of BP neural network, the most rapid descent method is used to optimize the weights, namely, the minimum value of formula (5.8). Thus, the weight change corresponding to the Lth output signal in the training process is obtained:

$$\begin{aligned}\Delta w_{1,L}(j) &= -\eta \frac{\partial E_D}{\partial w_{1,L}(j)} \\ &= -\frac{2\eta}{N}\left[\sum_{i=1}^{p} w_{2,L}(i)\sec^2\left(\sum_{j=1}^{m} w_{1,L}(j) I_j + \theta_1(j)\right) I_j\right] \\ &\times \left\{\left[\sum_{i=1}^{p} w_{2,L}(i)\tanh\left(\sum_{j=1}^{m} w_{1,L}(j) I_j + \theta_1(j)\right) + \theta_2\right] - O_L\right\},\end{aligned} \tag{5.9}$$

$$\begin{aligned}\Delta w_2(i) &= -\eta \frac{\partial E_D}{\partial w_2(i)} \\ &= -\frac{2\eta}{N}\tanh\left(\sum_{j=1}^{m} w_1(j) I_j + \theta_1(j)\right) \\ &\times \left\{\left[w_2(i)\tanh\left(\sum_{j=1}^{m} w_1(j) I_j + \theta_1(j)\right) + \theta_2\right] - O_L\right\},\end{aligned} \tag{5.10}$$

where η is the learning rate for the weight correction process, which is used for adjusting the step size for the weights in the gradient descent algorithm.

When the weights of BP neural network are modified by using the above model, the number of hidden layer neurons should be set in advance. If the number of neurons is too high, it will lead to a high degree of complexity and poor generalization ability. If the number of neurons is too small, it will lead to poor nonlinear fitting of the neural network. In order to adjust the number of neurons to the appropriate value, the attenuation function is added on the basis of the objective function in the training of neural network:

$$E_W = \sum_m \frac{1}{2} w_1^2 + \sum_p \frac{1}{2} w_2^2. \tag{5.11}$$

The complexity of the neural network can be effectively controlled by the value of formula (5.11). Therefore, the final objective function of the neural network training can be obtained:

$$E = \alpha E_W + \beta E_D = \alpha\left(\sum_m \frac{1}{2} w_1^2 + \sum_p \frac{1}{2} w_2^2\right) + \beta \sum_n \frac{1}{2}\left(\widehat{O_n} - O_n\right)^2, \tag{5.12}$$

where α and β are super-parameters, which are used to adjust the complexity of the neural network.

If $\alpha \gg \beta$, the training objective is to minimize the error between the actual output and the target output, then there is less control to the size of the network, leading to the overfitting of neural network. If $\alpha \ll \beta$, the training objective is to reduce the complexity of the neural network, and the nonlinear fitting ability of the neural network is poor.

By using the steepest descent algorithm to optimize the neural network, it is impossible to adjust the super-parameters α and β. Therefore, the Bayesian algorithm is introduced into the optimization process of the BP neural network, and the adaptive adjustment of the super-parameters is realized in the training process, which can balance the error of the neural network with the scale of the neural network.

Bayesian algorithm is a commonly used method to estimate the range of parameters. The essence of the algorithm is to predict the probability of occurrence in the future by using the occurrence probability in the past, which is to use the prior probability distribution to solve the posterior probability distribution. According to the Bayesian theorem, the amount of arbitrary demand can be regarded as random variables, which can be described by probability distribution.

Bayesian theorem is expressed by Bayesian formula:

$$p(\theta\,|\,x) = \frac{p(x\,|\,\theta)\,p(\theta)}{p(x)} = \frac{p(x\,|\,\theta)\,p(\theta)}{\int^{p}(x\,|\,\theta)\,p(\theta)\,d\theta}, \tag{5.13}$$

where θ is the random variable to be solved; x is the sample; $p(\theta|x)$ is the posterior distribution information of θ, that is, the distribution information of random variable to be solved; $p(\theta)$ represents the prior distribution information of θ, that is, the initial probability of the random variable to be solved; $p(x)$ is the marginal distribution of the sample x, which is usually regarded as a normalization factor that does not depend on the variable θ.

Formula (5.13) shows that the Bayesian algorithm solves the posterior distribution of variables by the prior distribution information of the random variable and the distribution probability of the sample. Since $p(x)$ is a normalization factor, the formula of Bayesian theorem is equivalent to

$$p(\theta|x) \propto p(x\,|\,\theta)\,p(\theta)\,. \tag{5.14}$$

From the above formula, the posterior distribution of the amount θ to be obtained reflects the knowledge of random variable θ after sampling, which is the effect of sampling information on the prior distribution.

BP neural network based on Bayesian algorithm adjusts the weights of the neural network through the Bayesian algorithm. The specific operational procedures are all parameters of the neural network as random variables. Based on the prior probability distribution of the parameters of the target and actual input and output signal of the sample data, range of neural network weights is further modified, and through the BP neural network model complexity control, the training time of neural network is saved and the overfitting possibility of the neural network is reduced, and the generalization ability of the BP neural network is improved.

A represents the neural network; according to the Bayesian theorem, the posterior distribution $p(w|D,\alpha,\beta,A)$ of the neural network weights with sample $D=(I_m,O_n)$ is estimated through the prior distribution $p(w|\alpha,A)$ of neural network weights without sample data. The posterior probability distribution function of the neural network weights is obtained:

$$p(w|D,\alpha,\beta,A)=\frac{p(D|w,\beta,A)\,p(w|\alpha,A)}{p(D|\alpha,\beta,A)}, \tag{5.15}$$

where $p(D\,|\,w,\beta,A)$ is the likelihood function, that is, the joint distribution of samples:

$p(D\,|\,\alpha,\beta,A)$ is the normalization factor

$$p(D\,|\,\alpha,\beta,A)=\int_{-\infty}^{\infty}P(D\,|\,\alpha,\beta,A)P(w\,|\,\alpha,H)\,dw. \tag{5.16}$$

The distribution of neural network samples can be seen as a Gauss distribution with zero mean and variance $\sigma^2=\frac{1}{\beta}$, then the likelihood function can be obtained by

$$p(D\,|\,w,\beta,A)=\frac{1}{Z_p(\beta)}\exp(-\beta E_D). \tag{5.17}$$

In the above formula, $Z_p(\beta)$ as the normalization factor can be written:

$$Z_p(\beta)=\left(\frac{2\pi}{\beta}\right)^{\frac{m}{2}}. \tag{5.18}$$

In the same way, it is assumed that the distribution of the neural network weights is a Gauss distribution with zero mean and variance $\sigma^2=\frac{1}{\alpha}$, so the prior distribution of the neural network weights can be obtained:

$$p(w\,|\,\alpha,A)=\frac{1}{Z_w(\alpha)}\exp(-\alpha E_w). \tag{5.19}$$

In the above formula, $Z_w(\alpha)$ is the normalization factor, which can be written as

$$Z_w(\alpha)=\left(\frac{2\pi}{\alpha}\right)^{\frac{m}{2}}. \tag{5.20}$$

Based on formulas (5.15), (5.17) and (5.19), the probability distribution function of the neural network weights can be obtained:

$$p(w\,|\,D,\alpha,\beta,A)=\frac{\frac{1}{Z_p(\beta)}\exp(-\beta E_D)*\frac{1}{Z_w(\alpha)}\exp(-\alpha E_w)}{p(D|\alpha,\beta,A)}=\frac{\exp(-E)}{Z_F(\alpha,\beta)}. \tag{5.21}$$

In the above framework, the objective function E is minimized, so the most possible neural network weight parameter w can be obtained.

On the premise of the given value w of the neural network weight, the posterior probability distribution of the super-parameters α and β of the neural network can be estimated. According to the Bayesian theorem, the posterior probability distribution of the super-parameter is obtained:

$$p(\alpha,\beta\,|\,D,A) = \frac{p(D\,|\,\alpha,\beta,A)\,p(\alpha,\beta\,|\,A)}{p(D\,|\,A)}. \tag{5.22}$$

Based on formulas (5.12) and (5.21), the normalization factor $p(D\,|\,\alpha,\beta,A)$ can be written as

$$p(D\,|\,\alpha,\beta,A) = \frac{Z_F(\alpha,\beta)}{Z_w(\alpha)\,Z_p(\beta)}. \tag{5.23}$$

The objective function E is carried out by Taylor series expansion in the network parameter w, and then the objective function is minimized to obtain the weight w_{MP} of the neural network that minimizes the objective function. And based on this, the most likely super-parameters α and β are obtained:

$$\alpha_{MP} = \frac{y}{2E_w(w_{MP})}, \tag{5.24}$$

$$\beta_{MP} = \frac{m-y}{2E_D(w_{MP})}, \tag{5.25}$$

where y can be expressed by the effective parameters of the neural network.

Based on the above analysis, the training process of BP neural network based on Bayesian algorithm is obtained, as shown in Figure 5.9.

The specific training process of BP neural network is as follows:

1. Input the training sample $D = (I_m, O_n)$ into the neural network.
2. To give the initial neural network structure, initialize the neural network's super-parameters α and β, and initialize the BP neural network weight w.
3. According to the principle of minimum error, minimize the target function E of neural network and calculate the optimal weight w of BP neural network.
4. On the basis of the optimal network weight w, the optimal network parameters α and β are calculated.
5. If the BP neural network optimization function E meets the set of error requirements, then ends the training process of neural network; if the optimization function E does not meet the requirements of the set error, and the neural network training number has not yet reached the upper limit, then repeat step 3 until the optimization function E meets the requirements; if the neural network training time limit is reached, the optimization function E still does not meet the requirement of error, which indicates that the current neural network does not converge.

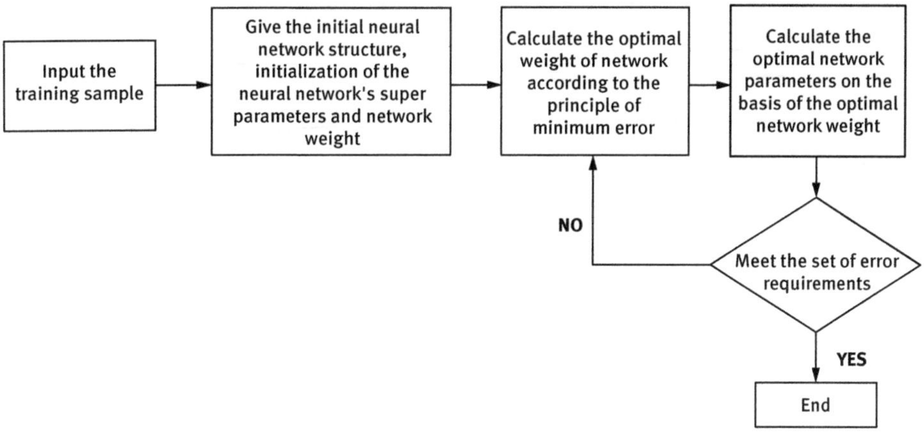

Figure 5.9: The training process of BP neural network based on Bayesian algorithm.

According to the training process of BP neural network, the nonlinear mapping relationship between the defect 3D geometry information and the MFL signal characteristics can be constructed.

In order to verify the effectiveness of the BP neural network based on Bayesian algorithm in reducing the training time of neural network, and the quantitative accuracy of the algorithm, the corresponding defect quantization test is needed. First, the BP neural network based on Bayesian algorithm is constructed, which uses the key characteristics of defect 3D MFL signal as input signals of BP neural network, the defect geometry as the output signal to obtain the mapping relationship between the defect geometry and the key characteristics of defect 3D MFL signal. Then, the network training time before and after optimization of the network using Bayesian algorithm is compared, and the test data is used to verify the accuracy of the trained neural network and the anti-interference ability.

For pits, features of 3D MFL signals are extracted in the sample of 90 defects. From these defects, 79 defects are selected to train the neural network, and the remaining 11 samples will be used as the test data to verify the accuracy of the neural network.

The mean square error of 0.001 is used as the target of neural network training. Figures 5.10 and 5.11 show the neural network training processes of the diameter and depth quantization of pits. Figures 5.10(a) and 5.11(a) show the training processes of the traditional BP neural network using the key characteristics of the defect 3D MFL signals to quantify the diameter and depth. Figures 5.10(b) and 5.11(b) show the training processes of the BP neural network based on the Bayesian algorithm, which are used to quantify the defect diameter and depth by using the key characteristics of the defect one-dimensional (1D) MFL signals. Figures 5.10(c) and 5.11(c) show the training processes of the BP neural network based on the Bayesian algorithm, which are used to quantify the defect diameter and depth by using the key characteristics of the defect 3D MFL signals. In the course of each training process, the broken line is used to identify the target error of neural network training.

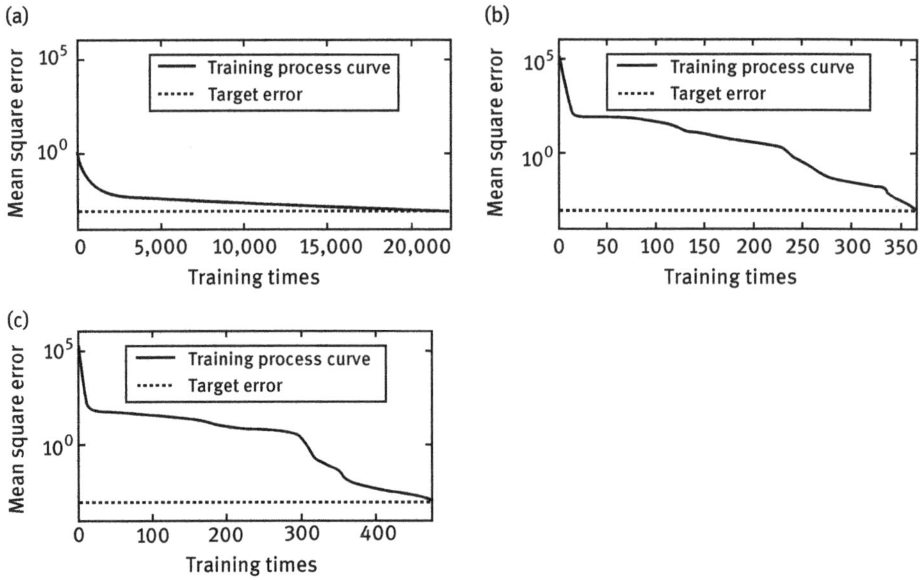

Figure 5.10: Training process of the neural network for the pit diameter quantization: (a) 3D MFL signal and traditional algorithm; (b) 1D MFL signal and Bayesian algorithm; and (c) 3D MFL signal and Bayesian algorithm.

Comparison of Figure 5.10(a) and 5.10(c) shows that in the neural network training process using the defect 3D MFL signal characteristics, traditional BP neural network after training 22,598 times reached expected goals, and the BP neural network based on the Bayesian algorithm after only 473 times training reached the expected target, whose convergence speed is faster, which can effectively reduce the training time. Comparing Figure 5.10(b) and 5.10(c), we can see that when using defect 1D MFL signal characteristics, the neural network reached expected goals after 365 times training, which is less than the number of the training of the 3D MFL signal neural network. This is consistent with the decrease of the input variables of the neural network. Comparison of Figure 5.11(a) and 5.11(c), and Figure 5.11(b) and 5.11(c) also shows the same conclusion. The above results show that the BP neural network based on Bayesian algorithm, compared to the traditional BP neural network, has the advantages of fast convergence speed and short training time.

For pits, the quantitative results of the diameters and depths of 11 test samples are shown in Figure 5.12. The horizontal coordinates represent the number of 11 samples, and the vertical coordinates represent the relative proportion of the quantitative result to the actual defect size. Table 5.11 shows the maximum relative error of the quantitative results of the pits under different conditions.

The quantitative results in Table 5.11 show in adopting the 3D MFL signal characteristics to quantify defects, the maximum relative errors of defect diameter and depth quantification using the traditional BP neural network are larger than that of the BP

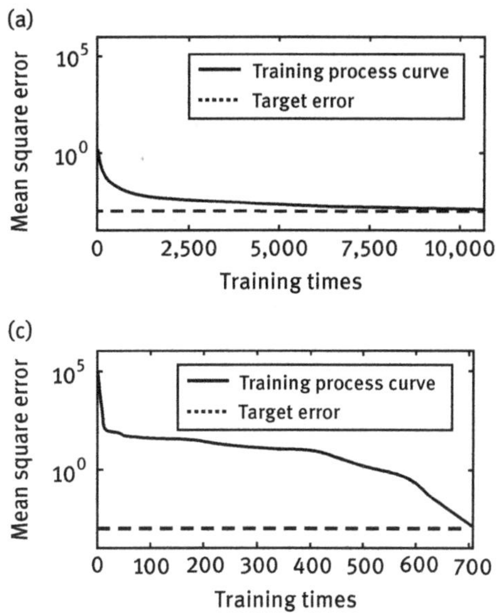

Figure 5.11: Training process of the neural network for the pit depth quantization: (a) 3D MFL signal and traditional algorithm; (b) 1D MFL signal and Bayesian algorithm; and (c) 3D MFL signal and Bayesian algorithm.

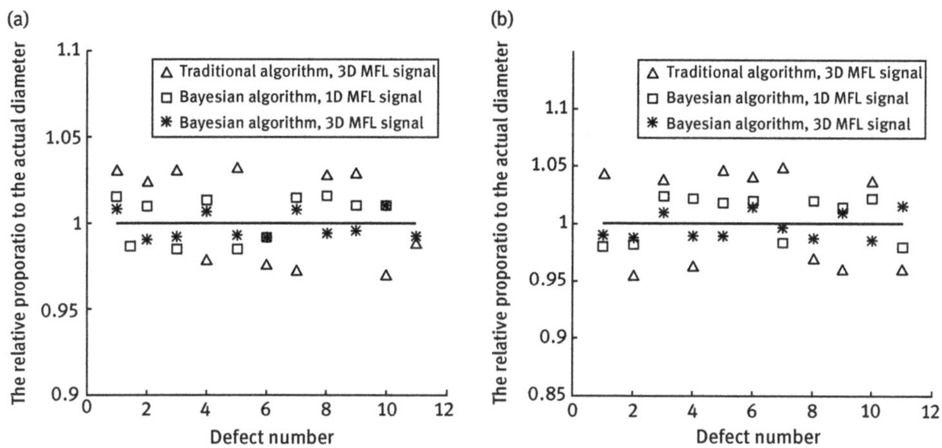

Figure 5.12: The relative proportions of the quantization results with the pit actual size: (a) diameter and (b) depth.

neural network based on Bayesian algorithm. As shown in Figure 5.12, in the quantitative results of the test samples, the quantization error of the traditional BP neural network is generally higher than that of the BP neural network based on the Bayesian algorithm. These results show that BP neural network based on Bayesian algorithm

Table 5.11: Maximum relative error of pit quantization.

Defect parameter	3D MFL signal traditional algorithm (%)	1D MFL signal Bayesian algorithm (%)	3D MFL signal Bayesian algorithm (%)
Diameter	3.12	1.59	0.97
Depth	4.78	2.26	1.54

can improve the quantitative accuracy of the pit by establishing a better mapping relationship between the defect 3D MFL characteristics and the defect geometry.

In addition, in the defect quantification of the BP neural network based on Bayesian algorithm, the maximum relative error of the defect diameter and depth using 3D MFL signal is smaller than that of 1D MFL signal. It is also found that the relative error of the 3D MFL signal quantization is generally lower than that of the 1D MFL signal.

In view of horizontal grooves, the key characteristics of 3D MFL are extracted by using 150 samples. In the samples, 130 samples are selected for training the neural network, and the remaining 20 samples are used to verify the accuracy of the trained neural network.

Taking 0.001 as the target error of network training, Figures 5.13–5.15 give the training process of the neural network to quantify the length, width and depth of the

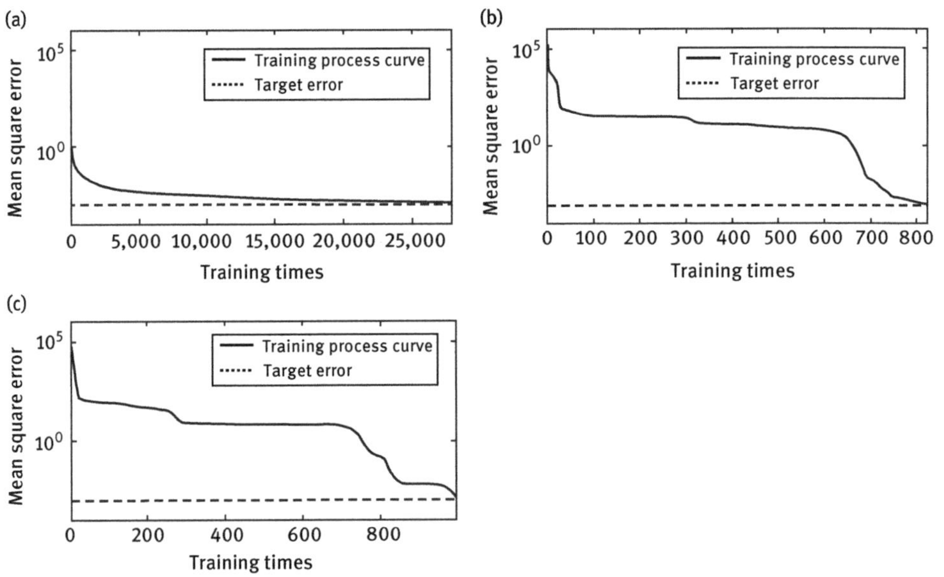

Figure 5.13: Training process of the neural network for the length quantization of the horizontal groove: (a) 3D MFL signal and traditional algorithm; (b) 1D MFL signal and Bayesian algorithm; and (c) 3D MFL signal and Bayesian algorithm.

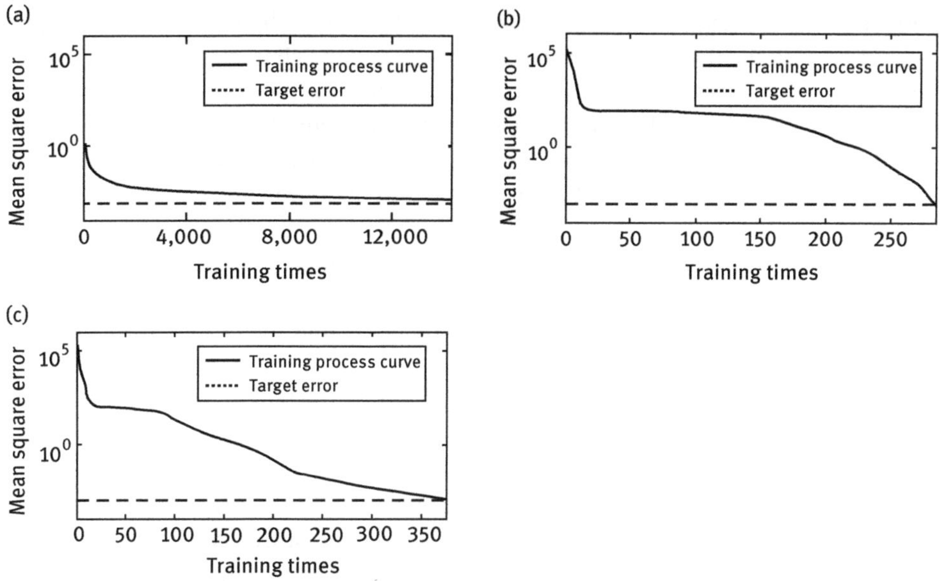

Figure 5.14: Training process of the neural network for the width quantization of the horizontal groove: (a) 3D MFL signal and traditional algorithm; (b) 1D MFL signal and Bayesian algorithm; and (c) 3D MFL signal and Bayesian algorithm.

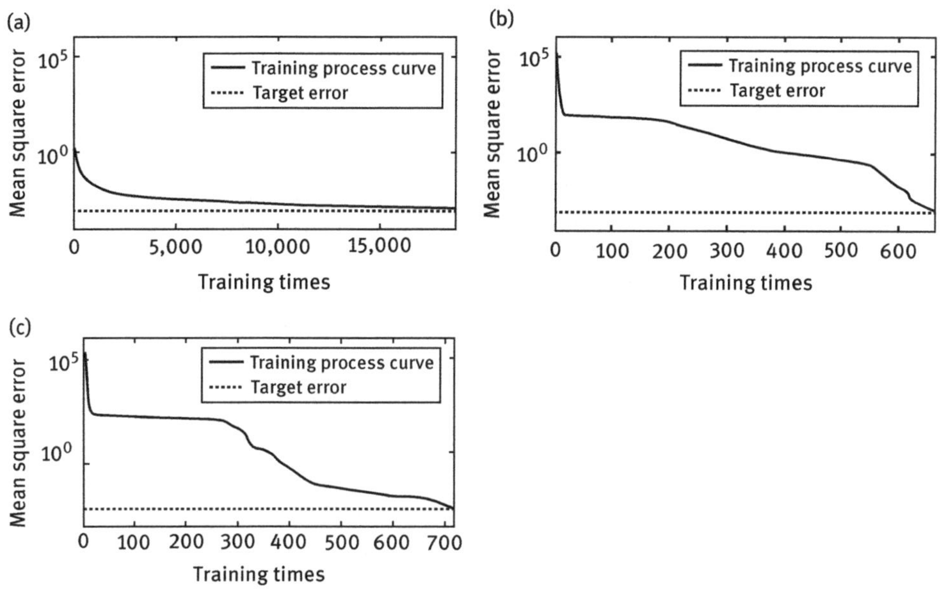

Figure 5.15: Training process of the neural network for the depth quantization of the horizontal groove: (a) 3D MFL signal and traditional algorithm; (b) 1D MFL signal and Bayesian algorithm; and (c) 3D MFL signal and Bayesian algorithm.

horizontal groove. Figures 5.13(a), 5.14(a) and 5.15(a) show the network training processes based on the traditional BP neural network and the characteristics of the defect 3D MFL signal. Figures 5.13(b), 5.14(b) and 5.15(b) show the network training processes of the BP neural network based on Bayesian algorithm and the characteristics of the defect 1D MFL signal. Figures 5.13(c), 5.14(c) and 5.15(c) show the network training processes of the BP neural network based on Bayesian algorithm and the characteristics of the defect 3D MFL signal. In the course of each training process, the broken line is used to identify the target error of neural network training.

The training processes of the traditional BP neural network and BP neural network based on the Bayesian algorithm can be compared using Figure 5.13(a), and 5.13(c), Figure 5.14(a) and 5.14(c), and Figure 5.15(a) and 5.15(c). In the training processes of the neural networks that are used to quantify the length, width and depth of the defect, the traditional BP neural network can achieve the expected goal by 28,931, 14,305 and 19,625 times, respectively, while BP neural network based on Bayesian algorithm achieves the expected goal only after 994, 375, 717 times trainings. This result verifies that the BP neural network based on Bayesian algorithm converges faster.

The training processes of neural network based on Bayesian algorithm using 1D MFL signals and 3D MFL signals can be compared by Figures 5.13(b) and 5.13(c), Figure 5.14(b) and 5.14(c), and Figure 5.15(b) and 5.15(c). In the training processes of neural networks that are used to quantify the length, width and depth of the defect, neural networks using 1D MFL signal characteristics reach the expected target after 824, 286, 666 times training, respectively, slightly less training times than that of neural networks using 3D MFL signal characteristics.

For horizontal grooves, Figure 5.16 shows the quantitative results of the length, width and depth of 20 test samples. Table 5.12 shows the maximum relative error of the quantitative results of the horizontal grooves.

From Table 5.12, we can see that the maximum quantitative relative errors of defect length, width and depth of the horizontal groove using the traditional BP neural network are larger than that of BP neural network based on Bayesian algorithm. As shown in Figure 5.16, in the quantitative results of the 20 test samples, the quantization error of the traditional BP neural network is generally higher than that of BP neural network based on the Bayesian algorithm.

In addition, in the defect quantification of the BP neural network based on Bayesian algorithm, the maximum relative error of the horizontal groove using 3D MFL

Table 5.12: Maximum relative error of horizontal groove quantification.

Defect parameter	3D MFL signal traditional algorithm (%)	1D MFL signal Bayesian algorithm (%)	3D MFL signal Bayesian algorithm (%)
Length	2.32	1.07	0.86
Width	7.98	4.52	2.13
Depth	5.98	2.69	3.03

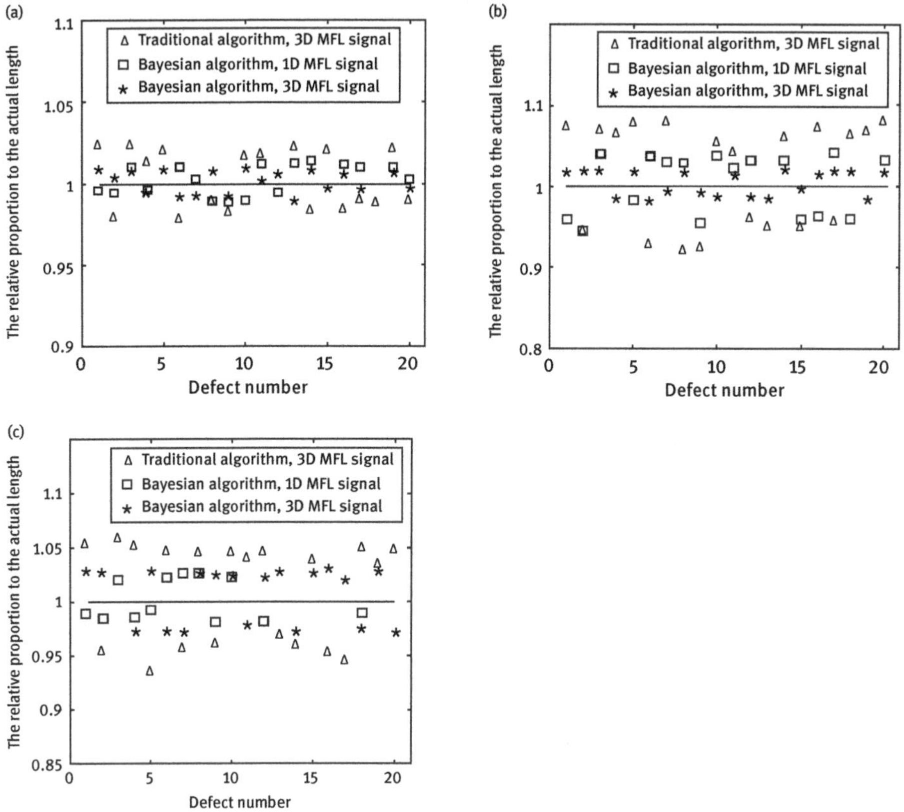

Figure 5.16: The relative proportions of the quantization results with the horizontal groove actual size: (a) length, (b) width and (c) depth.

signal is smaller than that of 1D MFL signal. From Figure 5.16, it is also found that the relative error of the 3D MFL signal quantization is generally lower than that of the 1D MFL signal because the 3D MFL signal contains more defects information.

In the 150 tangential groove samples, 130 are selected to train the neural network, and the remaining 20 are used to verify the quantitative accuracy of the trained neural network.

The training processes of the neural networks for the length, width and depth of the tangential grooves are presented in Figures 5.17–5.19. Figures 5.17(a), 5.18(a) and 5.19(a) show the training processes of traditional BP neural network using the characteristics of the 3D MFL signal. Figures 5.17(b), 5.18(b) and 5.19(b) show the training processes of BP neural network based on the Bayesian algorithm using the characteristics of the 1D MFL signal. Figures 5.17(c), 5.18(c) and 5.19(c) show the training processes of BP neural network based on the Bayesian algorithm using the characteristics of the 3D MFL signal.

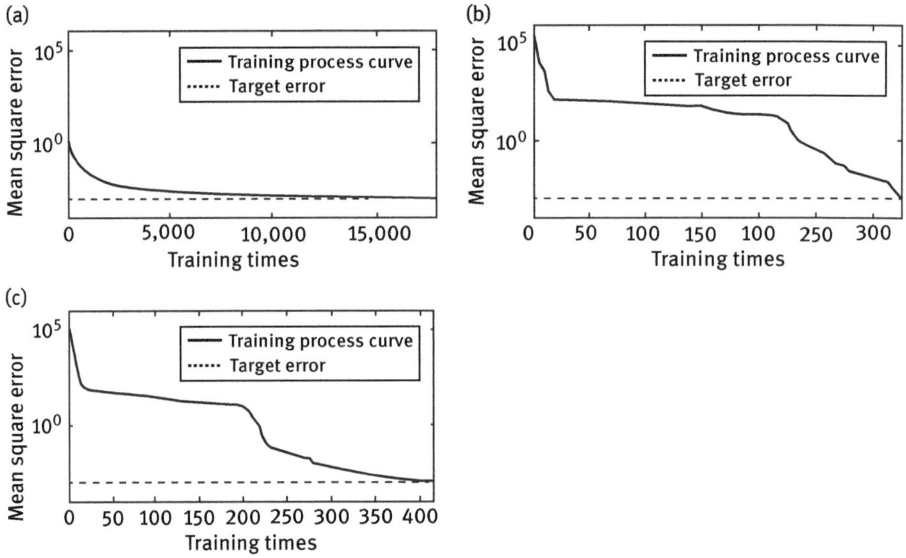

Figure 5.17: Training process of the neural network for the length quantization of the tangential groove: (a) 3D MFL signal and traditional algorithm; (b) 1D MFL signal and Bayesian algorithm; and (c) 3D MFL signal and Bayesian algorithm.

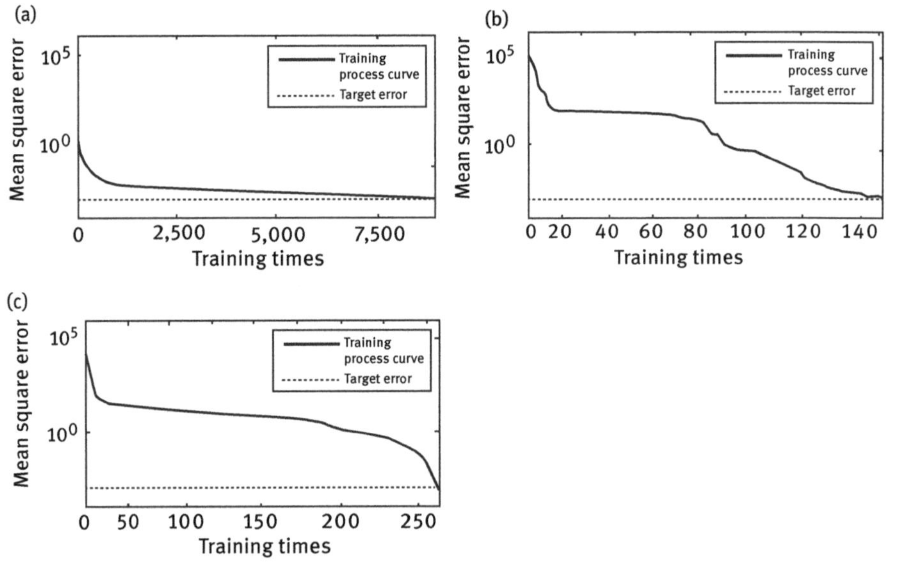

Figure 5.18: Training process of the neural network for the width quantization of the tangential groove: (a) 3D MFL signal and traditional algorithm; (b) 1D MFL signal and Bayesian algorithm; and (c) 3D MFL signal and Bayesian algorithm.

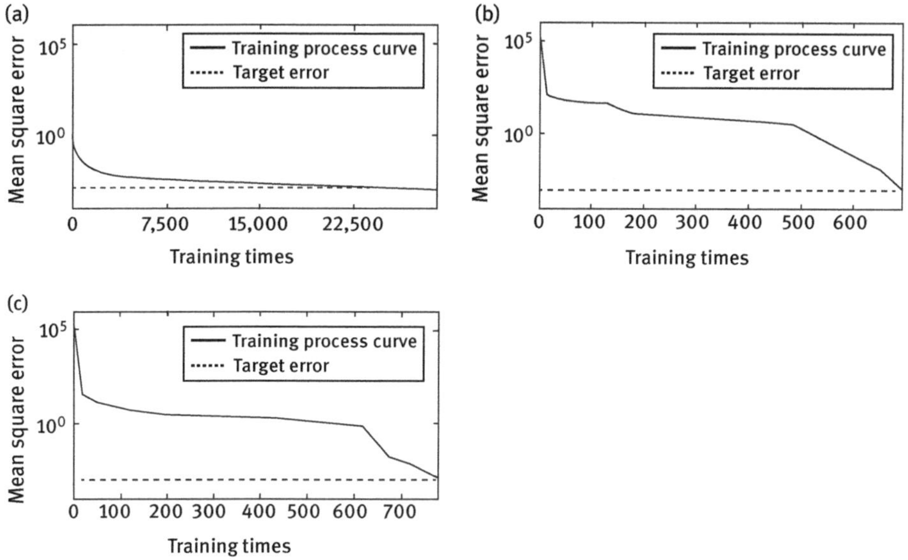

Figure 5.19: Training process of the neural network for the depth quantization of the tangential groove: (a) 3D MFL signal and traditional algorithm; (b) 1D MFL signal and Bayesian algorithm; and (c) 3D MFL signal and Bayesian algorithm.

Based on Figures 5.17(a), 5.17(c), 5.18(a), 5.18(c) and 5.19(a), 5.19(c), when the characteristics of the 3D MFL signals are adopted, the training time of traditional BP neural network and BP neural network based on Bayesian algorithm is compared. Traditional BP neural networks for the length, width and depth achieve the desired goal after 17,882, 9,185, 28,652 trainings, respectively. However, BP neural networks based on Bayesian algorithm achieve the desired goal after 416, 269, 779 trainings, respectively. This further verifies the fast convergence speed of BP neural network based on Bayesian algorithm. In addition, from Figures 5.17(b), 5.18(b) and 5.19(b), it can be seen that, in the training processes of neural networks that are used to quantify the length, width and depth of the tangential groove, neural networks using 1D MFL signal characteristics reach the expected target after 326, 146, 694 trainings, respectively, slightly less training times than that of neural networks using 3D MFL signal characteristics in Figures 5.17(c), 5.18(c) and 5.19(c). The reason is that the number of 3D MFL signal characteristics is more than that of 1D MFL signal characteristics, which increases the input variables of the neural network.

Figure 5.20 shows the quantitative results of lengths, widths and depths of 20 test samples. The horizontal coordinates represent the number of 20 samples, and the vertical coordinates represent the relative proportion of the quantitative result to the actual defect size. Table 5.13 shows the maximum relative error of the quantitative results of the tangential grooves under different conditions.

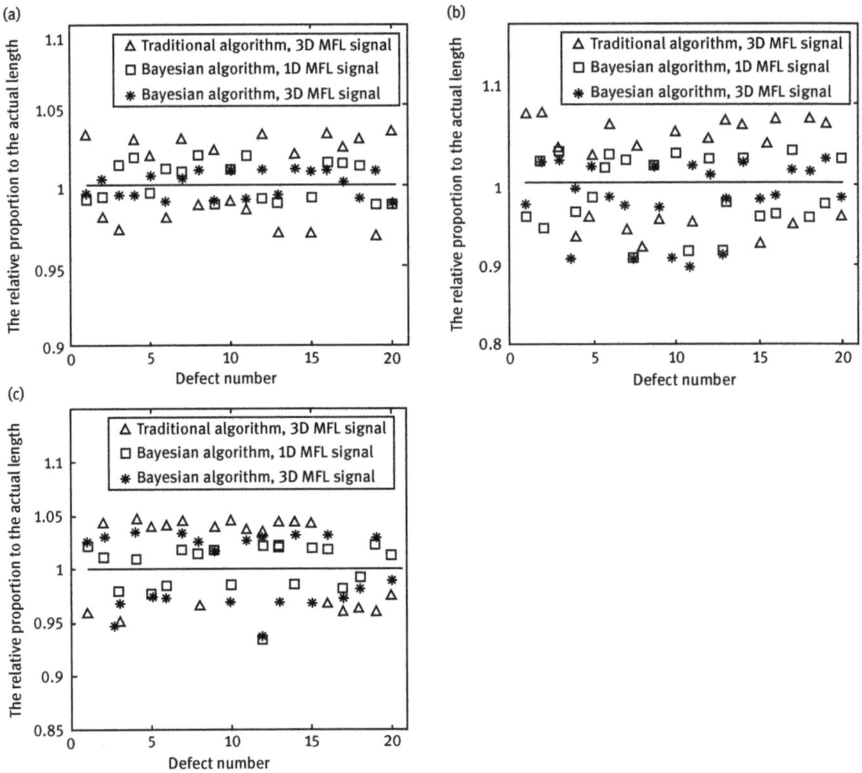

Figure 5.20: The relative proportions of the quantization results with the tangential groove actual size: (a) length, (b) width and (c) depth.

From the quantitative results in Figure 5.20 and Table 5.13, it can be seen that the results of tangential grooves are similar to those of the pits and horizontal grooves. The geometry quantization error of the traditional BP neural network is generally higher than that of BP neural network based on Bayesian algorithm. It is proved that the BP neural network based on Bayesian algorithm can effectively improve the quantitative accuracy of the tangential grooves.

Table 5.13: Maximum relative error of tangential groove quantification.

Defect parameter	3D MFL signal traditional algorithm (%)	1D MFL signal Bayesian algorithm (%)	3D MFL signal Bayesian algorithm (%)
Length	3.32	1.86	1.02
Width	8.65	5.49	3.01
Depth	4.48	2.45	3.51

When tangential grooves are quantified by using BP neural network based on Bayesian algorithm, the quantization errors of defect length and width using 1D MFL signals are higher than those of 3D MFL signals. It is verified that the 3D MFL signal contains more defect information, which can effectively improve the accuracy of the defect quantification.

5.3 Quantization and Display Method of Defect Under Incomplete Signal

In this section, we will use the detected 3D MFL signal to obtain the geometric size of the defect so as to realize the real-time display of the defect profile.

In the defect detection process, the data processing region may not include the whole-defect 3D MFL signal, which will not be able to accurately realize the classification and quantification of defects. In order to solve this problem, a fast defect contour display method is proposed in the case of the incomplete 3D MFL signal that has been obtained. Through the identification of the defect edge and the estimation of the defect depth, the defect real-time display under the condition of incomplete 3D MFL signal can be realized with a certain allowable error.

First, in order to identify the defect edge from the incomplete 3D MFL signal, a defect edge identification method based on Sobel discrete difference operator is proposed. By using the Sobel operator, the gradient of the 3D MFL signal is solved. Using the method of threshold interception, the defect edge points based on 3D MFL signal are obtained. And then, curve fitting is performed to obtain the final defect edge.

Second, in order to realize the estimation of defect depth under the incomplete 3D MFL signal, a fast estimation method of defect depth based on signal equivalent processing is proposed. By the equivalent treatment of defect-incomplete 3D MFL signals, an equivalent complete 3D MFL signal is obtained, and the key characteristic values are extracted as the input signal of the depth quantization neural network.

Finally, based on the defect edge and depth estimated from the incomplete 3D MFL signal, the real-time display method of the defect is summarized. The defect classification process is summarized, and for the pit, the horizontal groove and the tangential groove, the method for obtaining defect profile in defect real-time display is given. The feasibility of real-time display method is verified by real-time display experiment.

5.3.1 Defect Edge Recognition

Under the incomplete 3D MFL signal, the data processing area contains only part of the defect MFL signal. At the same time, every time the detector completes a scan, and the data queue in the data processing area is updated. In the process of defect inversion, the data of the static MFL field in the data processing area is still processed.

The real-time display of the defect testing results can be obtained by the real-time update after each scan.

In the MFL detection process, magnetic flux lines will mutate in the defect edge. Therefore, the mutation point of 3D MFL detection signal can reflect the edge information of the detected defect. According to the above, the defect edge can be detected by using the gradient value of the 3D MFL signal.

In view of the discontinuity of the signal in the data processing area, the gradient value of each point in the data processing area is calculated by using the Sobel discrete difference operator. First, according to the signal point (x, y) in the data processing area, the gradient values $G_x(x, y)$ and $G_y(x, y)$ along the horizontal direction and the tangential direction are obtained, respectively. The horizontal edge identification gradient value G_x can be calculated by the weighted difference of the left column points $(x-1, y-1)$, $(x-1, y)$ and $(x-1, y+1)$ and the right column points $(x+1, y-1)$, $(x+1, y)$ and $(x+1, y+1)$. Differential operation can be used to replace the first-order partial derivative operation. The tangential edge identification gradient value G_y can be calculated by the weighted difference of the top row points $(x-1, y-1)$, $(x, y-1)$ and $(x+1, y-1)$ and the bottom row points $(x-1, y+1)$, $(x, y+1)$ and $(x+1, y+1)$. From this, we can find out

$$\begin{aligned} G_x(x, y) = & (-1) \times f(x-1, y-1) + 0 \times f(x, y-1) + 1 \times f(x+1, y-1) \\ & + (-2) \times f(x-1, y) + 0 \times f(x, y) + 1 \times f(x+1, y) \\ & + (-1) \times f(x-1, y+1) + 0 \times f(x, y+1) + 1 \times f(x+1, y+1) \end{aligned} , \tag{5.26}$$

$$\begin{aligned} G_y(x, y) = & 1 \times f(x-1, y-1) + 2 \times f(x, y-1) + 1 \times f(x+1, y-1) \\ & + 0 \times f(x-1, y) + 0 \times f(x, y) + 0 \times f(x+1, y) \\ & + (-1) \times f(x-1, y+1) + (-2) \times f(x, y+1) + (-1) \times f(x+1, y+1) \end{aligned} . \tag{5.27}$$

According to the gradient values $G_x(x, y)$ and $G_y(x, y)$ along the horizontal direction and the tangential direction, the gradient $G(x, y)$ and the gradient direction $\theta(x, y)$ of the signal point (x, y) can be calculated:

$$G(x, y) = \sqrt{G_x(x, y)^2 + G_y(x, y)^2}, \tag{5.28}$$

$$\theta(x, y) = \arctan\left(G_y(x, y) / G_x(x, y)\right). \tag{5.29}$$

In the processing of all signal points **A** in the data processing area, Sobel convolution factors $\mathbf{G_x}$ and $\mathbf{G_y}$ along the horizontal direction and tangential direction perform convolution to achieve difference operation along the horizontal direction and the tangential direction:

$$\mathbf{G_x} = \begin{bmatrix} -1 & 0 & 1 \\ -2 & 0 & 0 \\ -1 & 0 & 1 \end{bmatrix}, \mathbf{G}_y = \begin{bmatrix} 1 & 2 & 1 \\ 0 & 0 & 0 \\ -1 & -2 & -1 \end{bmatrix}, \tag{5.30}$$

$$G_x(\mathbf{A}) = \mathbf{G_x} * \mathbf{A} = \begin{bmatrix} -1 & 0 & 1 \\ -2 & 0 & 0 \\ -1 & 0 & 1 \end{bmatrix} * \mathbf{A}, \tag{5.31}$$

$$G_y(\mathbf{A}) = \mathbf{A} * \mathbf{G}_y = \mathbf{A} * \begin{bmatrix} 1 & 2 & 1 \\ 0 & 0 & 0 \\ -1 & -2 & -1 \end{bmatrix}. \tag{5.32}$$

Thus, the gradient $G(\mathbf{A})$ and gradient direction $\theta(\mathbf{A})$ of all signal points in the data processing area can be calculated:

$$G(\mathbf{A}) = \sqrt{G_x(\mathbf{A})^2 + G_y(\mathbf{A})^2}, \tag{5.33}$$

$$\theta(\mathbf{A}) = \arctan\left(\frac{G_y(\mathbf{A})}{G_x(\mathbf{A})}\right). \tag{5.34}$$

Using the above method, the gradients of the 3D MFL signals of pits, horizontal grooves and tangential grooves are solved. Figures 5.21–5.23 show the gradient distributions of 3D MFL signals of the three types of defects, respectively.

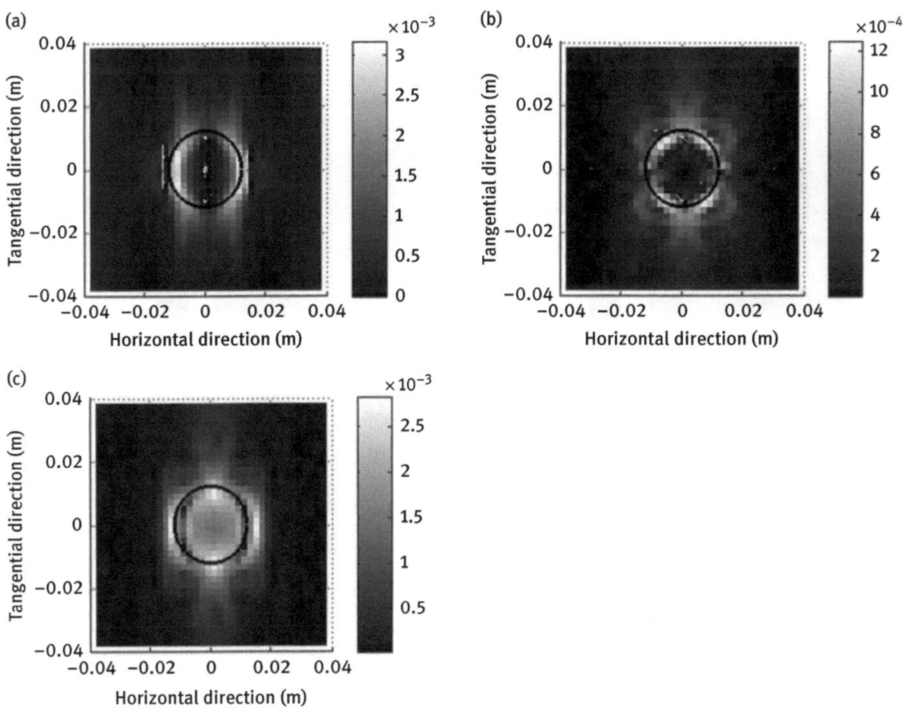

Figure 5.21: Gradient distribution of 3D MFL signal of pit 24 mm × 24 mm × 2.4 mm: (a) horizontal component; (b) tangential component; and (c) normal component.

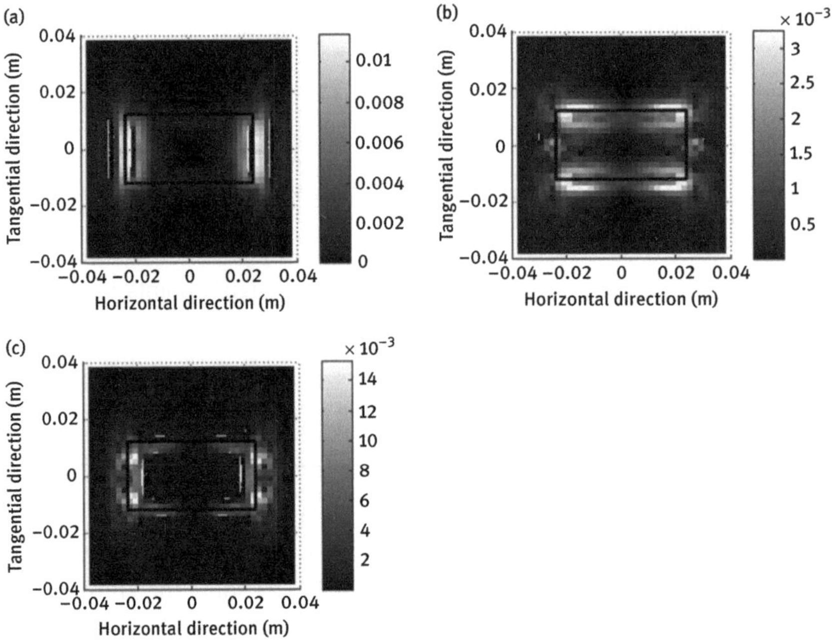

Figure 5.22: Gradient distribution of 3D MFL signal of horizontal groove 48 mm × 24 mm × 1.2 mm: (a) horizontal component; (b) tangential component; and (c) normal component.

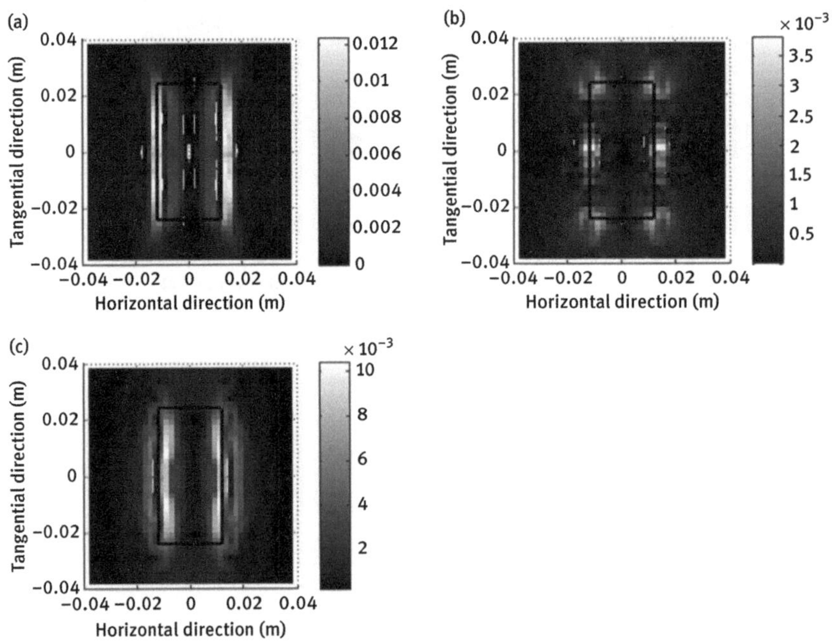

Figure 5.23: Gradient distribution of 3D MFL signal of tangential groove 24 mm × 48 mm × 1.2 mm: (a) horizontal component; (b) tangential component; and (c) normal component.

In Figures 5.21–5.23, the black lines are the actual edges of the defects. It can be seen that the gradient peak distribution region of the 3D MFL signal has a good agreement with the edge of the defect, but there is a certain difference between the horizontal, tangential and normal gradient peaks.

In order to obtain the final edge of the defect, the peak area distribution of the gradient is extracted according to the MFL signals in horizontal, tangential and normal directions. Setting 80 %, 65 % and 75 % of the gradient signal peak–valley values as the thresholds of the interception, the edge recognition results of the pit, the horizontal groove and the tangential groove are shown in Figures 5.24, 5.25 and 5.26, respectively.

It can be seen that the recognition results of the edge points in the three directions are basically in the actual edge, and the positions of the edge points in each direction are different. Therefore, the 3D edge points are synthesized, and the coverage of the synthesized edge points is larger than the actual defect edge. From Figures 5.24(d), 5.25(d) and 5.26(d), it can be seen that the synthesis recognition edge along the tangential direction is more accurate. The reason is that the magnetization direction is along the horizontal direction and the magnetic field changes in that direction is more obvious, so the tangential direction edge recognition will be clearer.

In order to obtain complete information about the defect edge, curve fitting of the outer edge points in the 3D synthetic edge recognition results is proceeded, which

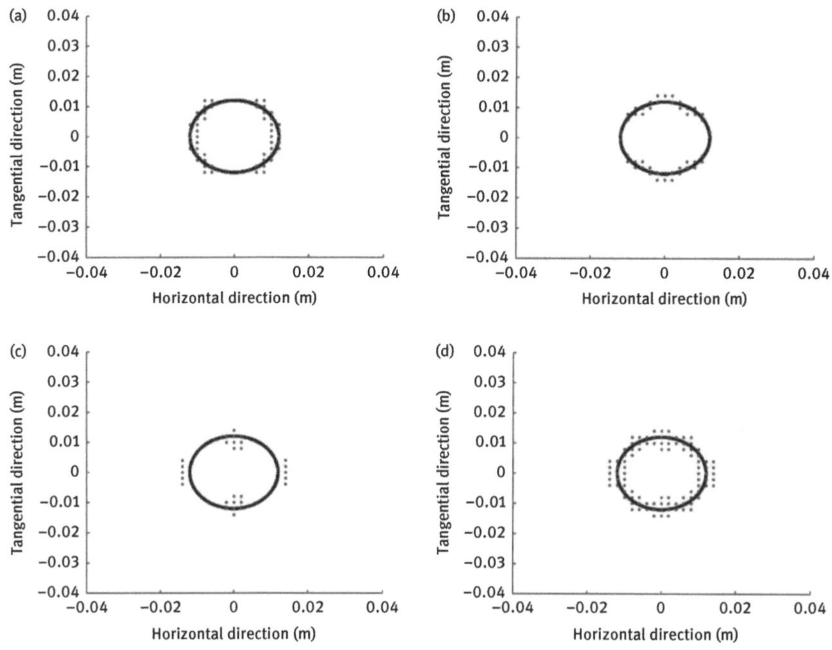

Figure 5.24: Edge point recognition of pit 24 mm × 24 mm × 2.4 mm: (a) horizontal component; (b) tangential component; (c) normal component; and (d) 3D synthesis.

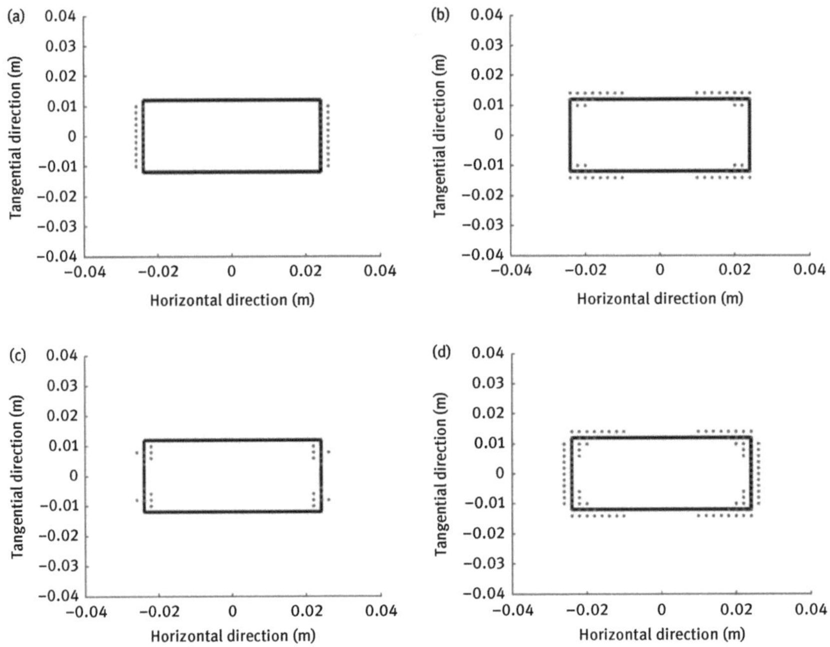

Figure 5.25: Edge point recognition of horizontal groove 48 mm × 24 mm × 1.2 mm: (a) horizontal component; (b) tangential component; (c) normal component; and (d) 3D synthesis.

gets a closed area as the recognized defect, as the gray areas shown in Figures 5.27(a), 5.28(a) and 5.29(a). Figures 5.27(b), 5.28(b) and 5.29(b) show the contrast between the recognized edge and the actual one.

From the comparison of recognized defect edge and actual defect edge, the defect edge recognition results based on 3D signal gradient are very similar to the actual defect edges. In addition, from Figures 5.27 to 5.29, we can see that the recognized edges are only 1 grid (2 mm) beyond the actual edges. Compared to the detector sampling interval, the error can be neglected.

In order to validate the proposed defect edge detection method for incomplete defect signals, the defect edge recognitions of pits, horizontal grooves and tangential grooves under incomplete defects MFL signals are tested, respectively. For each type of defects, there are 3D MFL signals of 10 %, 30 %, 60 %, 90 % defects in the data processing region, respectively. Under the condition, the defects edges are recognized. In Figure 5.30, the defect region identification results of 30 mm × 30 mm × 2 mm pit defect in 10 %, 60 %, 30 %, 90 % are presented. Figure 5.31 shows the comparison between the recognized edges and the actual edges.

From Figures 5.30 and 5.31, when the data processing region includes 10 %, 30 %, 60 %, 90 % incomplete defect 3D MFL signals, the recognized defect edge is consistent with the actual edge. This shows that the edge recognition method can accurately

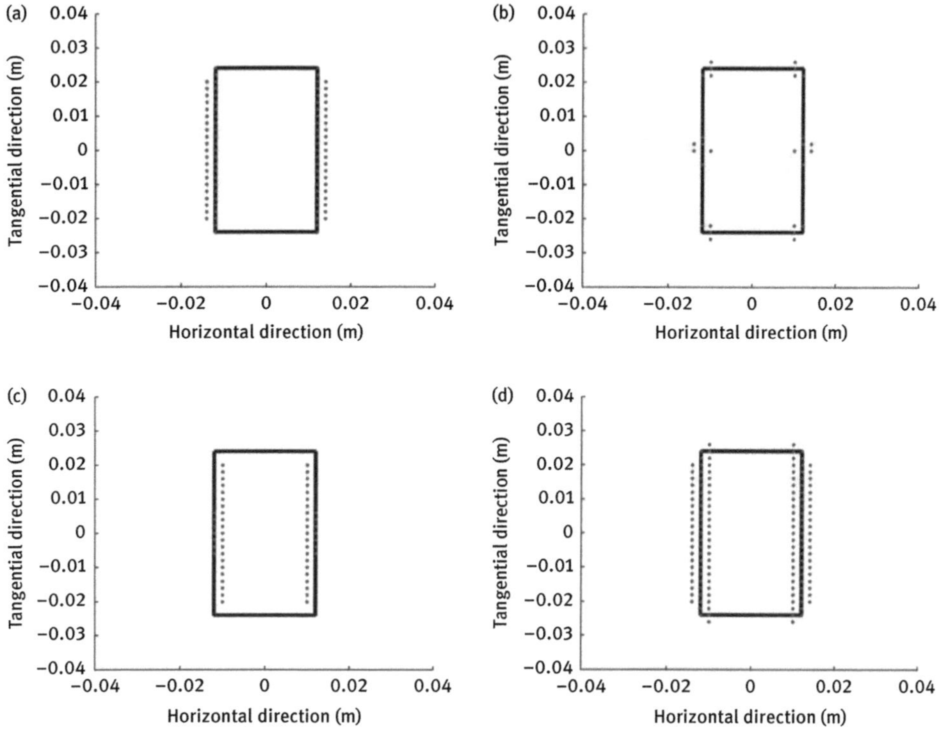

Figure 5.26: Edge point recognition of tangential groove 24 mm × 48 mm × 1.2 mm: (a) horizontal component; (b) tangential component; (c) normal component; and (d) 3D synthesis.

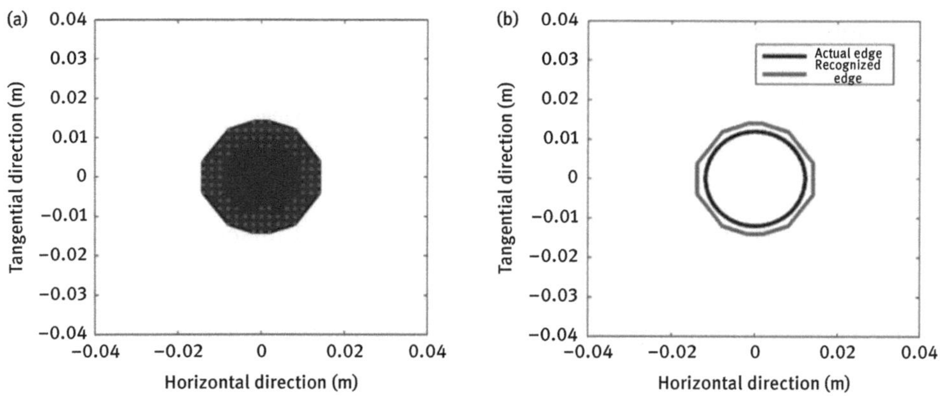

Figure 5.27: Edge recognition result of pit 24 mm × 24 mm × 2.4 mm: (a) defect area identification result and (b) comparison of the recognized edge and the actual edge.

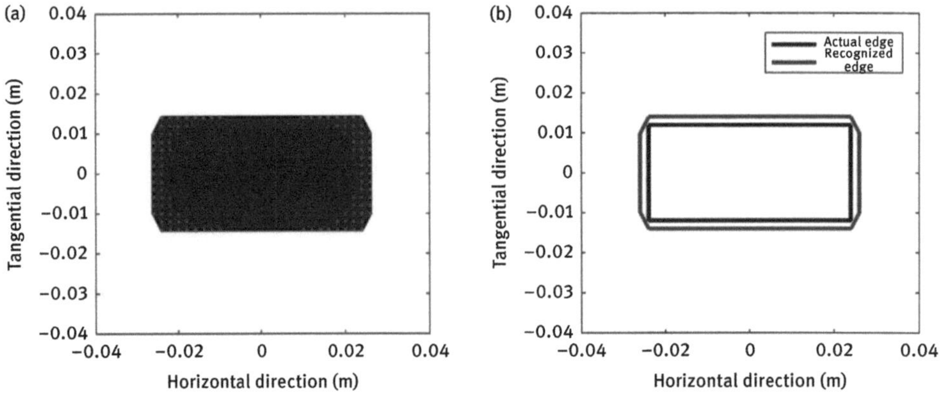

Figure 5.28: Edge recognition result of horizontal groove 48 mm × 24 mm × 1.2 mm: (a) defect area identification result and (b) comparison of the recognized edge and the actual edge.

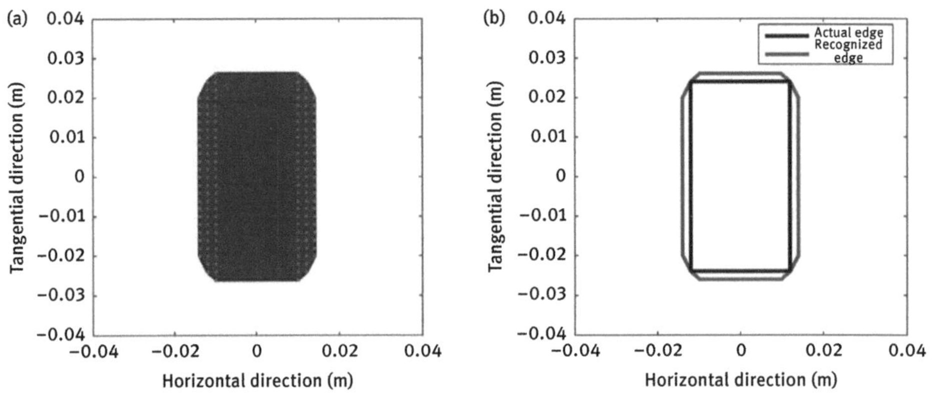

Figure 5.29: Edge recognition result of tangential groove 24 mm × 48 mm × 1.2 mm: (a) defect area identification result and (b) comparison of the recognized edge and the actual edge.

identify the edge of the corresponding part of the defect under the incomplete 3D MFL signal.

It is worth noting that there are two problems when the defect edge detection method is used to identify the defect edge under incomplete signal.

First, when the gradient signal at the edge does not reach the peak, the percentage of peak–valley value cannot be used as a threshold, such as the interception of the threshold directly, the nondefect location will be identified as a defect. If the interception threshold is directly used, the position of the nondefect can be identified as a defect. The solution for this problem is that a gradient threshold is set for the judgment of defect initiation in the gradient signal. When the gradient signal reaches the gradient threshold, the threshold intercepting method is adopted for edge recognition.

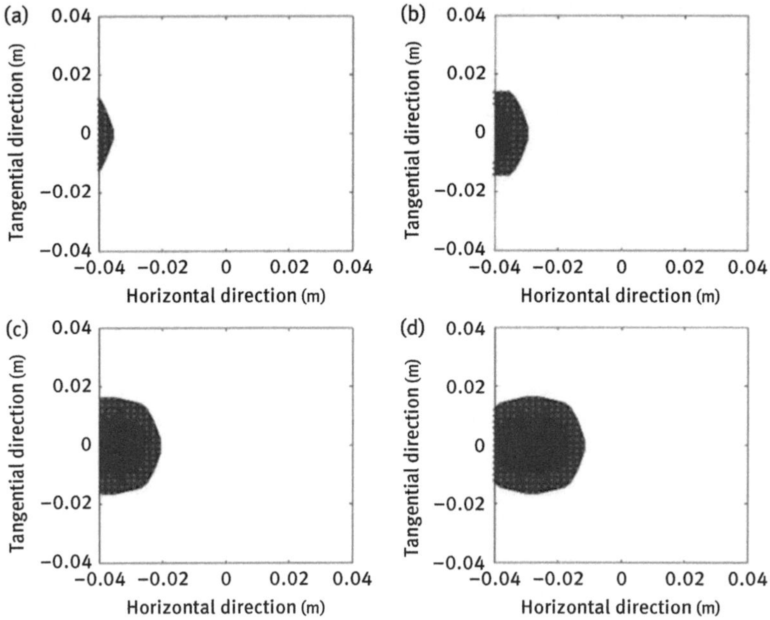

Figure 5.30: Defect region identification results of 30 mm × 30 mm × 2 mm pit: (a) 10 % 3D MFL signal; (b) 30 % 3D MFL signal; (c) 60 % 3D MFL signal; and (d) 90 % 3D MFL signal.

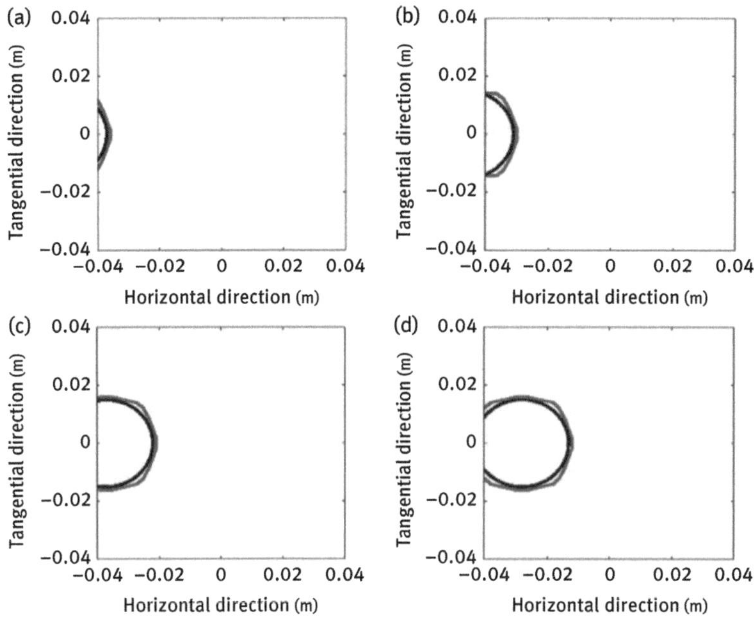

Figure 5.31: Comparison between recognized edge and actual edge of 30 mm × 30 mm × 2 mm pit: (a) 10 % 3D MFL signal; (b) 30 % 3D MFL signal; (c) 60 % 3D MFL signal; and (d) 90 % 3D MFL signal.

Because 3D MFL signal changes very fast at the edge of the defect, the gradient signal can reach the peak value in a short time. Experiments show that the method may cause the error of two grid points (4 mm) in the initial stage of edge recognition. Compared to the detector sampling interval, the error can be neglected.

Second, when the data processing area contains less than 50 % of the defect 3D MFL signal, the defect area cannot be determined by the two boundary points along the horizontal direction. The solution to solve the problem is that the defect edge within the 50 % defect 3D MFL signal will be used to restrict the defect width, and the edge points of the defect width within the data processing area are bound together to determine the region of the defect.

In order to identify the 3D MFL signal of 50 % defect, the horizontal component of defect 3D MFL signal is tested. The last six column points in the data processing area are taken. The centerline of the tangential direction is taken as the symmetry axis, and the relative error of the corresponding data points at both sides of the symmetry axis is compared. If the relative error of 90 % or more data points is less than 10 %, it is considered that the 3D MFL signal of 50 % defect is identified.

Figures 5.33–5.35 show the defect region recognition results and the comparison between recognized edge and actual edge of 40 mm × 20 mm × 2 mm horizontal groove and 20 mm × 40 mm × 2 mm tangential groove. As shown from Figures 5.32–5.35, under the different percentage of 3D MFL signals, the edge recognition of the corresponding part of the defect is highly consistent with the actual defect edge.

5.3.2 Defect Depth Estimation

Under the premise of the known edge of defects, in order to achieve the real-time display of the corresponding defect part in incomplete signal, the defect depth is estimated by using the 3D MFL signal in the area of data processing. Due to the incompleteness of the signal, it is inevitable that there will be some errors in the estimation of the defect depth.

In Section 5.2, the neural network method is used to train and to calculate the depth of the defect based on complete 3D MFL signal with high accuracy. In order to satisfy the requirement of defect real-time display, the neural network is used to estimate the defect depth, which is based on incomplete 3D MFL signal. However, the neural network is used to estimate the depth of defect, which usually requires complete 3D MFL signals of defects. In order to make use of the fast characteristic of neural network, an equivalent treatment method of the incomplete 3D MFL signal is proposed, which is used to estimate the defect depth.

Before the equivalent treatment of the signal, the key characteristics of the complete 3D MFL signal of pits, horizontal grooves and tangential grooves used in the depth quantification are analyzed.

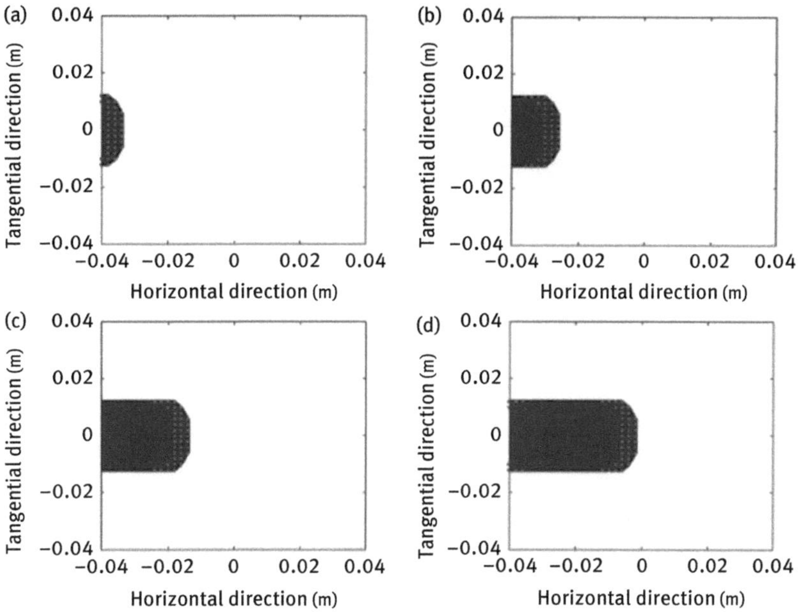

Figure 5.32: Defect area identification results of 40 mm × 20 mm × 2 mm horizontal groove: (a) 10 % 3D MFL signal; (b) 30 % 3D MFL signal; (c) 60 % 3D MFL signal; and (d) 90 % 3D MFL signal.

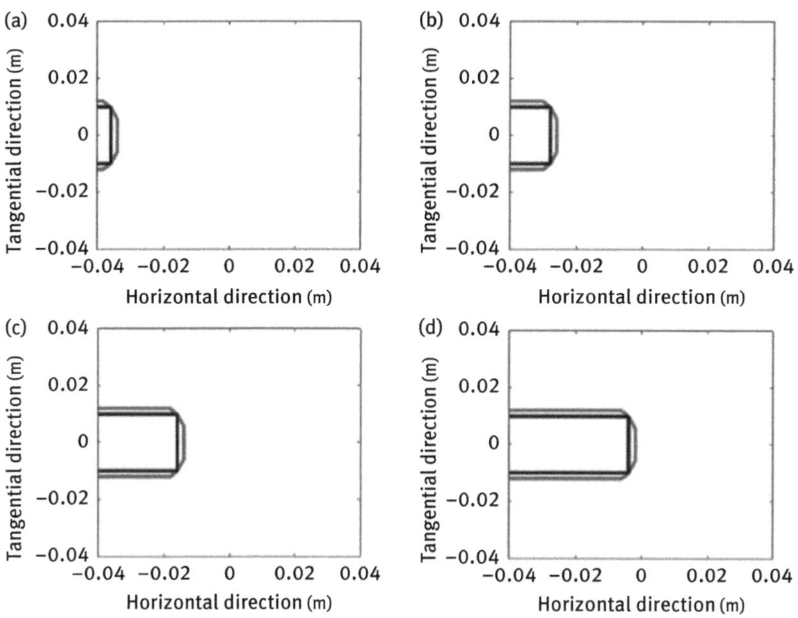

Figure 5.33: Comparison between recognized edge and actual edge of 40 mm ×20 mm × 2 mm horizontal groove: (a) 10 % 3D MFL signal; (b) 30 % 3D MFL signal; (c) 60 % 3D MFL signal; and (d) 90 % 3D MFL signal.

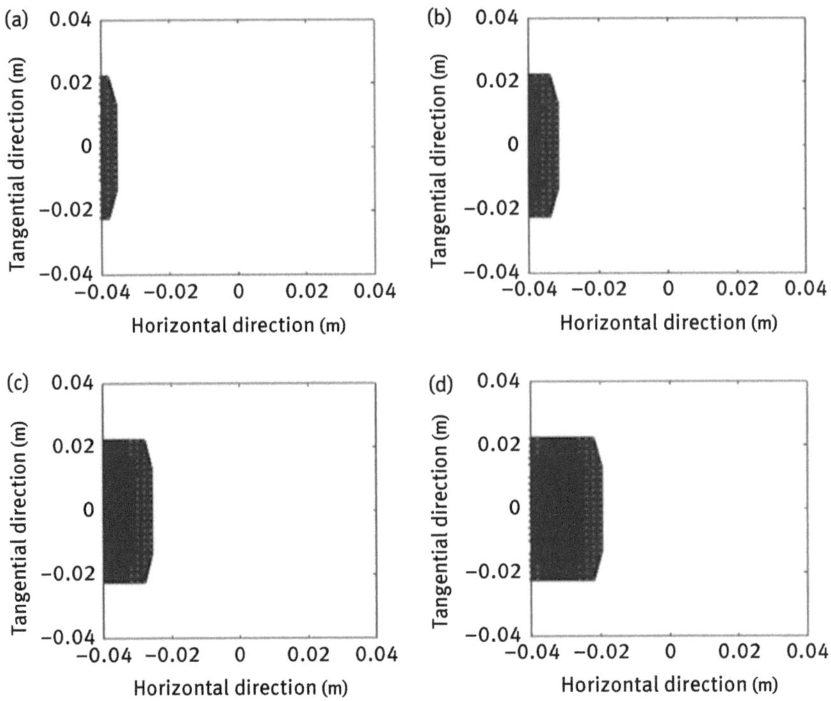

Figure 5.34: Defect area identification results of 20 mm × 40 mm × 2 mm tangential groove: (a) 10 % 3D MFL signal; (b) 30 % 3D MFL signal; (c) 60 % 3D MFL signal; and (d) 90 % 3D MFL signal.

In the depth quantification of the pit, the neural network input signals are the magnetic signal peak P_{h1} and the signal valley P_{h3} of the horizontal component, the intensity integral P_{t8} of the tangential component and the intensity integral P_{v5} of normal component of the complete 3D MFL signals. In the horizontal groove depth quantization process, neural network input signals are the magnetic signal peak H_{h1} and signal intensity integral H_{h9} of the horizontal component, the signal maxima H_{t4} of the tangential component, the signal intensity integral H_{v4} of the normal component and the horizontal first-order differential signal peak H_{v6} of the complete 3D MFL signals. In the tangential groove depth quantization, neural network input signals are the peak–valley T_{h5} and signal intensity integral T_{h10} of the horizontal component, the signal maxima T_{t1} of the tangential component, the signal intensity integral T_{v5} of the normal component and the horizontal signal peak T_{v7} of the complete 3D MFL signals.

For the input signals of the above depth quantization neural network, when the diameter of the defect, the length or width is constant, all the key characteristics, except the key characteristic P_{h3} of pit depth quantification, increase gradually with the increase of defect depth. The key characteristic P_{h3} of pit depth quantification decreases gradually with the increase of defect depth. In order to ensure the accuracy

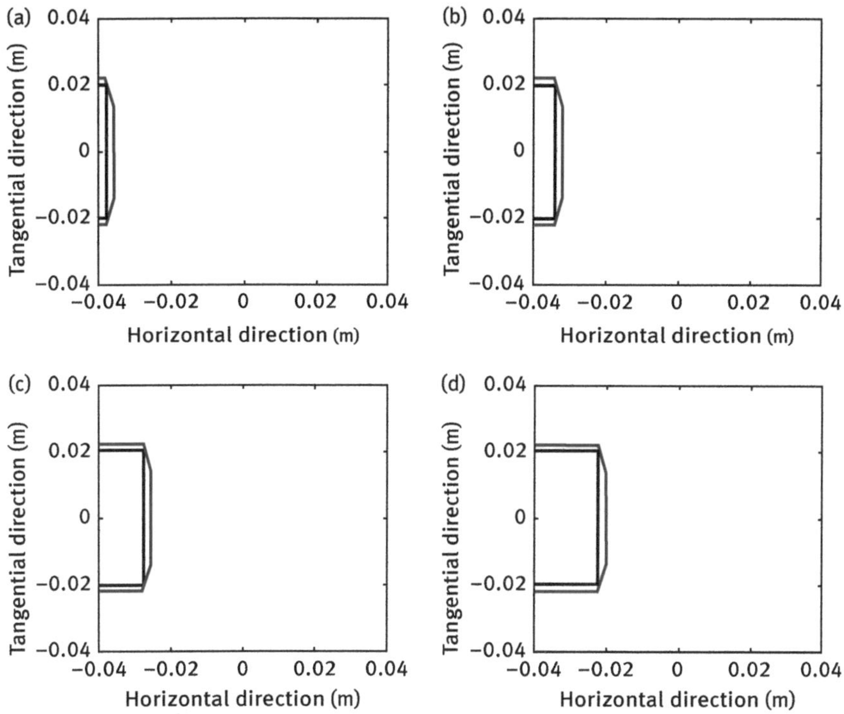

Figure 5.35: Comparison between recognized edge and actual edge of 20 mm × 40 mm ×2 mm tangential groove: (a) 10 % 3D MFL signal; (b) 30 % 3D MFL signal; (c) 60 % 3D MFL signal; and (d) 90 % 3D MFL signal.

of the estimation of defect depth increasing with the increase of the 3D MFL signal information, the 3D MFL signal equivalent processing for the incomplete 3D MFL signal should meet: With the increase of the amount of 3D MFL signal, all the key characteristics, except the key characteristic P_{h3} of pit depth quantification, should increase gradually.

Based on the above analysis, in the equivalent processing of the incomplete 3D MFL signal, when the defect signal is less than 50 % complete 3D MFL signal, the horizontal component of the incomplete 3D MFL signal will be processed with even symmetry along the tangential direction; The tangential component will be processed with even symmetry along the tangential plane, and then along the horizontal plane. The normal component will be processed with odd symmetry along the tangential plane.

Taking 24 mm × 24 mm × 2.4 mm pit for example, the horizontal component, the tangential component and the normal component of 10 %, 25 %, 40 % defect 3D MFL signals in the data processing region are equivalently processed, respectively. Figures 5.36–5.38 show the processing results.

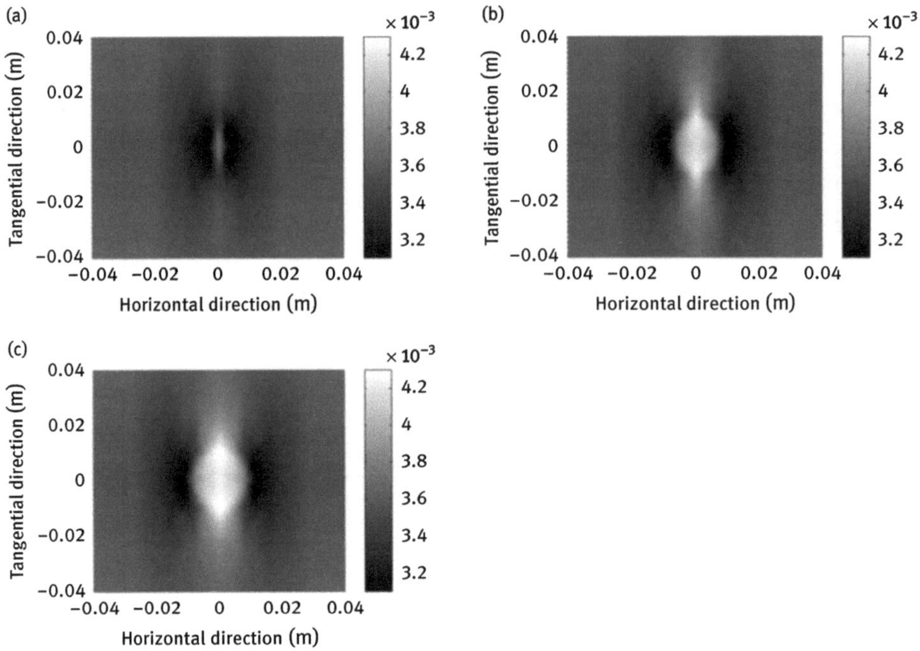

Figure 5.36: The equivalent processing results of the horizontal component of incomplete 3D MFL signal for 24 mm × 24 mm × 2.4 mm pit: (a) 10 % 3D MFL signal; (b) 25 % 3D MFL signal; and (c) 40 % 3D MFL signal.

These results show that, with the increase of defect 3D MFL signal information, the accuracy of the extracted key characteristics of 3D MFL signal also increases. When the data processing region includes 10 % defect 3D MFL signal, the peak and valley values of the tangential and normal components are equal to those of complete 3D MFL signal. With the gradual increase of the proportion of 3D MFL signal in the data processing area, the signal intensity integrals of the horizontal component, the tangential component and the normal component of the signal are increased. It can be seen that the estimation accuracy of the defect depth will be increased with the increase of the amount of information.

When the defect 3D MFL signal is more than 50 % of the complete defect signal, only 50 % of the horizontal component, tangential component and normal component of the defect-incomplete 3D MFL signal will be processed equivalently with even symmetry.

After the equivalent processing of the defect-incomplete 3D MFL signal, the key characteristics for the equivalent 3D MFL signals of pits, horizontal grooves and tangential grooves are extracted and inputted into the defect depth quantization neural network, which can obtain the corresponding defect depth.

In order to verify the effectiveness of the proposed 3D MFL signal equivalent treatment method, taking 30 mm × 30 mm × 2 mm pit as an example, when the data

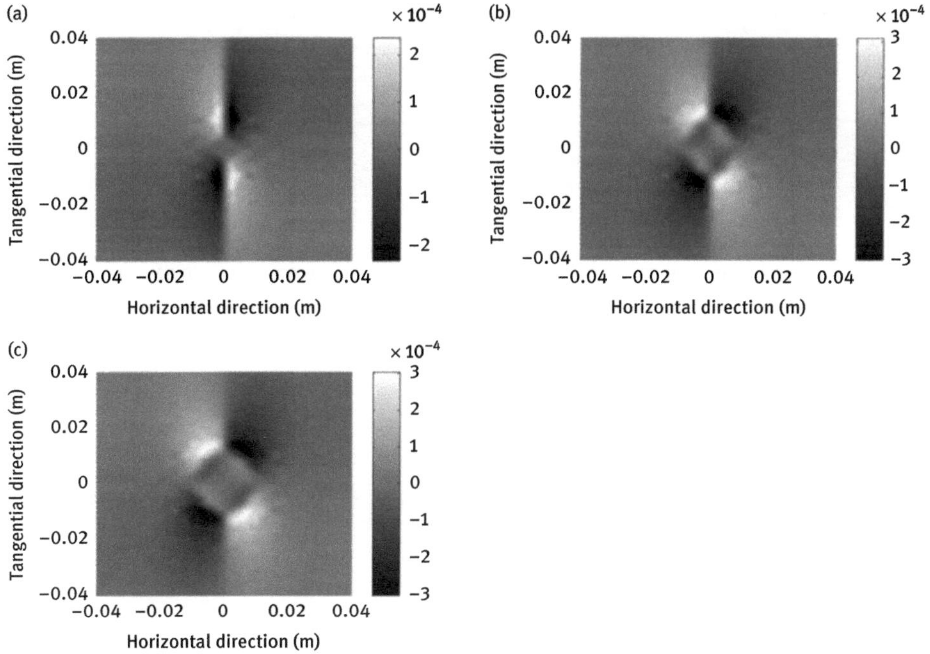

Figure 5.37: The equivalent processing results of the tangential component of incomplete 3D MFL signal for 24 mm × 24 mm × 2.4 mm pit: (a) 10 % 3D MFL signal; (b) 25 % 3D MFL signal; and (c) 40 % 3D MFL signal.

processing area contains different proportions of the complete 3D MFL signal, the signal is equivalently processed, and using the neural network, the defect depth of each stage is obtained. The variation of the error of the defect depth estimation in the real-time scanning process is shown in Figure 5.39.

As shown from Figure 5.39, with the gradual increase of the ratio to the complete MFL signal, the quantification error of the defect depth is gradually reduced. When the signal acquisition ratio reached 50 %, all the key features of the defect 3D MFL signal can be obtained and the depth quantification error will no longer change.

Figures 5.40 and 5.41 show the depth quantification error curves of 40 mm × 20 mm × 2 mm horizontal groove and 20 mm × 40 mm × 2 mm tangential groove, respectively. The data processing region contains different proportions of the complete 3D MFL signal.

From Figures 5.40 and 5.41, similar to the pit, the defect depth quantitative errors of horizontal grooves and tangential grooves decrease with the increase of the acquired ratio of 3D MFL signal by using equivalent processing of the defect-incomplete 3D MFL signal. When the signal ratio reaches 50 %, the defect depth quantification results with high accuracy are obtained. The results verify that the defect depth quantification method based on signal equivalent processing is feasible to quantify the defect depth of horizontal grooves and tangential grooves.

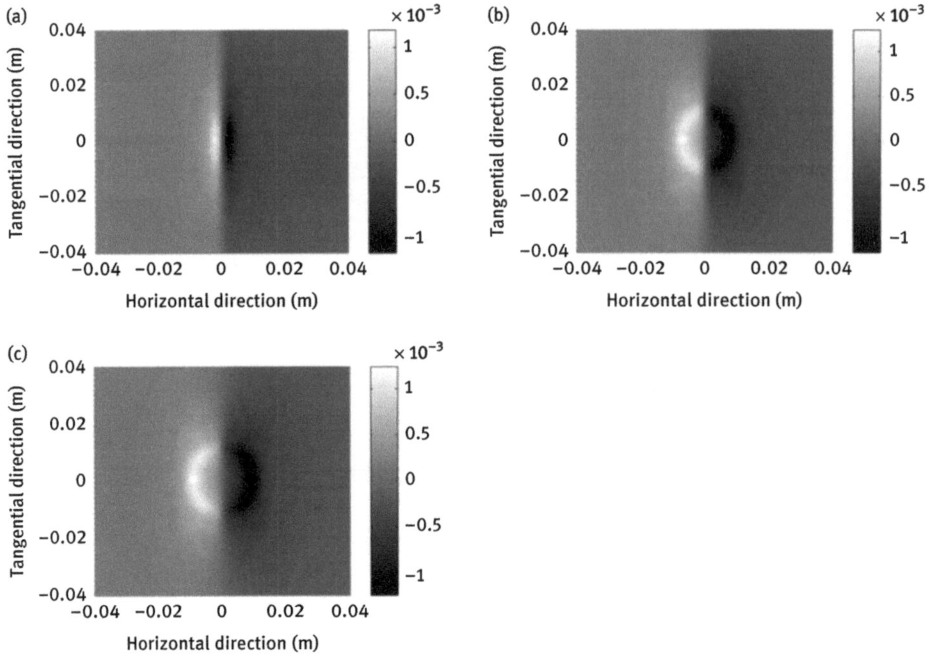

Figure 5.38: The equivalent processing results of the normal component of incomplete 3D MFL signal for 24 mm × 24 mm × 2.4 mm pit: (a) 10 % 3D MFL signal; (b) 25 % 3D MFL signal; and (c) 40 % 3D MFL signal.

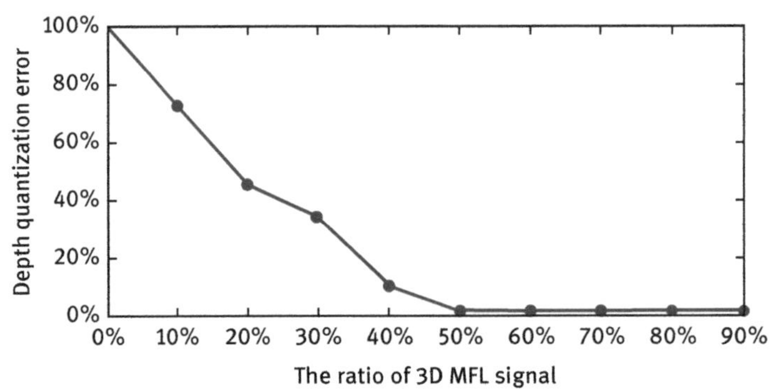

Figure 5.39: Depth quantization error of different ratio 3D MFL signal of 30 mm × 30 mm × 2 mm pit.

5.3.3 Real-Time Display of Defect Under Incomplete Signal

After the estimation results of defect edge recognition and the depth obtained from the incomplete signal, the specific defect real-time display method is given for the three types of defects.

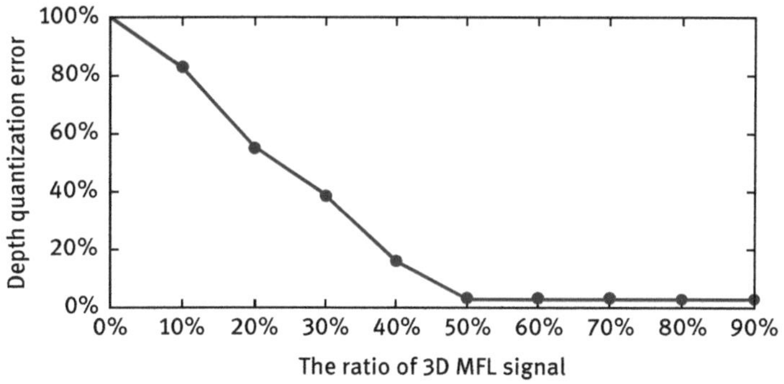

Figure 5.40: The depth quantification error curve of 40 mm × 20 mm ×2 mm horizontal groove with different proportions of the complete 3D MFL signal.

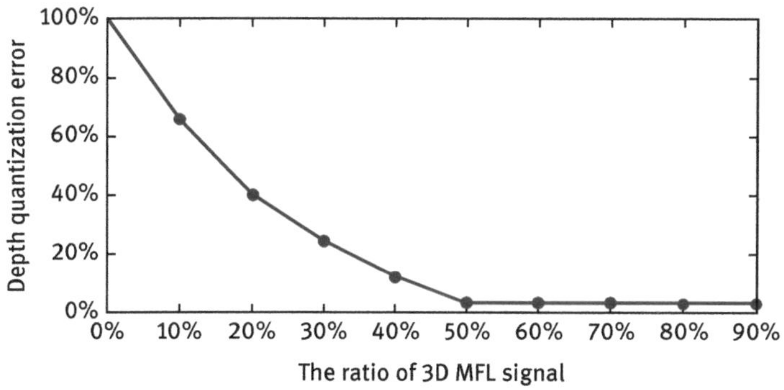

Figure 5.41: The depth quantification error curve of 20 mm × 40 mm ×2 mm tangential groove with different proportions of the complete 3D MFL signal.

Because the depth quantification neural networks for pits, horizontal grooves and tangential grooves are not the same, the types of the defects should be judged first before the defect depth is quantified. If the defect 3D MFL signal information is less enough, the defect types cannot be judged. So a defect judgment and display method in different amount of 3D MFL signal information is put forward.

The defect-type judgment is constantly updated with the scanning of the detector. Figure 5.42 shows the specific flowchart of the defect judgment. When the defect edge is identified for the first time in the data processing region, namely only the last column data containing the defect edge in the data processing area, defect depth quantification and defect display will begin as a pit defect. Along with the increase of the edge points in the data processing region, when there are three continuous columns of defect edge points not extended in the tangential direction, the defect will not be identified as a pit; instead the defect depth quantification and display will be

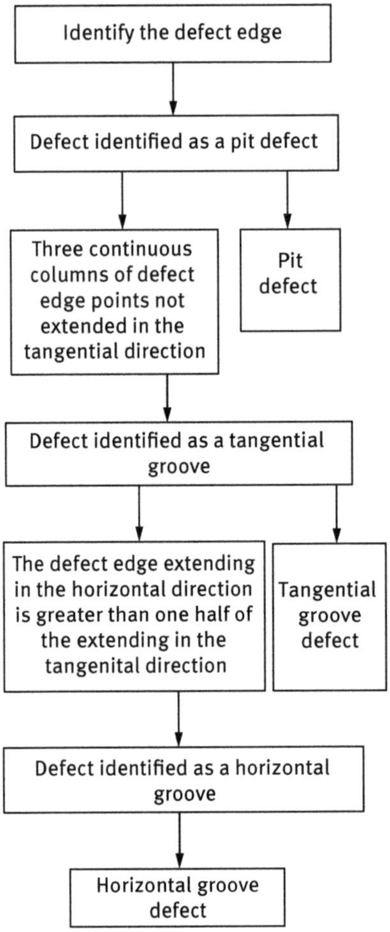

Figure 5.42: The flowchart of the defect judgment and display for incomplete MFL signal.

preceded as a tangential groove. Until the defect edge extending in the horizontal direction is greater than one-half of the extending in the tangential direction, the defect is identified as a horizontal groove and the defect depth quantification and display will proceed as a horizontal groove.

After the defect edge and depth are obtained, the outline of the defect will be calculated in accordance with the different defect types, respectively.

5.3.3.1 Outline of Pits

For the pit, the defect profile is obtained by solving the surface of the sphere determined by the defect edge and depth value. Figure 5.43 shows the sketch map for solving the relevant parameters of the spherical surface, where h is the depth of the defect, a is the distance between the two ends of the defect edge curve, b is the distance

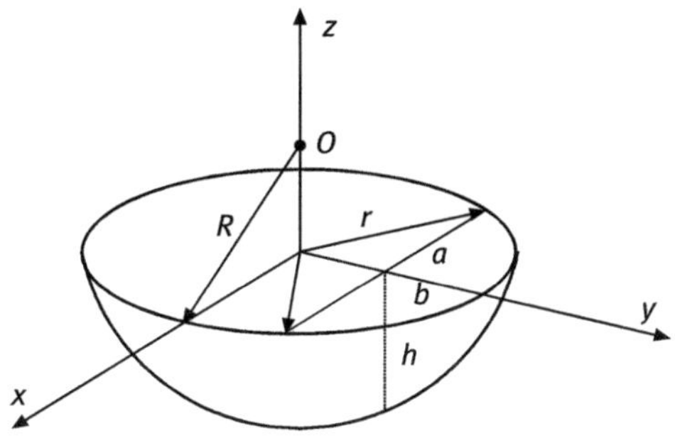

Figure 5.43: Solution of spherical-related parameters in the pit.

between the center of the edge curve and the center of two ends. Based on the above parameters, the radius r of the circle around the edge can be obtained:

$$r = \frac{b}{2} + \frac{a^2}{8b}. \tag{5.35}$$

Then, calculate the center position O and radius R of the spherical defect:

$$R = \sqrt{r^2 + \left(\frac{rb}{h} - \frac{b^2}{2h} - \frac{h}{2}\right)^2} = \sqrt{\left(\frac{b}{2} + \frac{a^2}{8b}\right)^2 + \left(\frac{a^2}{bh} - \frac{h}{2}\right)^2}. \tag{5.36}$$

The contour of the spherical defect can be calculated by the center position O and the radius R.

Figure 5.44 shows the schematic diagram of the display of the pit. Among them, Figure 5.44(a) shows the result when the 3D MFL signal is less than 50 %, and Figure 5.44(b) shows the result when the 3D MFL signal is more than 50 %.

5.3.3.2 Contours of Horizontal Grooves and Tangential Grooves

For the horizontal groove and the tangential groove, after the defect edge and the depth value are obtained, it will be directly displayed according to the corresponding

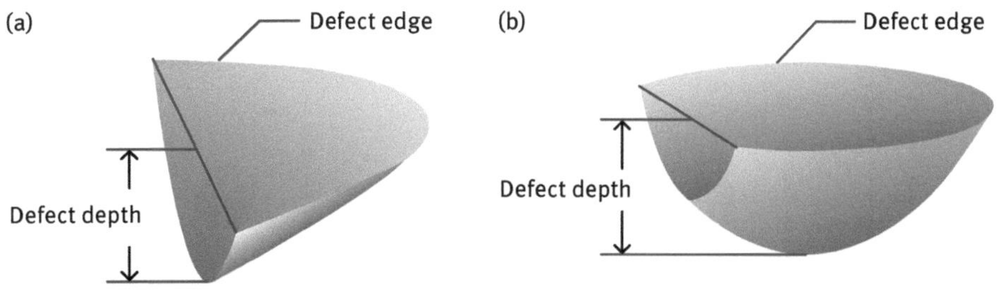

Figure 5.44: Schematic diagram of real-time display of the pit: (a) the defect 3D MFL signal is less than 50 % and (b) the defect 3D MFL signal is more than 50 %.

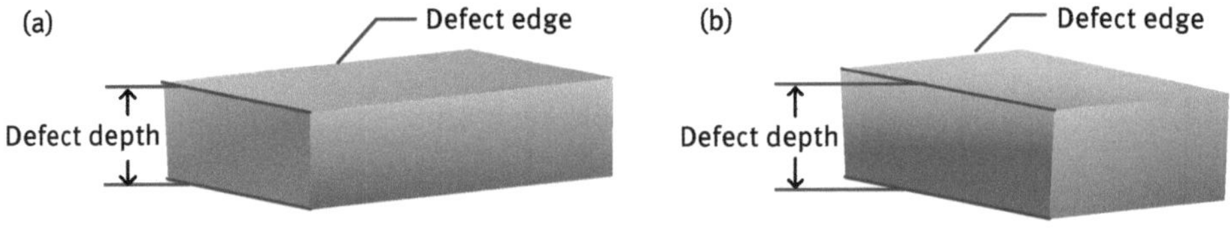

Figure 5.45: Schematic diagrams of the real-time display of the horizontal groove and the tangential groove: (a) horizontal groove and (b) tangential groove.

rectangular defect. Figure 5.45(a) and 5.45(b) shows the display defect diagrams of a horizontal groove and a tangential groove.

Based on the obtained defect edge and depth, the incomplete pit, horizontal groove and tangential groove are displayed in real time.

In order to validate the feasibility of the proposed pit defect display method based on the incomplete signal, Figures 5.46 and 5.47 show the actual defect and the real-time display results for 30 mm × 30 mm × 2 mm pit, whose signals are 10 %, 60 %, 30 %, 90 % of the complete MFL signal, respectively.

The above results show that the method can be used to display in real time the corresponding part of the defect with the known different ratio of the defect 3D MFL signal. At the same time, the real defect contours and the real-time display profiles in different proportions of the 3D MFL signal are basically the same, which verifies the defect quantification and real-time display method for the incomplete 3D MFL signal.

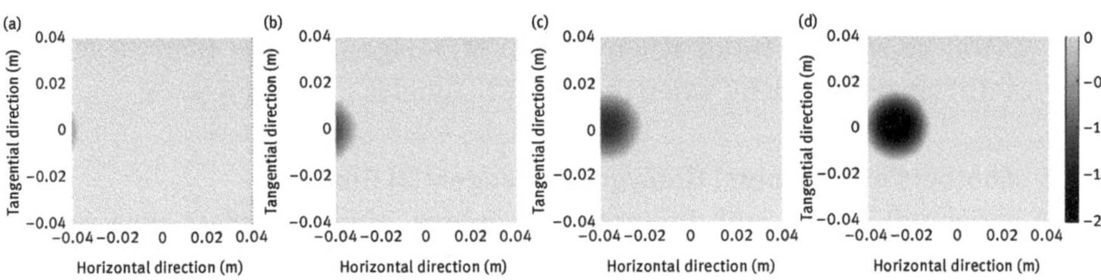

Figure 5.46: Actual defect of 30 mm × 30 mm × 2 mm pit: (a) 10 % MFL signal; (b) 30 % MFL signal; (c) 60 % MFL signal; and (d) 90 % MFL signal.

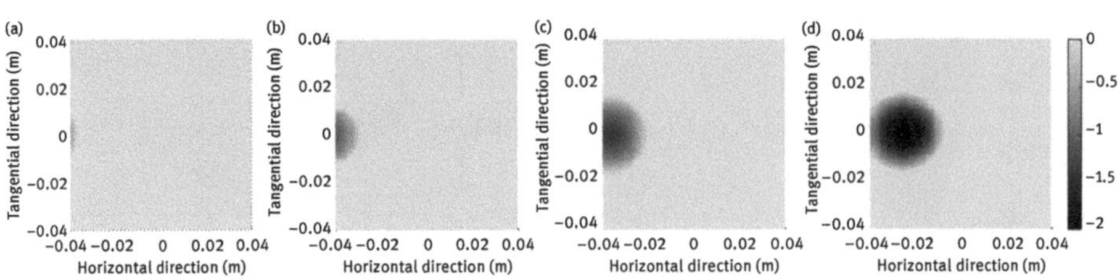

Figure 5.47: The real-time display results for 30 mm × 30 mm × 2 mm pit: (a) 10 % MFL signal; (b) 30 % MFL signal; (c) 60 % MFL signal; and (d) 90 % MFL signal.

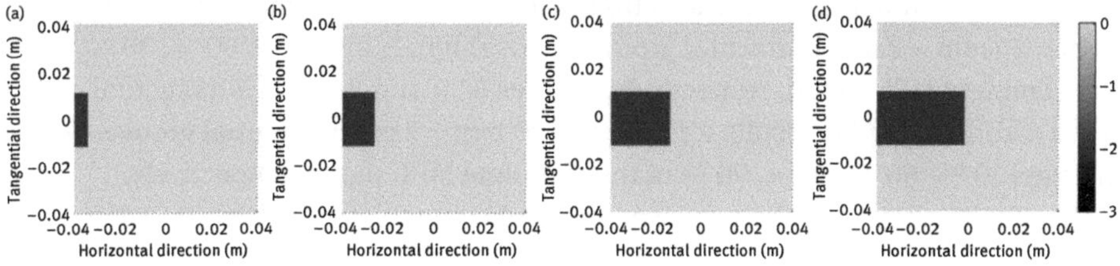

Figure 5.48: Actual defect of 40 mm × 20 mm × 2 mm horizontal groove: (a) 10 % MFL signal; (b) 30 % MFL signal; (c) 60 % MFL signal; and (d) 90 % MFL signal.

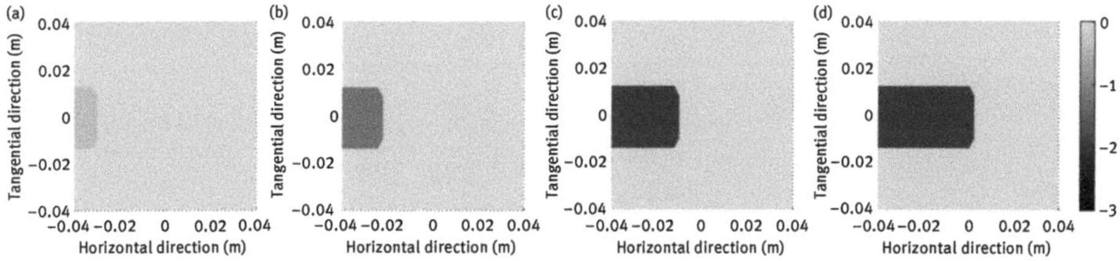

Figure 5.49: The real-time display results for 40 mm × 20 mm × 2 mm horizontal groove: (a) 10 % MFL signal; (b) 30 % MFL signal; (c) 60 % MFL signal; and (d) 90 % MFL signal.

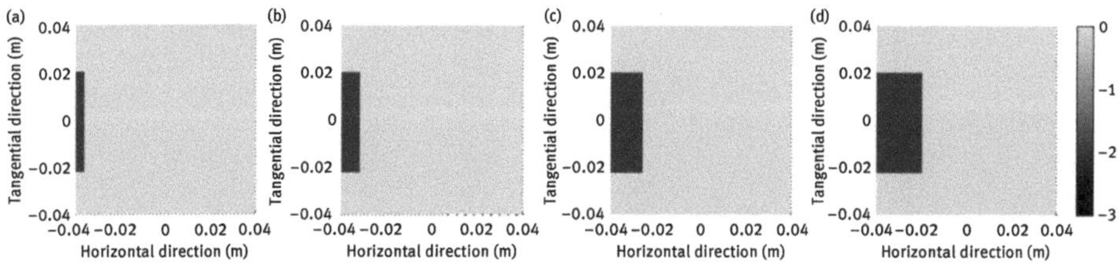

Figure 5.50: Actual defect of 20 mm × 40 mm × 2 mm tangential groove: (a) 10 % MFL signal; (b) 30 % MFL signal; (c) 60 % MFL signal; and (d) 90 % MFL signal.

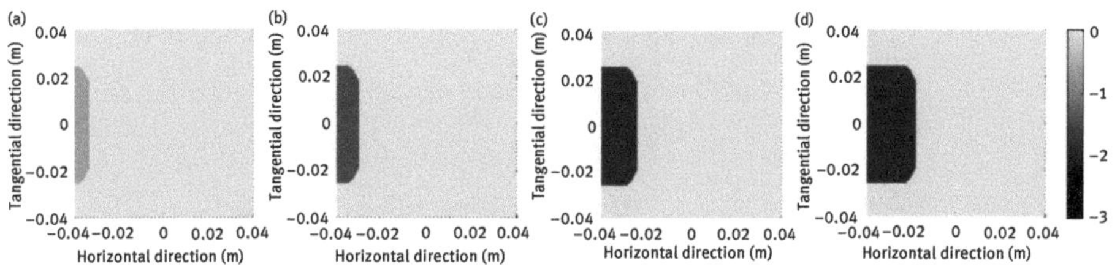

Figure 5.51: The real-time display results for 20 mm × 40 mm × 2 mm tangential groove: (a) 10 % MFL signal; (b) 30 % MFL signal; (c) 60 % MFL signal; and (d) 90 % MFL signal.

Figures 5.48 and 5.49 show the actual defect and the real-time display results for 40 mm × 20 mm × 2 mm horizontal groove, whose signals are 10 %, 60 %, 30 %, 90 % of the complete MFL signal, respectively. Figures 5.50 and 5.51 show the actual defect and the real-time display results for 20 mm × 40 mm × 2 mm tangential groove, whose signals are 10 %, 60 %, 30 %, 90 % of the complete MFL signal, respectively.

The results show that the proposed method can achieve the real-time display of the horizontal groove and the tangential groove defect according to the detected edge and depth. The comparison of actual defects and the real-time display results show that the real-time display defect profiles based on different proportions of defect 3D MFL signal coincide highly with the actual defect contour.

References

[1] Huang, S. The theory and application of magnetic internal inspection for defects of oil and gas pipeline. Beijing: Machinery Industry Press, 2013. (in Chinese)

[2] Huang S. New technology of electromagnetic nondestructive testing. Beijing: Tsinghua University Press, 2014. (in Chinese)

[3] Huang S, Zhao W. Theory and method of magnetic flux leakage imaging. Beijing: Tsinghua University Press, 2016. (in Chinese)

[4] Cui W. Study on the quantitative method of magnetic flux leakage detection for pipeline corrosion defects: [Ph.D Dissertation]. Beijing: Tsinghua University, 2006. (in Chinese)

[5] Kikuchi H, Kurisawa Y, Ara K, et al. Feasibility study of magnetic flux leakage method for condition monitoring of wall thinning on tube. International Journal of Applied Electromagnetics and Mechanics, 2010, 33(3): 1087–1094.

[6] Kikuchi H, Sato K, Shimizu I, et al. Feasibility study of application of MFL to monitoring of wall thinning under reinforcing plates in nuclear power plants. IEEE Transactions on Magnetics, 2011, 47(10): 3963–3966.

[7] Ding J F, Kang Y H, Wu X J. Tubing thread inspection by magnetic flux leakage. NDT & E International, 2006, 39(1): 53–56.

[8] Christen R, Bergamini A, Motavalli M. Three-dimensional localization of defects in stay cables using magnetic flux leakage methods. Journal of Nondestructive Evaluation, 2003, 22(3): 93–101.

[9] Song X C, Wu X J, Kang Y H. An inspection robot for boiler tube using magnetic flux leakage and ultrasonic methods. Insight-Non-Destructive Testing and Condition Monitoring, 2004, 46(5): 275–278.

[10] O'Connor S, Clapham L, Wild P. Magnetic flux leakage inspection of tailor-welded blanks. Measurement Science and Technology, 2002, 13(2): 157.

[11] Amos D M. Magnetic flux leakage as applied to aboveground storage tank flat bottom tank floor inspections. Materials Evaluation, 1996, 54(1): 26–28.

[12] Dobmann G, Walle G, Höller P. Magnetic leakage flux testing with probes: physical principles and restrictions for application. NDT International, 1987, 20(2): 101–104.

[13] Porter P C. Use of magnetic flux leakage (MFL) for the inspection of pipelines and storage tanks. NDT and E International, 1997, 30(1): 33–33.

[14] Tsukada K, Yoshioka M, Kiwa T, et al. A magnetic flux leakage method using a magnetoresistive sensor for nondestructive evaluation of spot welds. NDT & E International, 2011, 44(1): 101–105.

[15] Sharatchandra Singh W, Rao B P C, Mukhopadhyay C K, et al. GMR-based magnetic flux leakage technique for condition monitoring of steel track rope. Insight-Non-Destructive Testing and Condition Monitoring, 2011, 53(7): 377–381.

[16] Lei H M, Tian G Y. Broken wire detection in coated steel belts using the magnetic flux leakage method. Insight-Non-Destructive Testing and Condition Monitoring, 2013, 55(3): 126–131.

[17] Mandache C, Clapham L. A model for magnetic flux leakage signal predictions. Journal of Physics D: Applied Physics, 2003, 36(20): 2427.

[18] Melikhov Y, Lee S J, Jiles D C, et al. Analytical approach for fast computation of magnetic flux leakage due to surface defects. Digests of the IEEE International Magnetics Conference, 2005. NTERMAG Asia 2005. 2005: 1165–1166.

[19] Sun Y H, Kang Y H. Magnetic mechanisms of magnetic flux leakage nondestructive testing. Applied Physics Letters, 2013, 103(18): 184104.

[20] Wu J, Sun Y, Kang Y, et al. Theoretical analyses of MFL signal affected by discontinuity orientation and sensor-scanning direction. IEEE Transactions on Magnetics, 2015, 51(1): 1–7.

[21] Dutta S M, Ghorbel F H, Stanley R K. Dipole modeling of magnetic flux leakage. IEEE Transactions on Magnetics, 2009, 45(4): 1959–1965.

[22] Mandal K, Dufour D, Krause T W, et al. Investigations of magnetic flux leakage and magnetic Barkhausen noise signals from pipeline steel. Journal of Physics D: Applied Physics, 1997, 30(6): 962.

[23] Mandal K, Atherton D L. A study of magnetic flux leakage signals. Journal of Physics D: Applied Physics, 1998, 31(22): 3211.

[24] Mandal K, Dufour D, Atherton D L. Use of magnetic Barkhausen noise and magnetic flux leakage signals for analysis of defects in pipeline steel. IEEE Transactions on Magnetics, 1999, 35(3): 2007–2017.

[25] Snarskii A A, Zhenirovskyy M, Meinert D, et al. An integral equation model for the magnetic flux leakage method. NDT & E International, 2010, 43(4): 343–347.

[26] Kovács G, Kuczmann M. Simulation and measurement of magnetic based nondestructive tester. Przeglad Elektrotechniczny, 2009, 85(12): 76–79.

[27] Altschuler E, Pignotti A. Nonlinear model of flaw detection in steel pipes by magnetic flux leakage. NDT & E International, 1995, 28(1): 35–40.

[28] Al-Naemi F I, Hall J P, Moses A J. FEM modelling techniques of magnetic flux leakage-type NDT for ferromagnetic plate inspections. Journal of Magnetism and Magnetic Materials, 2006, 304(2): e790–e793.

[29] Zhang Z, Udpa L, Udpa S S, et al. An equivalent linear model for magnetostatic nondestructive evaluation. IEEE Transactions on Magnetics, 1996, 32(3): 718–721.

[30] Usarek Z, Augustyniak B, Augustyniak M. Separation of the effects of notch and macroresidual stress on the MFL signal characteristics. IEEE Transactions on Magnetics, 2014, 50(11): 1–4.

[31] Leonard S, Atherton D L. Calculations of the effects of anisotropy on magnetic flux leakage detector signals. IEEE Transactions on Magnetics, 1996, 32(3): 1905–1909.

[32] Leonard S, Atherton D L. Calculations of the effects of anisotropy on magnetic flux leakage detector signals. NDT and E International, 1997, 30(1): 31–31.

[33] Babbar V, Shiari B, Clapham L. Mechanical damage detection with magnetic flux leakage tools: modeling the effect of localized residual stresses. IEEE Transactions on Magnetics, 2004, 40(1): 43–49.

[34] Babbar V, Bryne J, Clapham L. Mechanical damage detection using magnetic flux leakage tools: modeling the effect of dent geometry and stresses. NDT & E International, 2005, 38(6): 471–477.

[35] Ivanov P A, Zhang V, Yeoh C H, et al. Magnetic flux leakage modeling for mechanical damage in transmission pipelines. IEEE Transactions on Magnetics, 1998, 34(5): 3020–3023.

[36] Du Z Y, Ruan J J, Peng Y, et al. 3-D FEM simulation of velocity effects on magnetic flux leakage testing signals. IEEE Transactions on Magnetics, 2008, 44(6): 1642–1645.

[37] Li Y, Tian G Y, Ward S. Numerical simulation on magnetic flux leakage evaluation at high speed. NDT & E International, 2006, 39(5): 367–373.

[38] Wang P, Gao Y, Tian G Y, et al. Velocity effect analysis of dynamic magnetization in high speed magnetic flux leakage inspection. NDT & E International, 2014, 64: 7–12.

[39] Norouzi E, Ravanbod H. Optimization of the flux distribution in magnetic flux leakage testing. Materials Evaluation, 2010, 68(3): 360–364.

[40] Park G S, Park S H. Analysis of the velocity-induced eddy current in MFL type NDT. IEEE Transactions on Magnetics, 2004, 40(2): 663–666.

[41] Cheng W, Chen L, Wang K, et al. Magnetic flux leakage data lossless compression based on Context Tree Weighting. IEEE 2011 International Conference on Consumer Electronics, Communications and Networks (CECNet), 2011: 1376–1379.

[42] Afzal M, Udpa S. Advanced signal processing of magnetic flux leakage data obtained from seamless gas pipeline. NDT & E International, 2002, 35(7): 449–457.

[43] Park G S, Park E S. Improvement of the sensor system in magnetic flux leakage-type nondestructive testing (NDT). IEEE Transactions on Magnetics, 2002, 38(2): 1277–1280.
[44] Pang H, Chen D, Pan M, et al. A new calibration method of three axis magnetometer with nonlinearity suppression. IEEE Transactions on Magnetics, 2013, 49(9): 5011–5015.
[45] Nguyen T T, Bouillault F, Daniel L, et al. Finite element modeling of magnetic field sensors based on nonlinear magnetoelectric effect. Journal of Applied Physics, 2011, 109(8): 084904.
[46] Nguyen T T, Mininger X, Daniel L, et al. Influence of mechanical boundary conditions on magnetoelectric sensors. IEEE Transactions on Magnetics, 2013, 49(5): 2009–2012.
[47] Kejik P, Bourdelle P F, Reymond S, et al. Offset compensation based on distributed hall cell architecture. IEEE Transactions on Magnetics, 2013, 49(1): 105–108.
[48] Pelkner M, Nenbaner A, Blome M, et al. Flux leakage measurements for defect characterization using NDT adapted OMR sensors. Studies in Applied Electromagnetics and Mechanics, 2011, 35: 217.
[49] Pelkner M, Blome M, Reimund V, et al. Flux leakage measurements for defect characterization using a high precision 3-axial GMR magnetic sensor. AIP Conference Proceedings-American Institute of Physics. 2011, 1335(1): 380.
[50] Tehranchi M M, Hamidi S M, Eftekhari H, et al. The inspection of magnetic flux leakage from metal surface cracks by magneto-optical sensors. Sensors and Actuators A: Physical, 2011, 172(2): 365–368.
[51] Tehranchi M M, Ranjbaran M, Eftekhari H. Double core giant magneto-impedance sensors for the inspection of magnetic flux leakage from metal surface cracks. Sensors and Actuators A: Physical, 2011, 170(1): 55–61.
[52] Xu J, Wu X, Cheng C, et al. A magnetic flux leakage and magnetostrictive guided wave hybrid transducer for detecting bridge cables. Sensors, 2012, 12(1): 518–533.
[53] Angani C S, Kim J, Le M, et al. Magnetic leakage testing using linearly integrated hall and GMR sensor arrays to inspect inclusions in cold-rolled strips. IEEE Transactions on Magnetics, 2014, 50(11): 1–4.
[54] Nara T, Takanashi Y, Mizuide M. A sensor measuring the Fourier coefficients of the magnetic flux density for pipe crack detection using the magnetic flux leakage method. Journal of Applied Physics, 2011, 109(7): 07E305.
[55] Nara T, Fujieda M, Gotoh Y. Non-destructive inspection of ferromagnetic pipes based on the discrete Fourier coefficients of magnetic flux leakage. Journal of Applied Physics, 2014, 115(17): 17E509.
[56] Jun J W, Choi M K, Lee J Y. Nondestructive evaluation of austenitic stainless steel using CIC-MFL and LIHaS. IEEE Transactions on Magnetics, 2011, 47(10): 3959–3962.
[57] Sun Y H, Kang Y H. The feasibility of MFL inspection for omni-directional defects under a unidirectional magnetization. International Journal of Applied Electromagnetics and Mechanics, 2010, 33(3): 919–925.
[58] Sun Y H, Kang Y H. High-speed magnetic flux leakage technique and apparatus based on orthogonal magnetization for steel pipe. Materials Evaluation, 2010, 68(4): 452–458.
[59] Sun Y H, Kang Y H. A new MFL principle and method based on near-zero background magnetic field. NDT & E International, 2010, 43(4): 348–353.
[60] Sophian A, Tian G Y, Zairi S. Pulsed magnetic flux leakage techniques for crack detection and characterisation. Sensors and Actuators A: Physical, 2006, 125(2): 186–191.
[61] Deng Y M, Liu X. Electromagnetic imaging methods for nondestructive evaluation applications. Sensors, 2011, 11(12): 11774–11808.
[62] Khodayari-Rostamabad A, Reilly J P, Nikolova N K, et al. Machine learning techniques for the analysis of magnetic flux leakage images in pipeline inspection. IEEE Transactions on Magnetics, 2009, 45(8): 3073–3084.

[63] Walters W T. Pipeline inspection apparatus for detection of longitudinal defects by flux leakage inspection of circumferential magnetic field. US Patent 3,539,915. 1970.
[64] Proctor N B. Printed circuit foils for use in magnetic flux leakage flow detection. US Patent 3,504,276. 1970.
[65] Veach W D, Lloyd T S, Ludlow J C. ID-OD discrimination sensor concept for a magnetic flux leakage inspection tool. US Patent 6,847,207. 2005.
[66] Huang Z Y, Que P W, Liang C. 3D FEM analysis in magnetic flux leakage method. NDT & E International, 2006, 39(1): 61–66.
[67] Zhang G G. Defect quantitative recognition technology of circumferential magnetic flux leakage inspection in pipeline. 2010 International Conference on Computer Application and System Modeling (ICCASM), 2010: 311–314.
[68] Zhang G G, Li P H. Signal processing technology of circumferential magnetic flux leakage inspection in pipeline. IEEE 2011 3rd International Conference on Measuring Technology and Mechatronics Automation (ICMTMA), 2011, 3: 229–232.
[69] Saranya R, Daniel J, Abudhahir A, et al. Comparison of segmentation techniques for detection of defects in non-destructive testing images. IEEE 2014 International Conference on. Electronics and Communication Systems (ICECS), 2014: 1–6.
[70] Qi J. Research on quantification technology and application of pipeline magnetic flux leakage testing: [Ph.D Dissertation]. Tianjin: Tianjin University, 2003. (in Chinese)
[71] Saha S, Mukhopadhyay S, Mahapatra U, et al. Empirical structure for characterizing metal loss defects from radial magnetic flux leakage signal. NDT & E International, 2010, 43(6): 507–512.
[72] Schifini R, Bruno A C. Experimental verification of a finite element model used in a magnetic flux leakage inverse problem. Journal of Physics D: Applied Physics, 2005, 38(12): 1875.
[73] Mukhopadhyay S, Srivastava G P. Characterisation of metal loss defects from magnetic flux leakage signals with discrete wavelet transform. NDT & E International, 2000, 33(1): 57–65.
[74] Reimund V, Blome M, Pelkner M, et al. Fast defect parameter estimation based on magnetic flux leakage measurements with GMR sensors. International Journal of Applied Electromagnetics and Mechanics, 2011, 37(2): 199–205.
[75] Ma Z, Liu H. Pipeline defect detection and sizing based on MFL data using immune RBF neural networks. IEEE Congress on Evolutionary Computation, 2007. CEC 2007, 2007: 3399–3403.
[76] Abe M, Biwa S, Matsumoto E. Three dimensional evaluation of parallelepiped flaw using amorphous MI sensor and neural network in biaxial MFLT. 3rd IEEE International Conference on Sensing Technology, 2008 (ICST 2008), 2008: 238–241.
[77] Amineh R K, Koziel S, Nikolova N K, et al. A space mapping methodology for defect characterization from magnetic flux leakage measurements. IEEE Transactions on Magnetics, 2008, 44(8): 2058–2065.
[78] Amineh R K, Nikolova N K, Reilly J P, et al. Characterization of surface-breaking cracks using one tangential component of magnetic leakage field measurements. IEEE Transactions on Magnetics, 2008, 44(4): 516–524.
[79] Liu S, Zhang Q C. LS-SVM method for 2-D reconstruction of the oil pipeline defect based on PSO algorithm. IEEE 2014 33rd Chinese Control Conference (CCC), . 2014: 7263–7267.
[80] Yang L J, Liu G, Zhang G G, et al. Oil-gas pipeline magnetic flux leakage testing defect reconstruction based on support vector machine. Second IEEE International Conference on Intelligent Computation Technology and Automation, 2009. ICICTA'09. 2009, 2: 395–398.
[81] Yuan X C, Wang C L, Zuo X Z, et al. A method of 2D defect profile reconstruction from magnetic flux leakage signals based on improved particle filter. Insight-Non-Destructive Testing and Condition Monitoring, 2011, 53(3): 152–155.
[82] Mukherjee D, Saha S, Mukhopadhyay S. Inverse mapping of magnetic flux leakage signal for defect characterization. NDT & E International, 2013, 54: 198–208.

[83] Han W H, Shen X H, Xu J, et al. Fast estimation of defect profiles from the magnetic flux leakage signal based on a multi-power affine projection algorithm. Sensors, 2014, 14(9): 16454–16466.
[84] Priewald R H, Magele C, Ledger P D, et al. Fast magnetic flux leakage signal inversion for the reconstruction of arbitrary defect profiles in steel using finite elements. IEEE Transactions on Magnetics, 2013, 49(1): 506–516.
[85] Hari K C, Nabi M, Kulkarni S V. Improved FEM model for defect-shape construction from MFL signal by using genetic algorithm. IET Science, Measurement & Technology, 2007, 1(4): 196–200.
[86] Han W H, Xu J, Wang P, et al. Defect profile estimation from magnetic flux leakage signal via efficient managing particle swarm optimization. Sensors, 2014, 14(6): 10361–10380.
[87] Xu C, Wang C L, Ji F Z, et al. Finite-element neural network-based solving 3-D differential equations in MFL. IEEE Transactions on Magnetics, 2012, 48(12): 4747–4756.
[88] Ramuhalli P, Udpa L, Udpa S S. Electromagnetic NDE signal inversion by function-approximation neural networks. IEEE Transactions on Magnetics, 2002, 38(6): 3633–3642.
[89] Ramuhalli P, Udpa L, Udpa S S. Neural network-based inversion algorithms in magnetic flux leakage nondestructive evaluation. Journal of Applied Physics, 2003, 93(10): 8274–8276.
[90] Han W H, Que P W. 2-D defect reconstruction from MFL signals based on genetic optimization algorithm. IEEE International Conference on Industrial Technology, 2005. ICIT 2005, 2005: 508–513.
[91] Yuan X C, Wang C L, Ji F Z, et al. A magnetic flux leakage analysis model based on finite element neural network. Insight-Nondestructive Testing and Condition Monitoring, 2011, 53(9): 482–486.
[92] Joshi A, Udpa L, Udpa S, et al. Adaptive wavelets for characterizing magnetic flux leakage signals from pipeline inspection. IEEE Transactions on Magnetics, 2006, 42(10): 3168–3170.
[93] Joshi A. Wavelet transform and neural network based 3D defect characterization using magnetic flux leakage. International Journal of Applied Electromagnetics and Mechanics, 2008, 28(1): 149–153.
[94] Zhang Y. Research on some key technologies of magnetic flux leakage testing: [Ph.D Dissertation]. Hefei: University of Science and Technology of China, 2007. (in Chinese)
[95] Li M, Lowther D A. The application of topological gradients to defect identification in magnetic flux leakage-type NDT. IEEE Transactions on Magnetics, 2010, 46(8): 3221–3224.
[96] Ravan M, Amineh R K, Koziel S, et al. Sizing of 3-D arbitrary defects using magnetic flux leakage measurements. IEEE Transactions on Magnetics, 2010, 46(4): 1024–1033.
[97] Ravan M, Amineh R K, Koziel S, et al. Sizing of multiple cracks using magnetic flux leakage measurements. IET Science, Measurement & Technology, 2010, 4(1): 1–11.
[98] Siebert M A, Sutherland J E. Application of the circumferential component of magnetic flux leakage measurement for in-line inspection of pipelines. BJ Pipeline Inspection Services, Calgary, Alberta (CA), 1999.
[99] Wilson J W, Tian G Y. 3D magnetic field sensing for magnetic flux leakage defect characterisation. Insight-Non-Destructive Testing and Condition Monitoring, 2006, 48(6): 357–359.
[100] Li Y, Wilson J, Tian G Y. Experiment and simulation study of 3D magnetic field sensing for magnetic flux leakage defect characterization. NDT&E International, 2007, 40(2): 179–184.
[101] Dutta S M, Ghorbel F H, Stanley R K. Simulation and analysis of 3-D magnetic flux leakage. IEEE Transactions on Magnetics, 2009, 45(4): 1966–1972.
[102] Ma F. Research on data compression and quantitative identification of defects in pipeline magnetic flux leakage testing: [Ph.D Dissertation]. Shenyang: Shenyang University of Technology, 2006. (in Chinese)

[103] Fuxiang W, Qingshan F, Hailiang Z. et al. Pipeline feature recognition based on three axis magnetic flux leakage testing technology. Nondestructive Testing, 2011, 33(1): 79–84. (in Chinese)

[104] Kim J, Lee J, Jun J, et al. Integration of Hall and giant magnetoresistive sensor arrays for real-time 2-D visualization of magnetic field vectors. IEEE Transactions on Magnetics, 2012, 48(11): 3708–3711.

[105] Chen Z M, Yusa N, Miya K. Some advances in numerical analysis techniques for quantitative electromagnetic nondestructive evaluation. Nondestructive Testing and Evaluation, 2009, 24(1–2): 69–102.

[106] Huang S, Tong Y, Zhao, W. An adaptive compression algorithm for pipeline EMAT inspection data. International Journal of Applied Electromagnetics Mechanics, 2010,33(3,4): 1095–1100.

[107] Huang S, Tong Y, Zhao W. A denoising algorithm for an electromagnetic acoustic transducer (EMAT) signal by envelope regulation. Measurement Science & Technology, 2010, 21(8): 1–6.

[108] Chen J, Huang S, Zhao W. Equivalent MFL model of pipelines for 3-D defect reconstruction using simulated annealing inversion procedure. International Journal of Applied Electromagnetics and Mechanics, 2015,47(2): 551–561.

[109] Chen J, Huang S, Zhao W. Three-dimensional defect reconstruction from magnetic flux leakage signals in pipeline inspection based on a dynamic taboo search procedure. Insight-Non-Destructive Testing and Condition Monitoring, 2014, 56(10):535–540.

[110] Chen J, Huang S, Zhao W, Wang S. Reconstruction of 3-D defect profiles from MFL signals using radial wavelet basis function neural network. International Journal of Applied Electromagnetics and Mechanics. 2014, 45(14): 465–471.

[111] Chen J, Huang S, Zhao W. Three-dimensional defect inversion from magnetic flux leakage signals using iterative neural network. IET Science, Measurement and Technology, 2015, 9(4): 418–426. DOI: 10.1049/iet-smt.2014.0173.

[112] Chen J. Study on inversion methods for three-dimensional profiles of defects on oil and gas pipelines using magnetic flux leakage testing: [Ph.D Dissertation]. Beijing: Tsinghua University, 2015. (in Chinese)

[113] Liu X. Study on method of real-time quantification and display of defects on tank floor using three-dimensional magnetic flux leakage testing: [Master Dissertation]. Beijing: Tsinghua University, 2015. (in Chinese)

Index

Bei Fragen zur Produktsicherheit wenden Sie sich bitte an:
If you have any questions regarding product safety,
please contact:

Walter de Gruyter GmbH
Genthiner Straße 13
10785 Berlin
productsafety@degruyterbrill.com